ADVANCES IN
ASYMMETRIC SYNTHESIS

Volume 2 • 1997

ADVANCES IN ASYMMETRIC SYNTHESIS

Editor: ALFRED HASSNER
Department of Chemistry
Bar-Ilan University
Ramat-Gan, Israel

VOLUME 2 • 1997

JAI PRESS INC.

Greenwich, Connecticut *London, England*

55 Old Post Road, No. 2
Greenwich, Connecticut 06836

JAI PRESS LTD.
38 Tavistock Street
Covent Garden
London WC2E 7PB
England

ISBN: 1-55938-797-1

Transferred to digital printing in 2009.

CONTENTS

LIST OF CONTRIBUTORS

Claudia Borm	Institut für Organische Chemie Universität Hannover Hannover, Germany
Patrizia Ferraboschi	Dipartimento di Chimica e Biochimica Medica Università degli Studi di Milano Milano, Italy
André Guingant	Laboratoire de Synthèse Organique CNRS Université de Nantes Nantes, France
H. B. Kagan	Institut de Chimie Moléculaire d'Orsay Laboratoire de Synthèse Asymétrique Université Paris-Sud Paris, France
Frank Nerenz	Institut für Organische Chemie Universität Hannover Hannover, Germany
Lendon N. Pridgen	Synthetic Chemistry Department Chemical Development SmithKline Beechman Pharmaceuticals King of Prussia, Pennsylvania
O. Riant	Institut de Chimie Moléculaire d'Orsay Laboratoire de Synthèse Asymétrique Université Paris-Sud Paris, France

Stanley M. Roberts
Department of Chemistry
University of Exeter
Exeter, England

Enzo Santaniello
Dipartimento di Chimica e Biochimica Medica
Università degli Studi di Milano
Milano, Italy

Nicholas J. Turner
Department of Chemistry
University of Exeter
Exeter, England

Ekkehard Winterfeldt
Institut für Organische Chemie
Universität Hannover
Hannover, Germany

PREFACE

In the last few years we have witnessed a meteoric rise in methods leading to enantiopure compounds containing various functional groups. Some advances have been stimulated by the need for pharmaceutical compounds as single enantiomers. Advances in the field have been rapid and have led to asymmetric syntheses of many natural products. Among the methods used are chiral reagents, chiral catalysts (synthetic or bio), and kinetic resolution.

In the six chapters of this volume, the second in the series on *Advances Asymmetric Synthesis*, the authors, each an authority in his field, have demonstrated effective pathways leading to syntheses of a variety of functionalized molecules with good to very high enantiomeric excess.

Two of the chapters deal with enzymatic processes, demonstrating that enzymes are very effective biocatalysts capable of creating a single stereogenic center. For instance, Santaniello and Ferraboschi describe biocatalysts as powerful synthetic tools. In this chapter it is shown how effective baker's yeast can be in enantioselective reduction not only of carbonyl groups but also of olefinic double bonds, leading to the synthesis of a variety of nonracemic functionalized molecules. Other examples of enzymatic catalysis are provided by lipases.

In a related chapter, Roberts and Turner show the application of lipases and esterases for the synthesis of chiral molecules on a large scale. Examples also include other enzymatic transformations such as stereocontrolled Baeyer–Villiger oxidation.

Chiral catalysts are able to catalyze diastereoselective transformations. When one thinks of alkylation of β-dicarbonyl compounds, chiral alkylation does not immediately come to mind. Yet in Chapter 3, Guingant shows how one can prepare α-α-disubstituted β-diketones and β-keto esters with moderate to high enantiomeric excess. This can be achieved via enols or enamines with the help of chiral metal catalysts.

Recent studies have shown how it is possible to prepare nonracemic amines via diastereoselective reactions involving chiral amino alcohol derivatives (oxazolidines). This is discussed in a chapter by Pridgen.

In a most interesting chapter, Winterfeldt and co-workers describe their synthetic approaches to many types of functionalized cyclic molecules with very high ee. This was accomplished by means of highly effective diastereoselective Diels–Alder reactions of chiral cyclopentadienes followed ultimately by retro Diels–Alder thermolysis.

Finally the chapter by Kagan demonstrates how asymmetric syntheses of ferrocene and other metallocene derivatives can be achieved by chemical or enzymatic methods, taking advantage of lateral or planar chirality of these systems.

I am indebted to the authors for their splendid cooperation and excellent presentations. This volume is devoted with love to my wonderful wife Cyd who has always stood by me come rain or shine.

Alfred Hassner
Editor

PREPARATION AND APPLICATION OF CHIRAL CYCLOPENTADIENES

Ekkehard Winterfeldt, Claudia Borm, and

Frank Nerenz

Advances in Asymmetric Synthesis
Volume 2, pages 1–53

ISBN 1-55938-797-1

I. INTRODUCTION

Chiral reagents and catalysts are currently the most popular tools for the preparation of enantiopure compounds. They range from pure enzymes and enzyme mixtures such as lipases via enantiopure natural products, e.g. tartrate, quinine, and proline including their derivatives, to purely synthetic molecules including chiral bisnaphthols and their phosphine analogs. These compounds either catalyze or exercise diastereoselective transformations, which after separation from the catalyst or reagent, give rise to pure enantiomers. However, they also have the ability to discriminate between enantiomers and thus lead to kinetic resolution and finally they may be able to differentiate between the enantiotopic groups of a prochiral substrate and transform the starting material into one enantiomer completely.

There have been various attempts to realize this concept in the field of Diels–Alder cycloadditions[1–7] and there are a number of reports on the preparation and the use of chiral 4π-systems in the literature.[8–23] One has to realize, however, that the demands on this tool are very high, as far as selectivity is concerned. As Scheme 1 clearly demonstrates, satisfactory results in kinetic resolution and differentiation of enantiotopic groups can only be expected if an extremely high degree of regioselectivity, face-selectivity, and *endo*-selectivity is guaranteed for all cycloaddition processes to be dealt with.

Since conformationally directed diastereoselective transformations have been shown to be generally more reliable and efficient than configurationally directed ones,[24] we concentrated our efforts right from the beginning on configurationally and conformationally well-defined 4π-systems. Since for cyclopentadienes substantially higher reaction rates are well documented,[5] a conformationally rigid bicyclic or polycyclic chiral cyclopentadiene was considered the candidate of choice for directed cycloadditions.

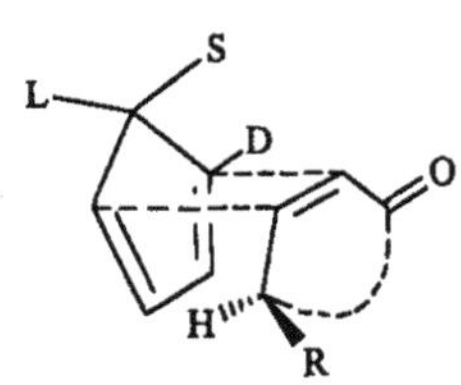

1

Scheme 1.

II. STEROID DERIVATIVES

Having the above in mind we were particularly intrigued by a series of papers by A.J. Solo and his colleagues from the Buffalo School of Pharmacy.[25–28] These workers, aiming at new 17α-substituted steroid derivatives, launched a very thorough investigation of cycloaddition reactions using 20-keto-cyclopentadiene **2** and revealed a very high face-selectivity along with excellent regio- and *endo*-selectivity for all Diels–Alder reactions studied. This chemistry was later extended by Bull[29–35] and his colleagues to the enol acetate of dehydro-estrone **3**; again a very remarkable set of selectivities was observed and it was also shown that various subsequent transformations of the cycloadducts proceeded with extremely high diastereoselectivity.

However, of particular importance for the understanding and the predictability of the Diels–Alder chemistry involved is the fact that these additions in contrast to a great number of diastereoselective transformations in the steroid field, which are all governed by the directing power of the β-oriented angular methyl groups, do not take place from the α-face as usual but turn out to operate via a very reliable and highly diastereoselective β-attack. Certainly, this quite unusual steroid–molecule behavior occurred due to the quite rigid concave–convex conformation as portrayed in structure **6** (Scheme 2).

The very remarkable degree of selectivity observed with these dienes invited further investigations and obliged us to look for an inert donor substituent on the 4π-system which would survive metalloorganic chemistry as well as hydride reductions. Hence we introduced a phenyl substituent by treating dehydro-estrone **7** with phenyllithium (Scheme 3). Spontaneous dehydration took place on work up and gave rise to cyclopentadiene **9** which in subsequent cycloadditions afforded the corresponding Diels–Alder products, again with excellent face-, regio-, and *endo*-selectivity (see **11** and **12**). The same was true for diene **10**, which is also easily prepared from 15α-hydroxandrostenedione (**8**).[8] According to expectations all chemical transformations of these adducts showed very high selectivity which is of course due to the cage-like structure of this subunit. To our great delight, adducts with symmetric dienophiles (see **11** and **12**) also showed a high degree of regioselectivity since nucleophilic attack took place with remarkable preference at the carbonyl group neighboring the phenyl group. Obviously the sp^3-centers of ring B in the steroid system exercise a higher steric hindrance than the sp^2-centers of the phenyl ring. Thus hydroxy lactone **13** turned out to be the main product from Grignard reactions and the lactones **14** that were generated in a highly diastereoselective silane reduction of these intermediates afforded pure (*S*)-butenolides in the final Kugelrohr distillation at 250 °C. HPLC analysis proved the enantiomeric excess to be higher than 99.5% and their absolute con-

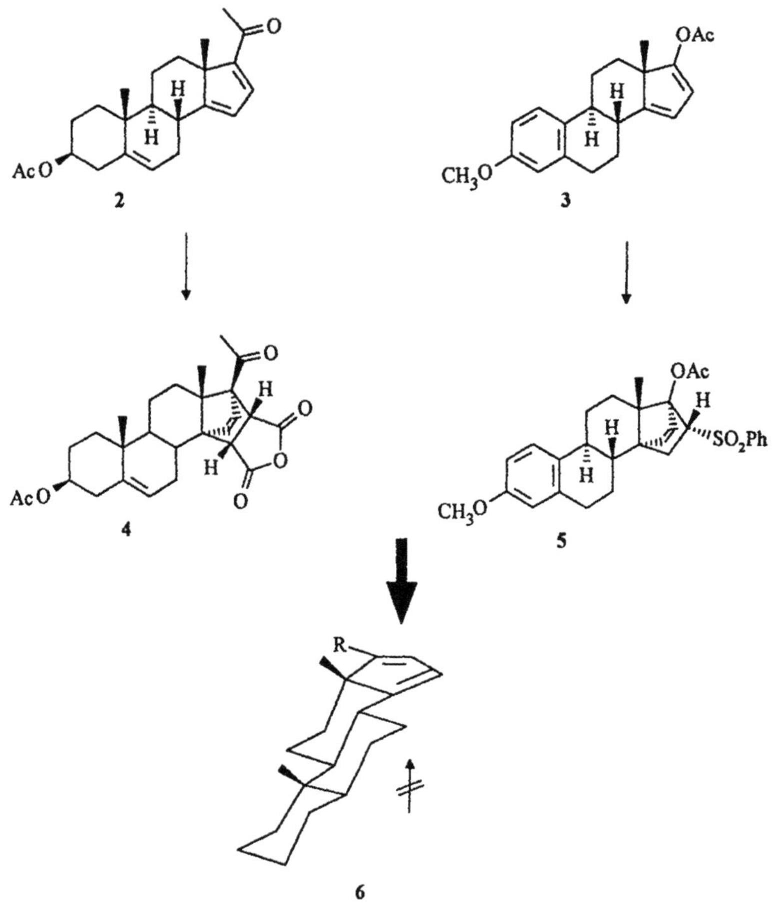

Scheme 2.

figuration was rigorously proven by an unequivocal synthesis of the corresponding methylderivative (**15**, R = CH_3) from methyl lactate.[36]

Although several Diels–Alder adducts were prepared, which in subsequent transformations performed with impressive and predictable diastereoselectivity, there remained the drawback that only one enantiomer is available from the steroid series; here we saw a severe limitation if one were to aim at kinetic resolution and at chiral recognition phenomena in general. Additionally there were good reasons for broader applications apart from the very special and also quite expensive steroid precursors, and to address a quite simple system which should have the additional advantage of providing both absolute configurations.

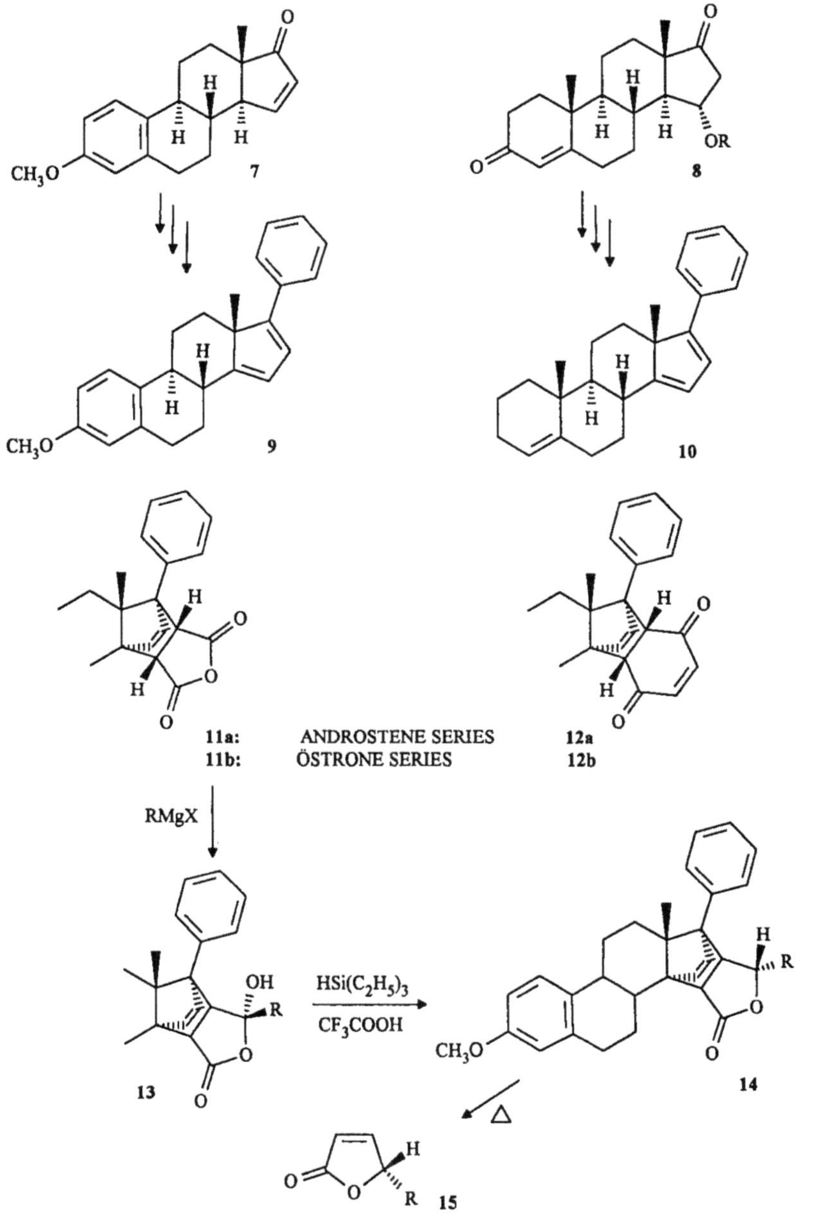

7
8
9
10
11a: ANDROSTENE SERIES 12a
11b: ÖSTRONE SERIES 12b
RMgX
13
HSi(C2H5)3
CF3COOH
14
15

Scheme 3.

III. HYDRINDANE DERIVATIVES

Since the CD ring system of the steroid molecule provides the 4π-system as well as the important directing groups, we decided to reduce the chiral cyclopentadiene to compound **18** which has the additional advantage of easy access to both enantiomers from the corresponding enantiomerically pure Hajos–Wiechert ketone **16**. This compound can easily be prepared in a proline-directed catalytic asymmetric synthesis with very high enantiomeric excess (Scheme 4).[37,38]

Treatment with borohydride under acidic conditions followed by reoxidation of the remaining hydroxy group, either using the Swern procedure or Bobbit reagent, generated the unsaturated ketone **17** which can be treated with various

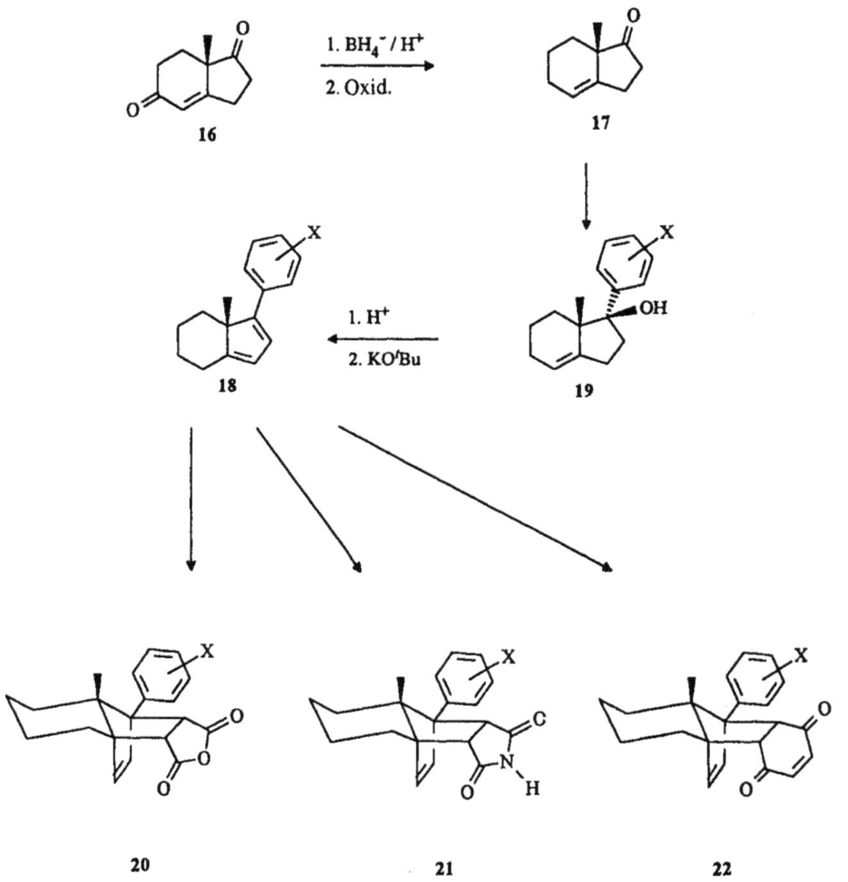

Scheme 4.

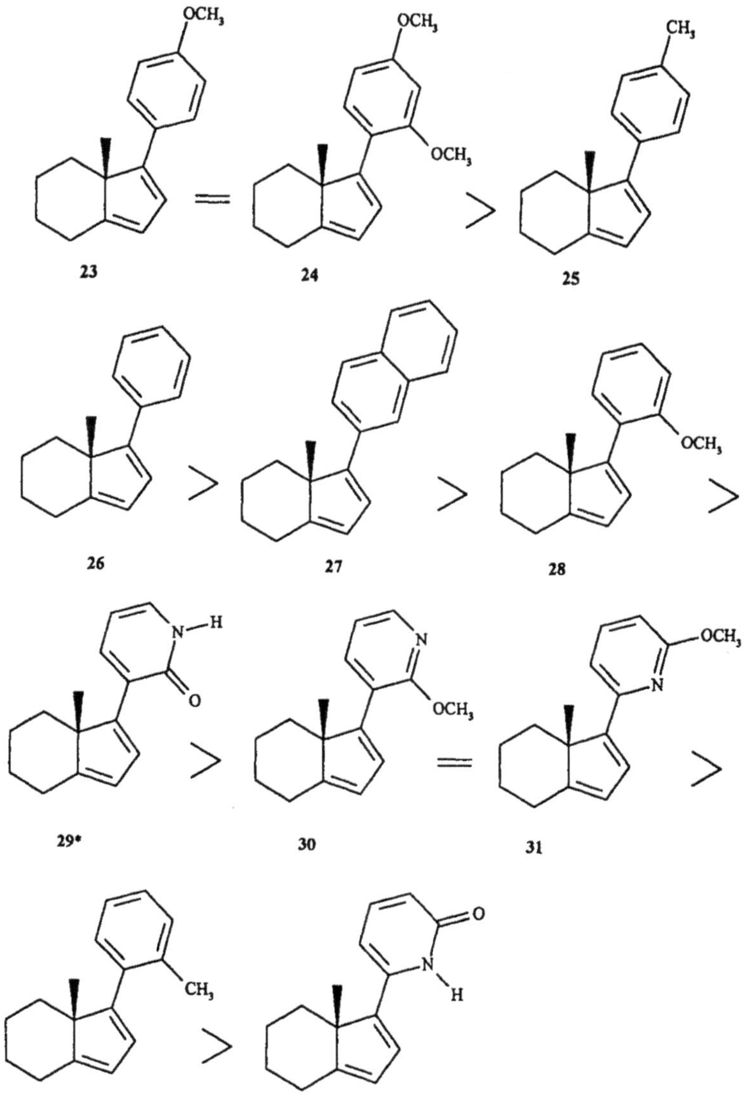

*$ZnCl_2$ catalysis

**No reaction at 6.5 kbar in the presence of $ZnCl_2$

Scheme 5.

aryllithium derivatives to provide the very acid-labile alcohol **19**. Proton-catalyzed elimination followed by base-catalyzed isomerization finally gave rise to the desired dienes in high yield, which again having excellent face- and *endo*-selectivity afforded the Diels–Alder adducts **20–22**.[39]

This very flexible route offers the possibility to manipulate the electronic as well as the stereochemical situation at the diene system by introducing electron-rich or electron-poor phenyl groups and by placing a very space-demanding group in the close neighborhood of the cyclopentadiene. With these aims in mind the dienes **23–33** were prepared (see Scheme 5) and their reaction rates with maleic imide (see **21**) were chosen for a rough comparison of their Diels–Alder reactivity.[39] According to these data the *p*-methoxy compound **23** is, as far as stability and reactivity are concerned, certainly the best choice. In the *o*-methoxy derivative **28**, steric hindrance leads to much lower reaction rates comparable to the also quite crowded β-naphthyl compound **27**. In the *o,p*-dimethoxy derivative **24**, steric hindrance of the *o*-substituent seems to counterbalance its electronic contribution, leading to reaction rates comparable to the *p*-methoxy compound **23**. These data additionally indicate that the pyridine or pyridone derivatives **29**, **30**, **31**, and **33**, which had mainly been prepared to look at attractive interactions between diene and dienophile, are quite generally characterized by rather low reaction rates. Therefore, for a closer inspection of the phenomena of chiral recognition, we did not focus on attractive forces like hydrogen bridges and chelation but instead focused exclusively on repulsive forces due to the simple space demand.

IV. SYMMETRIC DIENOPHILES

Since the space demand of the steroid ring B was previously observed to provide a directing influence in the regioselective attack of Grignard reagents to the symmetric anhydride moiety of adduct **11b**, we selected adducts **20**, **21**, and **22** (symmetric dienophiles) to initially probe into the regioselectivity problem. Anhydride **20** was first chosen to check the same metalloorganic transformation in the hydrindane series. Since ring B is completely missing here, one is not at all surprised to notice in this field a preference of the other carbonyl group (see Scheme 6) to obtain intermediate **34** which after reduction to **35** and thermolysis gave rise to the butenolide (*R*)-**15**.[36] It has to be stressed at this stage that with adducts like **11b** and **20** a change in regioselectivity in connection with subsequent diastereoselective reactions is translated into a change in enantioselectivity, thus giving access to both enantiomers of a racemic mixture by simply changing the regioselectivity of the initial step.

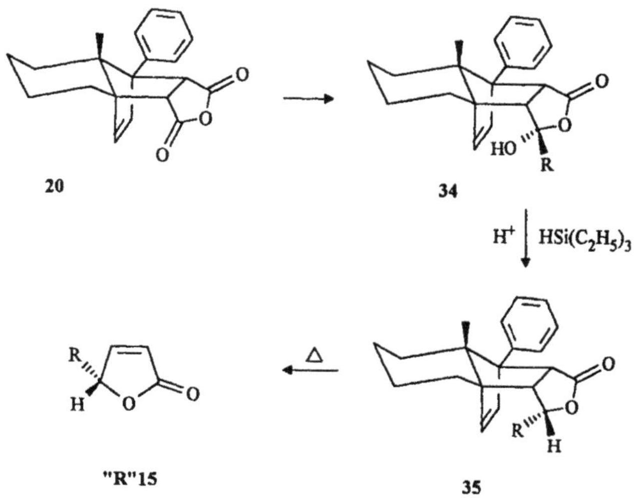

Scheme 6.

The same type of regioselectivity was noticed in the selectride reduction of imides **36** and **37** which were easily obtained in a Mitsunobu alkylation of **21**.[40] The acetylene derivatives **37a,b** obtained this way after regioselective reduction were transformed into the ethyl ether **38** or the thioether **40**, respectively (Scheme 7). While the former underwent the usual immonium salt cylization to form ketone **41**, which on subsequent thermolysis provided a 89% yield of the enantiopure indolizidone derivative **44**, the latter was taken through the Hart cyclization procedure to generate a 1:5 mixture of the epimeric aldehydes **46a,b** which were separated by chromatography. The main product **46b** was heated and provided the pyrrolizidone aldehyde **47** in 83% yield.

It should be noted at this stage that **37a** in a well-established sequence can also be converted into the thiophenol ether **48** which on treatment with formic-acid cyclizes in the usual way to generate the thioester **50**, which after borohydride reduction and thermolysis gave rise to the enantiopure pyrrolizidone alcohol **51**. The same compound was obtained from aldehyde **47b** by reduction thus establishing the configuration for this material (Scheme 8).

Although bromo compound **36** also served very well in cyclization reactions, the behavior of this material deserves a detailed inspection. First, nucleophilic attack to the carbonyl group of the imide triggered an immediate ether formation (see **39**) leading after acid treatment and Mannich reaction with malonate to alcohol **42** (see Scheme 7). When this intermediate was taken through a Mitsunobu

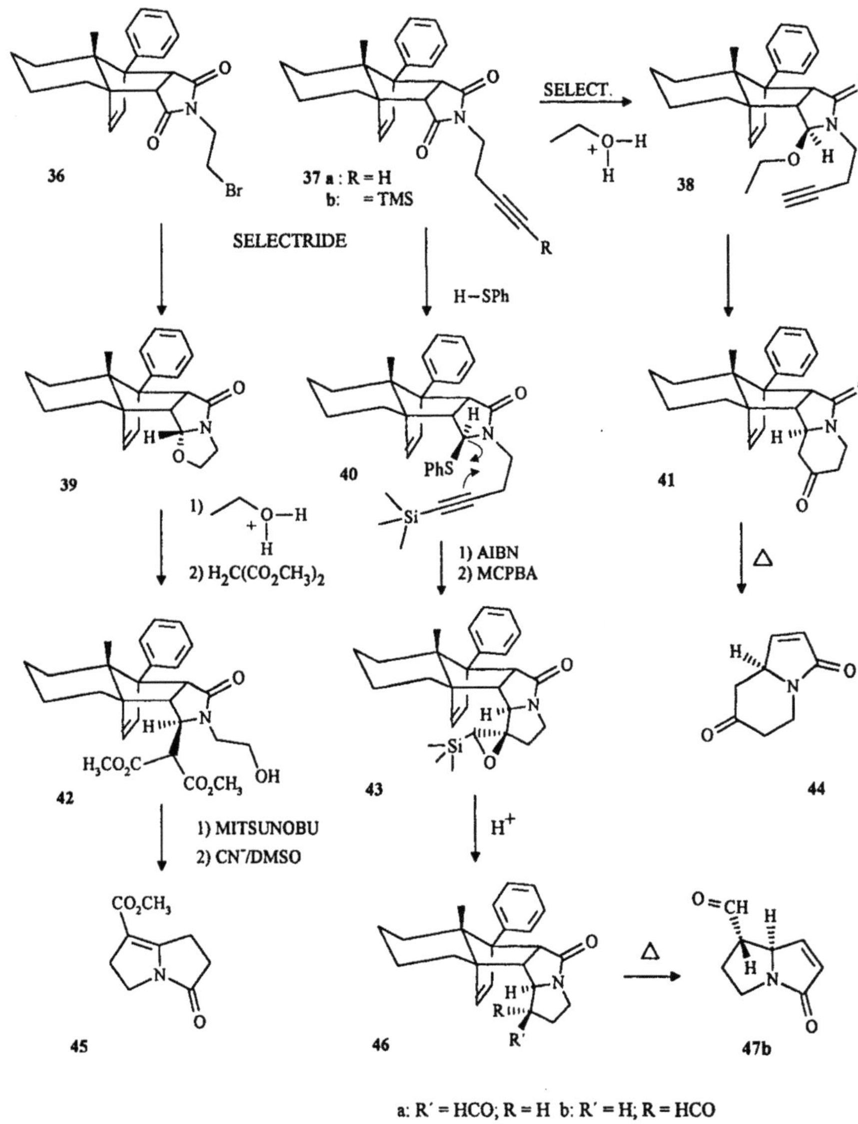
36
37 a : R = H
b: = TMS
SELECTRIDE
SELECT.
H–SPh
38
39
40
41
1) AIBN
2) MCPBA
2) $H_2C(CO_2CH_3)_2$
42
43
44
1) MITSUNOBU
2) CN⁻/DMSO
H^+
45
46
47b
a: R′ = HCO; R = H b: R′ = H; R = HCO

Scheme 7.

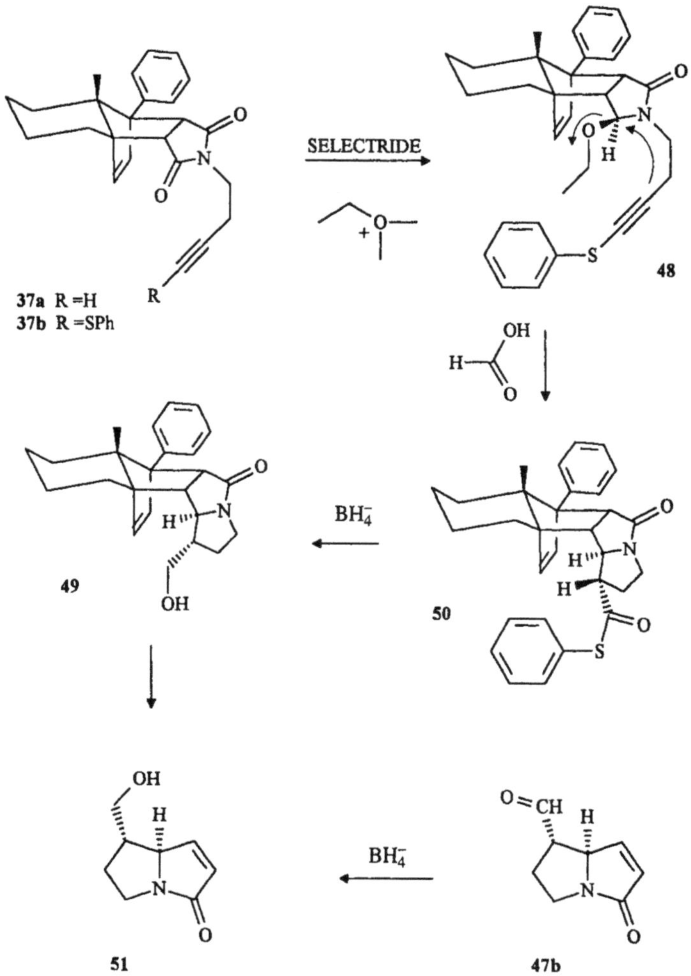

Scheme 8.

cyclization–decarboxylation sequence it unexpectedly generated the achiral dehydropyrrolizidone ester **45** spontaneously.[40] For an explanation (see Scheme 9) we assume diester **52** forms the dehydropyrrolizidone **53** in a retro-Diels–Alder process and undergoes an enolization–deconjugation sequence under the reaction conditions. The unsaturated lactam-acid **54** formed in a Krapcho process should then decarboxylate as indicated and generate **45**.

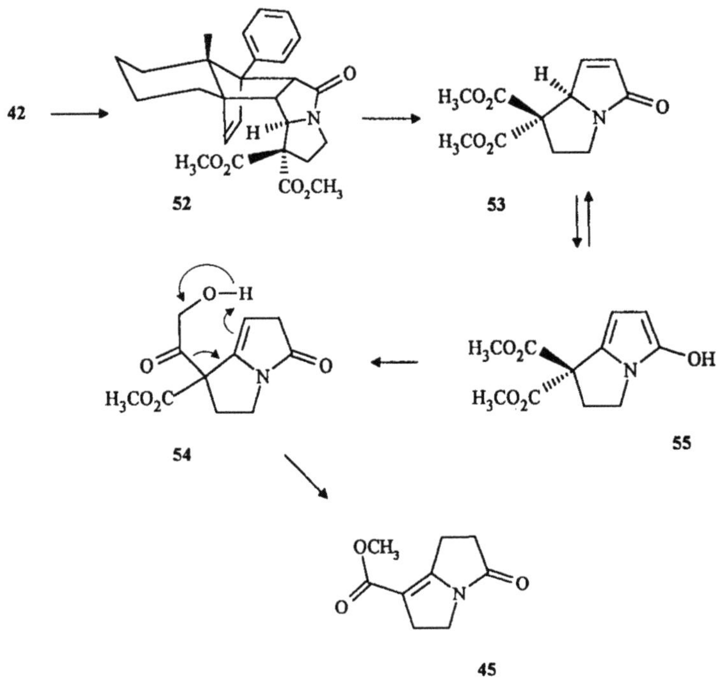

Scheme 9.

In connection with 5-membered-ring dienophiles, such as maleic anhydride and its imide, the all-carbon analog cyclopentene-1,3-dione **56** deserves special treatment here since, in contrast to the anhydride and the imide, this compound is comparatively unstable. Bases as well as acids lead to quick decomposition and temperatures higher than 60 to 70 °C are also unacceptable. This molecule is therefore a very special candidate for high-pressure cycloadditions (6–7 kbars). While the anhydrides and imides discussed so far can easily be added under purely thermal conditions and sometimes even at room temperature, this particular diketone does fail completely on heating. The high-pressure technique, however, provides a 98% yield of adduct **57** which behaves, of course, like a vinologous acid as equilibrium **57** ⇔ **58** indicates (see Scheme 10).[41]

This practically quantitative room-temperature cycloaddition of a quite unstable dienophile is an excellent example for the merits of the high-pressure technique in this field. We shall see further applications with quinones and quinone derivatives and very spectacular advantages will emerge in the area of chiral recognition including the differentiation of enantiotopic groups. With the vinolo-

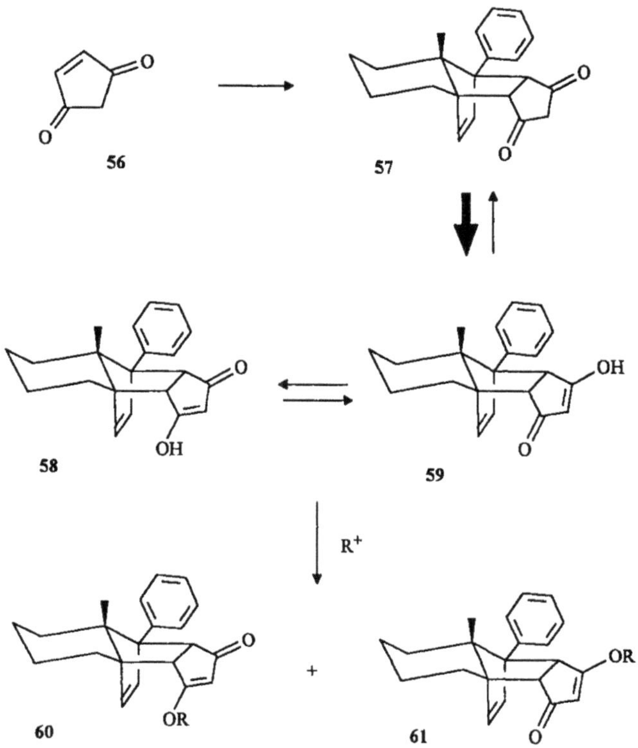

a) = CH_3	66	:	34
b) = Tos	30	:	70
c) = Piv	95	:	5

Scheme 10.

gous acid **58** ⇔ **59**, regioselectivity of ester formation is of course a major problem and as Scheme 10 demonstrates it does very much rely on the structure of the electrophile. Since the reaction with diazomethane, which gives rise to the methyl esters **60a/61a**, is a very fast transformation, the ratio given here may reflect the equilibrium of enols, while with other electrophiles one may well end up with a product controlled ratio of regioisomers. This is certainly indicated with the pivalates **60c/61c**.

If one remembers the regioselectivity of Grignard reactions with anhydrides and of selectride reductions in the imide series, pivalate **60c** should be the main product from the treatment with a bulky electrophile, and since the phenyl residue had been held responsible for this outcome, we checked some differently substituted phenyl residues in this reaction and arrived at a quite interesting result (see Scheme 11). While the *p*-methoxy derivative (see **60e/61e**) and the *o*-methyl compound (see **60f/61f**) behaved according to expectations, the former just indicated that the electron density of the aromatic ring is of no importance here.

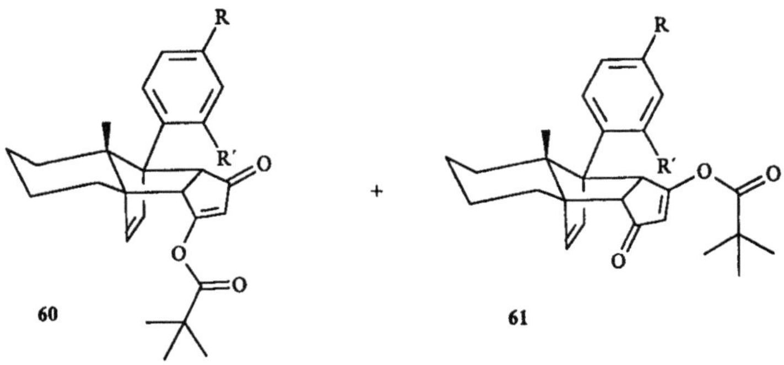

60 **61**

d: R = R' = H	95	:	5
e: R = OCH_3; R' = H	95	:	5
f: R = H; R' = CH_3	100	:	0
g: R = H; R' = OCH_3	26	:	74
h: R = R' = OCH_3	20	:	80

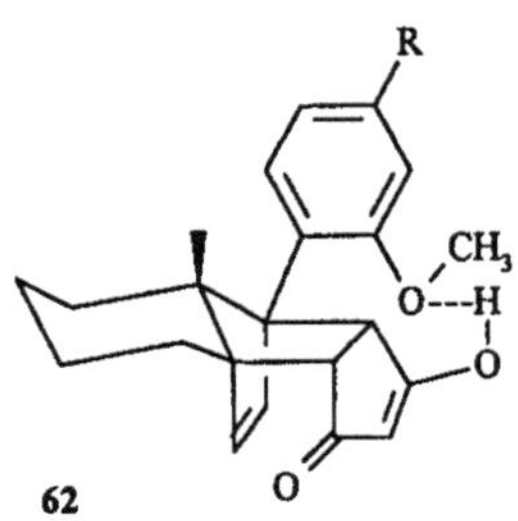

62

Scheme 11.

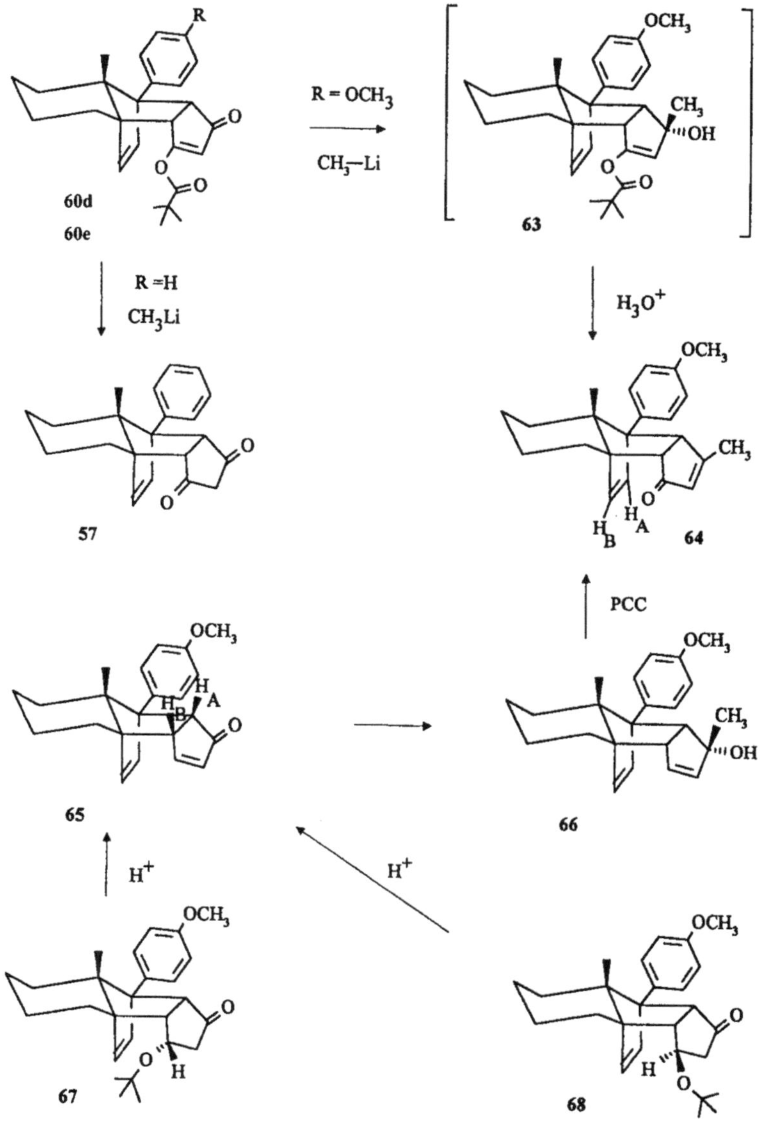
R
O
O
O
60d
60e
R = OCH3
CH3–Li
OCH3
CH3
OH
O
O
63
R =H
CH3Li
H3O+
O
O
57
OCH3
CH3
O
HB
HA
64
PCC
OCH3
HA
HB
O
65
OCH3
CH3
OH
66
H+
H+
OCH3
O
O
H
67
OCH3
O
H
O
68

Scheme 12.

Thus, a higher space demand favors isomer **60f**, as may be predicted and the methoxy derivatives **60g/60h** both strongly favored **61**.[42] The explanation of this result is by no means simple. There could of course be a precomplexation of the pivaloyl chloride by the *o*-methoxy group thus directing the electrophile to the formation of **61**, but one may also speculate about hydrogen bond formation as given in **62** which could lead to **61** preferentially. Anyway, having good access to **60d,e**, we expected both to be useful intermediates in the formation of enantiomerically pure cyclopentenones (Scheme 12) which are of value in their own right but could additionally provide final proof for the regiochemistry defined in Scheme 11, which had been assigned with the help of NMR data. Again a quite surprising and unexpected result emerged from the treatment of both compounds with lithium–alkyl derivatives (see Scheme 12).[41]

In contrast to our expectations methyllithium did attack the phenyl compound **60d** exclusively at the pivalate–carbonyl group, thus regenerating simply the starting diketone **57**. The *p*-methoxy derivative **60e**, however, did undergo attack at the other carbonyl group with excellent regioselectivity to generate cyclopentenone **64**. The intermediate **63** could not be isolated since acid workup converted it immediately into the corresponding unsaturated ketone. To explain this result one has to retreat again to precomplexation of the reagent by the electron-rich aromatic ring or the methoxy group itself. The correct orientation of the cyclopentenone moiety can easily be proven by NOE measurements as irradiation of the protons of the methyl group at the double bond leads to substantial Overhauser effects with proton H_A, while no effect is observed with proton H_B.

To provide a chemical proof too, we used a second independent route to **64** which started from the well-assigned cyclopentenone **65**, which is easily available from either 4-acetoxy cyclopentenone in a Diels–Alder-elimination sequence or from the 4-*tert*-butoxycyclopentenone addition products **67** and **68** by acid treatment (Scheme 13).[43] The multiplicity of protons H_A and H_B in **65** as well as in adducts **67** and **68** leaves no doubt about their structures, thus we were not surprised to note that **65** on treatment with methyllithium and subsequent PCC oxidation in a well-established route was also transformed into the substituted cyclopentenone **64**. Thanks to the high-pressure cycloaddition one has easy access to adduct **57** which can be transformed into cyclopentenone **64** in two simple and high-yield steps; hence, we investigated the potential of compounds of this type for the predictable preparation of enantiopure-substituted cyclopentenones.

Following the same lines as described for the methyl derivative **64**, its analogs **64a** and **64b** were prepared with butyllithium and phenyllithium, respectively. All these compounds were uneventfully treated with cuprates to provide the disubstituted reaction products **69** (Scheme 13); one has to realize that the configu-

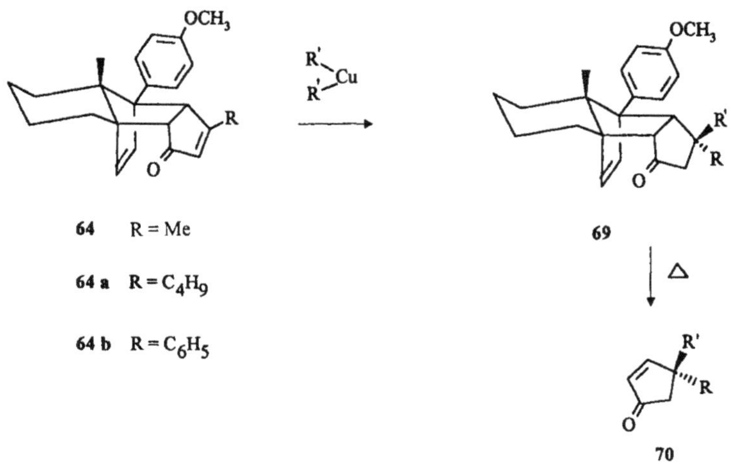

Scheme 13.

ration at the quaternary carbon atom is set in a predictable way by the sequence of reagents used in the process.[41] Since thermolysis of **69** gave rise to enantiomerically pure cyclopentenones of type **70** in 90–95% yield, this represents a constitutionally as well as configurationally highly flexible route to optically pure cyclopentenones, showing some similarity to the procedure described recently by Meyers.[44]

To evaluate the synthetic potential of the substituted cyclopentenones of type **64** the diastereoselectivity of reductions and oxidations was investigated with the methyl derivative **64**. Nucleophilic epoxidation provided a quantitative yield of epoxide **71** which easily underwent reductive ring opening with samarium diiodide to generate alcohol **73**. Selectride reduction unfortunately was not completely chemoselective and did even under very mild conditions give rise to about 10% of the overreduced compound **72**, but the mixture was efficiently focused into ketone **74** with the Bobbit reagent.[41] Finally the oxaziridine technique reported by Davis and his colleagues[45] turned out to be the method of choice for the formation of α-hydroxyketone **75** (yield: 90%). With the exception of epoxide **71**, which according to Zwanenburg[46,47] rearranges to form α-pyrones on heating, the ketones described in Scheme 14 are useful precursors for optically pure cyclopentenones.

Similarly to cyclopentenedione **56**, which operates as a master key for various highly functionalized, enantiomerically pure cyclopentenones, we expected quinone to be of the same value for a broad range of cyclohexenones and were pleased to note that the cycloadducts **76** were formed in 95% yield at 6.5 kbars

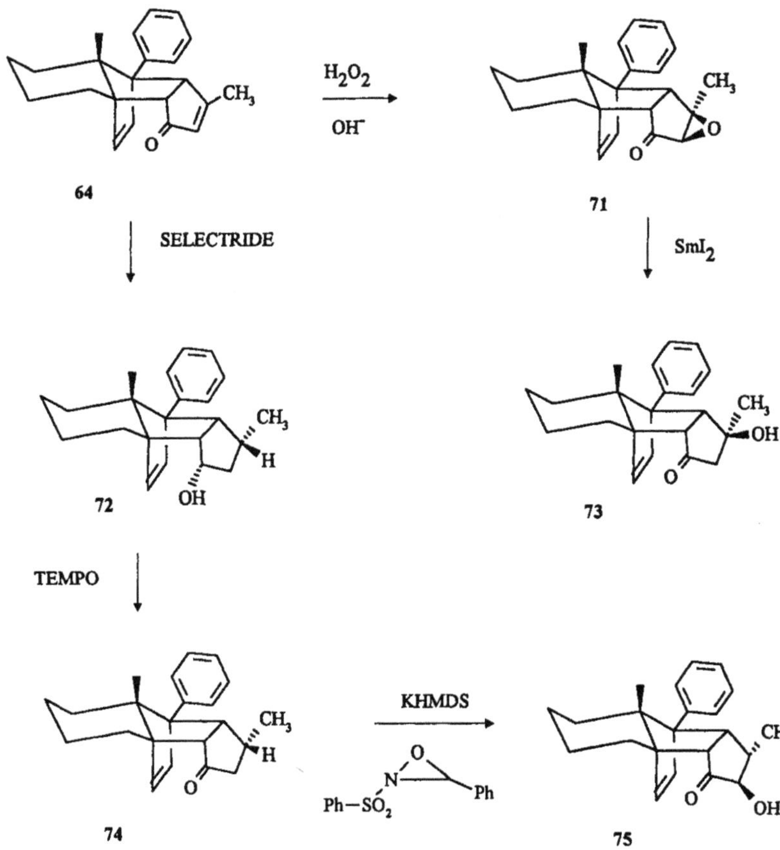

Scheme 14.

and room temperature, thus making large amounts of this material easily available for regioselective and diastereoselective transformations. As far as regioselectivity went, we again expected regioselective attack of selectride as had been noticed with imides. To our great delight, hydrogenation[48] as well as nucleophilic epoxidation[49] were indeed followed by a highly regioselective reduction to generate ketoalcohols **78** and **79**, which both on heating gave rise to optically pure cyclohexenones **80** and **81** in 70% and 90% yield, respectively with ee values higher than 98% (Scheme 15) (NMR data from Mosher's ester).

Since the regioselectivity of the reduction of all quinone adducts had proven to be very high, we became interested in the oxidation of the corresponding diols (see **82**). If oxidation turned out to be regioselective too, thus giving rise to hydroxyketone **84**, this result would correspond to a change in regioselectivity of the diketone reduction (cf. **79**). Since we had already recognized in the anhydride series that a change in regioselectivity does translate directly into a change of absolute configuration, this should open the road to the enantiomers of **80** and **81** (see Scheme 15).

Diol **82** was easily obtained in a low-temperature DIBAH reduction in excellent yield and the desired hydroxyketone **84** was indeed generated with perfect regioselectivity when the preformed oxoammonium salt from the Bobbit reagent was used. This process turned out to be highly dependent on the reaction conditions, however, and when the Bobbit reagent was added to the diol directly a rad-

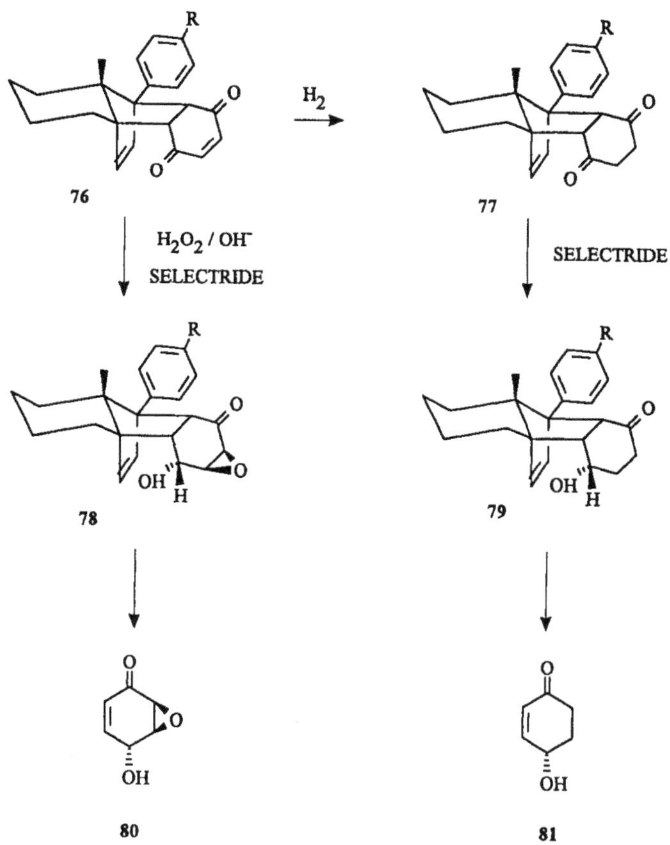

Scheme 15.

ical pathway initiated the addition of the hydroxy group to the strained double bond forming the less interesting cyclic ether **83** (Scheme 16).[41]

Having disclosed good regioselectivity for reductions as well as oxidations we next focused on the diastereoselectivity and regioselectivity of enolate alkylations and, since we had easy access to the saturated diketone **77** (see Scheme 15), we chose this material for a deprotonation sequence employing a very space-demanding proton acceptor. Since so far the carbonyl group away from the phenyl residue had been attacked preferentially, we expected alkylations to occur at this site also.

Using the lithium salt of hexamethyldisilazane as the deprotonating agent, the alkylation with various alkyl halides did, however, give rise to mixtures of

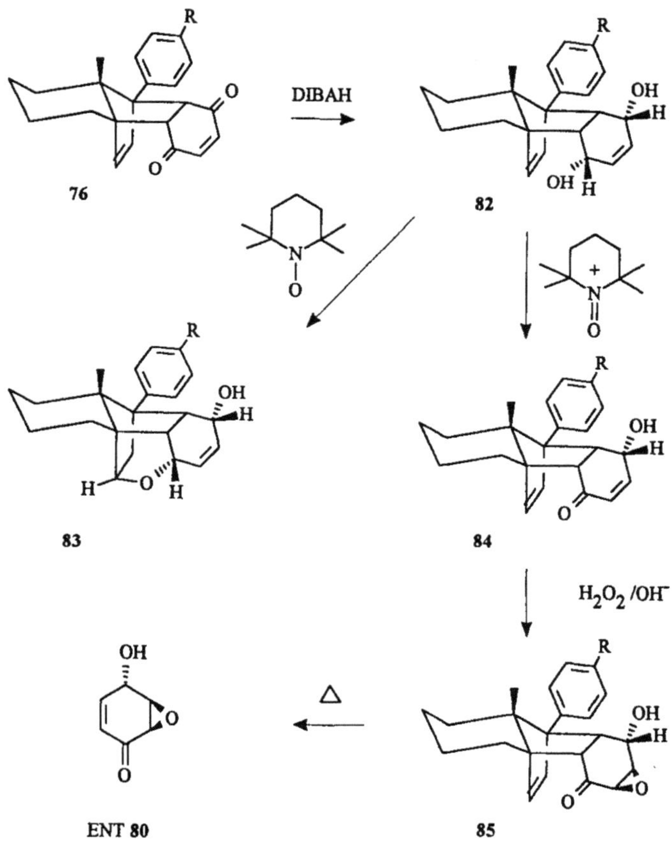

Scheme 16.

monoalkylation products with one isomer predominating very clearly (for ratio see Scheme 17). Bearing in mind the regioselectivity of earlier transformations we tentatively assigned structure **86** as the main product,[41] but a rigorous proof of these structures was of course mandatory and looking for an unambiguous experiment we were attracted by the quinone monoketal **92** for various reasons. First, we were interested in its behavior as a dienophile in general, particularly as far as regioselectivity and *endo*-selectivity are concerned. Second, the monoprotected diketone promised even higher synthetic flexibility than the quinone adduct **76**. Third, the alkylation of the corresponding dihydro derivative leaves no doubt about the structure of the product and additionally grants the possibility to study the configurational stability of the alkylation products, which boils down to the question: Will the sterically more hindered α-alkylation products **91**, which are not formed in the alkylation process, result from acid- or base-catalyzed epimerization reactions?

Cycloaddition reactions employing the monoketal **92** proved to be extremely efficient.[50] This cyclohexadienone **92** in contrast to the corresponding disubstituted cyclohexenones already provided a quantitative yield of cycloadduct **90** after 2 days at 6.5 kbars; one should mention at this stage that even for monosubstituted cyclohexenones the presence of a Lewis acid ($ZnCl_2$) in cycloaddition reactions is mandatory and still leads to only moderate yields[50] (vide infra). Since the subsequent Luche reduction to generate ketone **88** as well as the formation of the monoalkylated compounds **89** proceeded with high yields and excellent diastereoselectivity, the monoketal route is certainly the method of choice to prepare a wide variety of enantiopure cyclohexanones and cyclohexenones. As expected, the hydrolysis of the ketal proved to be the crucial experiment and since 6*N* HCl and comparatively long reaction times were needed, one was not surprised to notice epimerization. However, since one of these epimers turned out to be identical to the major regioisomer of the diketone alkylation (see **87**), these experiments prove our earlier tentative assignment of the diketone regioisomers to be wrong and leave no doubt that the diketone alkylation provides the compounds of type **87** as the major regioisomers. This result is in contrast to the observations made with nucleophilic attack at the corresponding diketones, but since in lithium amide mediated deprotonation experiments the interaction of the lithium cation with the oxygen atom of the carbonyl group determines the regioselectivity of enolate formation, a comparison of these mechanistically very different processes may be unacceptable anyway. Additionally one may again speculate about an enolate stabilizing effect from the π-cloud of the phenyl ring as indicated in **93** (see Scheme 18).

To avoid epimerization during ketal hydrolysis we decided to reduce the carbonyl group to the corresponding α-carbinol. Since the bicyclic structure

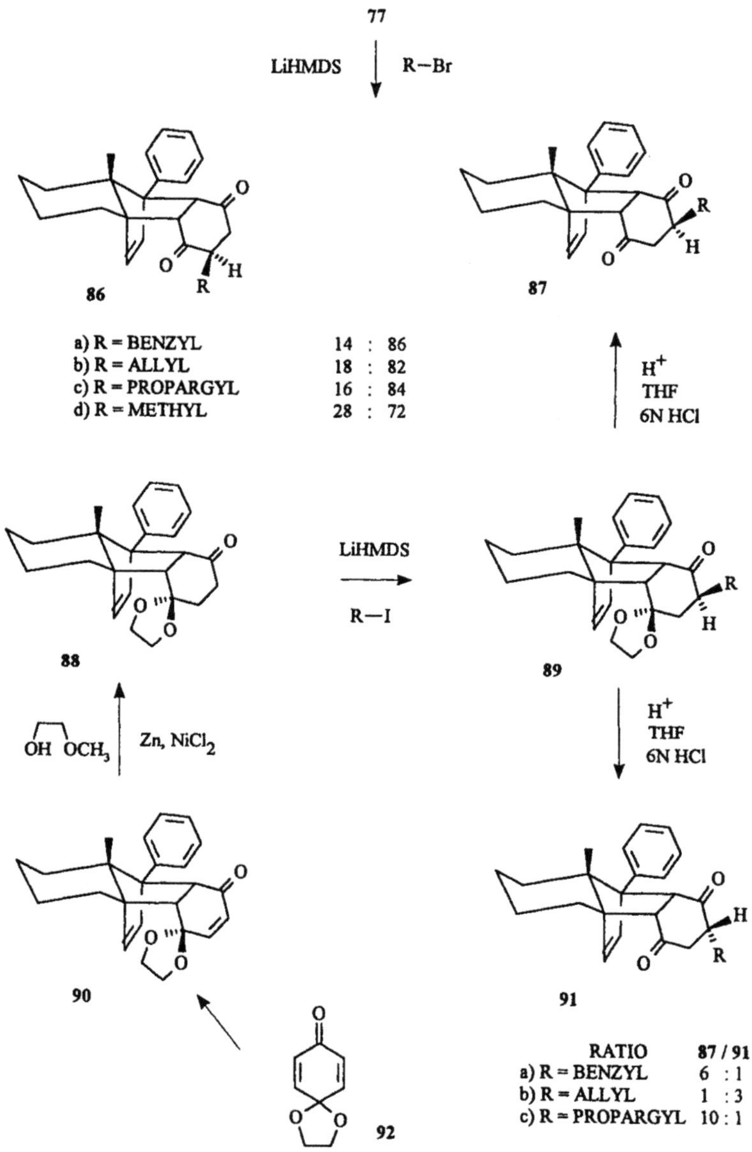
77
LiHMDS
R—Br
86
87
a) R = BENZYL
b) R = ALLYL
c) R = PROPARGYL
d) R = METHYL
14 : 86
18 : 82
16 : 84
28 : 72
H+
THF
6N HCl
88
LiHMDS
R—I
89
OH OCH3
Zn, NiCl2
H+
THF
6N HCl
90
91
92
RATIO 87 / 91
a) R = BENZYL 6 : 1
b) R = ALLYL 1 : 3
c) R = PROPARGYL 10 : 1

Scheme 17.

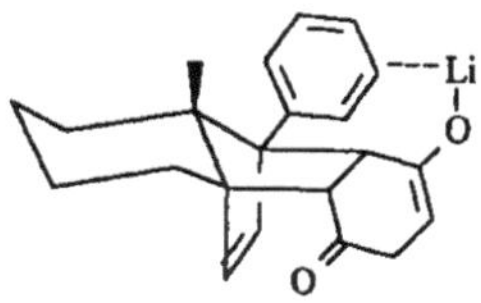

93

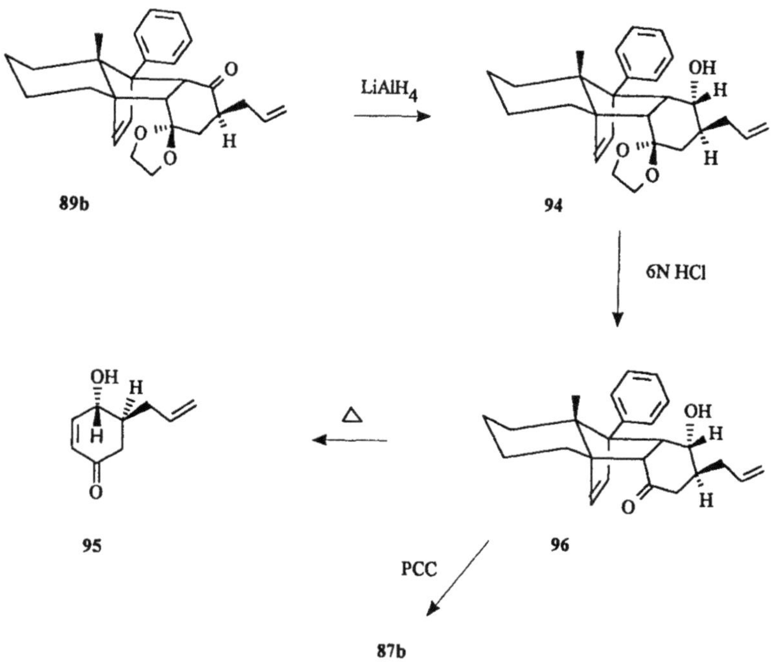

Scheme 18.

demands β-attack of the hydride reagent and since this very side is shielded by the alkyl residue, we were not surprised at all to notice a remarkable resistance against hydride attack. When the ketones of type **89** were treated with lithium aluminum hydride, however, smooth and diastereoselective reduction to the corresponding α-alcohol **94** took place and this material nicely yielded to ketal hydrolysis to form hydroxy ketone **96**[51] without any epimerization, thus opening the road to various optically pure substituted cyclohexenones. Since compound **95** had been prepared and used by Overman[52] we took advantage of this high-yield sequence for the preparation of this particular enantiomer, which proved to be identical in every detail to the sample kindly provided by Professor Overman.

Because epimerization on ketal hydrolysis of **89** had thwarted the stereoselective preparation of the β-alkylation products **87**, we investigated the PCC oxidation of hydroxy ketone **96** and were pleased to note the formation of a single β-alkyl diketone, e.g. **87b**.

Although the quinone–monoketal used in these final experiments is actually a nonsymmetric dienophile, the cycloadditions exercised with this cyclohexadienone proved to be absolutely regioselective and since this regioselectivity was quite in line with prediction, we decided to extend our investigations into other nonsymmetric dienophiles to check the generality of this result.

V. NONSYMMETRIC DIENOPHILES

Propargylic aldehyde, which already had served very well with other cycloadditions,[53–55] gave rise to a high yield of adduct **97** on heating in dichloromethane and subsequent reduction with samarium iodide provided aldehyde **98**, which proved to be identical to the adduct obtained from acrolein, thus proving both cycloadditions to be absolutely regioselective again. Of particular interest from the synthetic point of view are cuprate additions to the unsaturated aldehyde **97**, since these (1) all proved to be highly diastereoselective, and (2) as is demonstrated with the phenyl derivative **100**, they led to diastereoselective formation of secondary alcohols (see **101**) on treatment with Grignard reagents, which in a thermal retro-process provided the enantiomerically pure allylic alcohols of type **102** (Scheme 19).[56] Contrary to expectations the zinc dichloride-catalyzed addition of *trans*-croton aldehyde afforded the *cis*-substituted aldehyde **99** which was in every detail identical to the product from dimethylcuprate addition to the unsaturated aldehyde **97**. This result demands that zinc dichloride also catalyzes the *trans* ⇔ *cis* isomerization of crotonic aldehyde and additionally triggers the comparatively fast addition of *cis*-crotonaldehyde to the diene. Since the cycloaddition of *trans*-crotonaldehyde to cyclopentadiene is known to be comparatively slow, the *cis*-disubstituted compound **99** is the only product from this addition.[56]

Very similar results were observed with butynone, pentynone, and hexynone and in the case of butynone the very smooth formation of the 1:2 adduct **104** was noticed with an excess of the ketone, a result that may also be obtained with propargylic aldehyde. While samarium iodide provided the saturated ketone **105** in the usual way, one was a little bit surprised to find that lithium aluminum hydride as well as butyllithium also underwent conjugate addition exclusively to generate **105** and **106** as the sole reaction product (Scheme 20).[56] The conjugate reduction to give rise to *cis*-compound **109** is noticed with the pentynone adduct **107a** too, but with this special ketone one should mention the thermal isomerization (1,5-

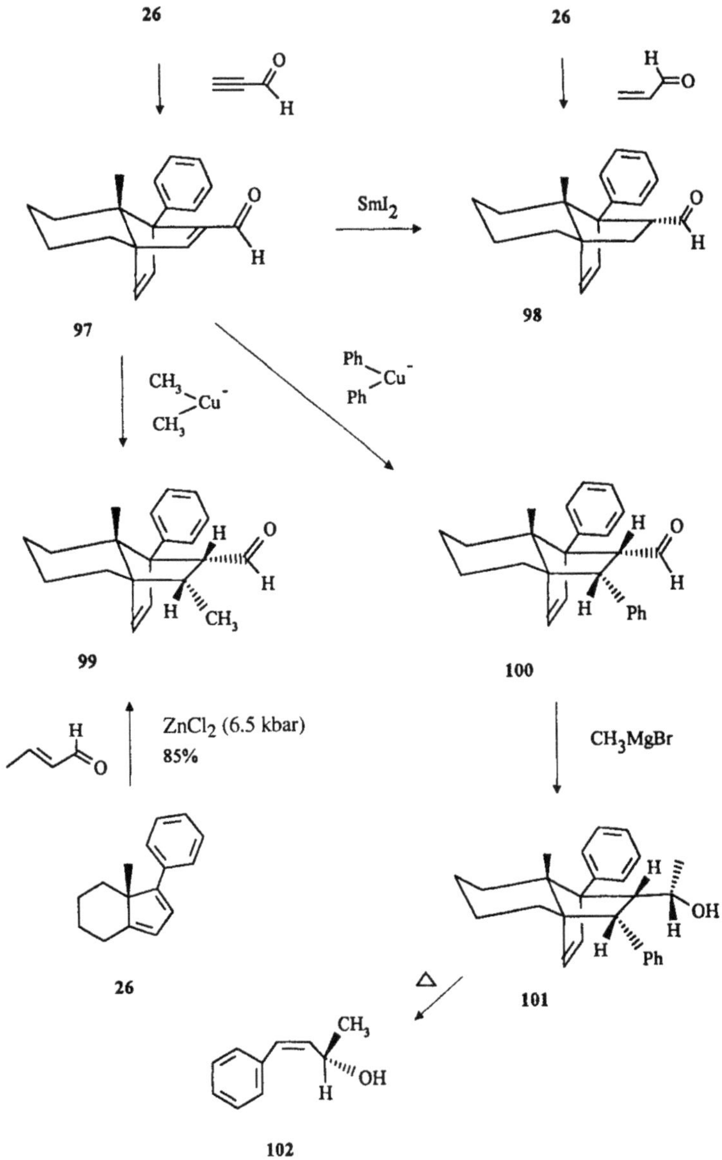
26
26
SmI2
97
98
CH3
Cu-
CH3
Ph
Cu-
Ph
99
100
ZnCl2 (6.5 kbar)
85%
CH3MgBr
26
101
102

Scheme 19.

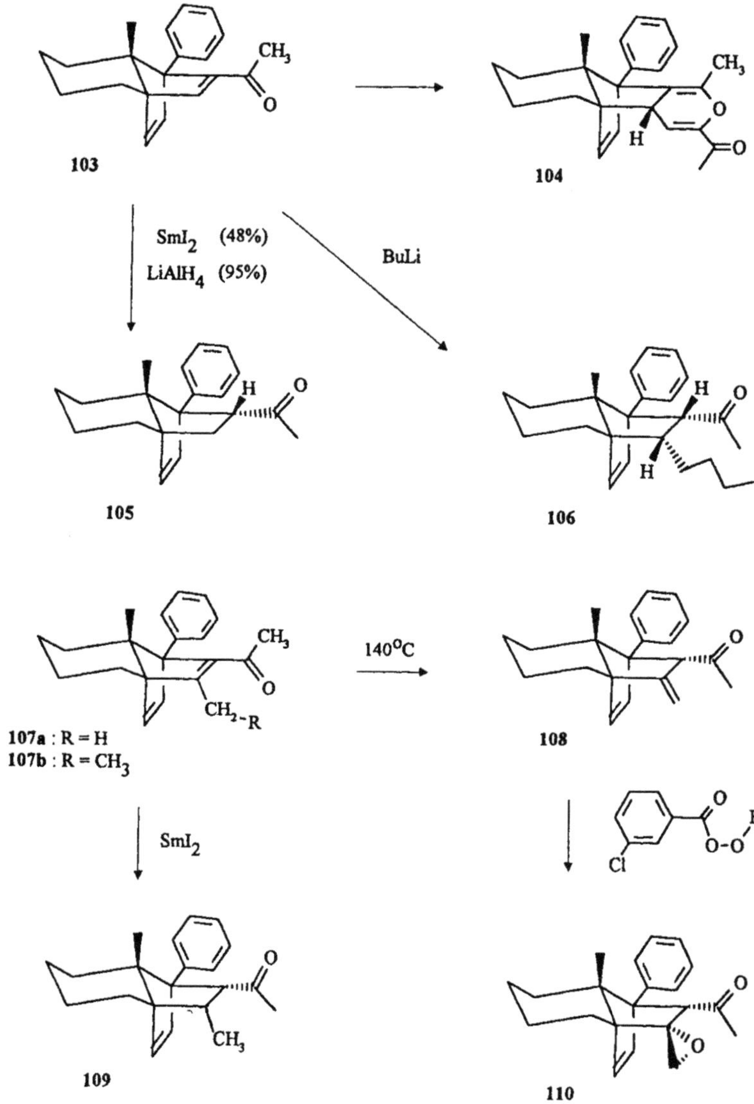
CH3
O
103
CH3
O
H
O
104
SmI2 (48%)
LiAlH4 (95%)
BuLi
H
O
105
H
O
H
106
CH3
O
CH2-R
107a : R = H
107b : R = CH3
140°C
O
108
SmI2
O
H
O-O
Cl
O
CH3
109
O
O
110

Scheme 20.

hydride shift) which gives rise to the deconjugated ketone **108** in quantitative yield. The exocyclic double bond of the latter proved to be quite reactive since epoxidation provided monoepoxide **110** with excellent chemoselectivity as a single stereoisomer.[56] Although many reactions of hexynone adduct **107b** parallel those of **107a**, this thermal isomerization providing an exocyclic double bond does not take place here. The results obtained so far indicate a high degree of regioselectivity for nonsymmetric dienophiles and as adducts formed from five- and six-membered rings additionally underwent a number of absolutely diastereoselective transformations we felt encouraged to investigate the Diels–Alder

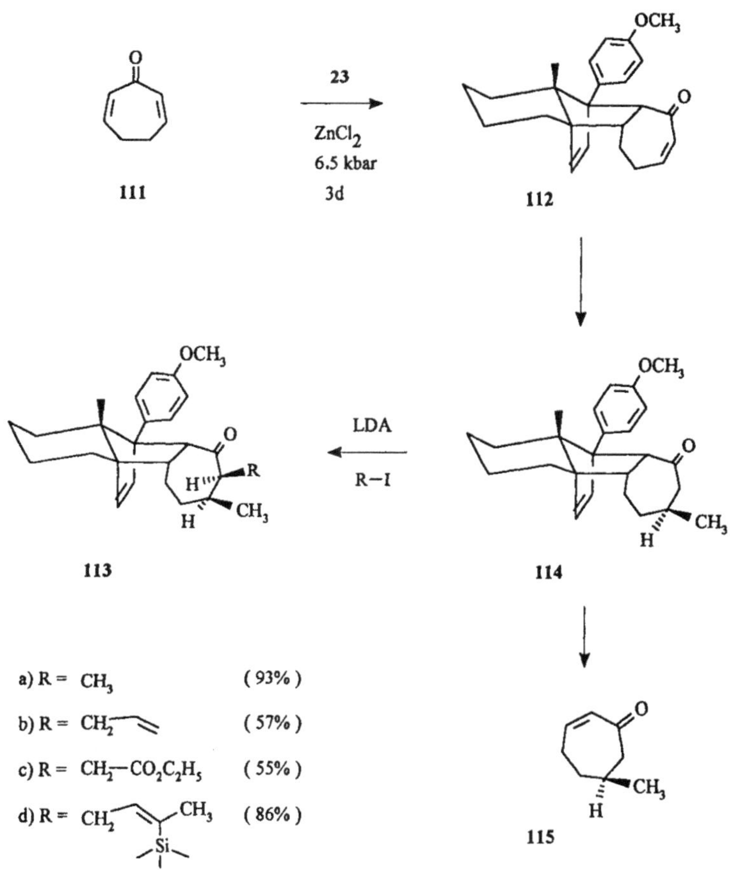

Scheme 21.

chemistry of cycloheptadienone **111** as well. This dienophile was selected for various reasons. First, one should expect regioselectivity for the addition process. Second, since the adduct remains an unsaturated ketone one would be in the position to study addition reactions to the electron-poor double bond in adduct **112** (see Scheme 21).

Since the seven-membered ketone is characterized by higher conformational mobility the diastereoselectivity of these transformations would be an important issue. Finally there is quite a number of natural products containing either a seven-membered ring or an eight-membered ring lactone, which could be synthetically connected to the corresponding cycloheptanones via the Baeyer–Villiger oxidation.

Although the cycloaddition with the cross-conjugated ketone **111** had to be catalyzed with $ZnCl_2$, a nearly quantitative yield of adduct **112** was obtained at 6.5 kbars and room temperature and the subsequent addition of dimethylcuprate in the presence of trimethylchlorosilane proceeded with very high yield as well as excellent diastereoselectivity to provide the β-substituted ketone **114** exclusively

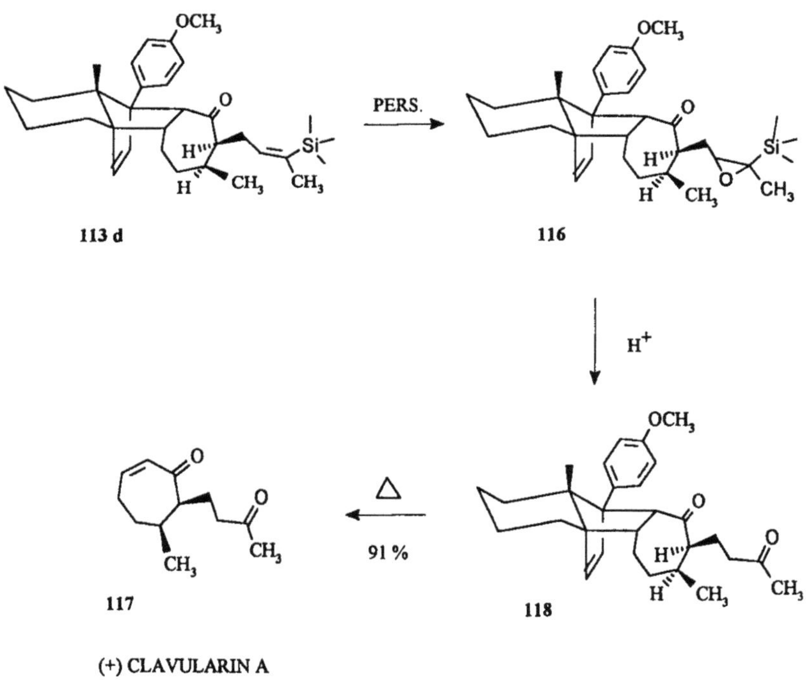

Scheme 22.

after treatment with tetrabutylammonium fluoride.[57] This result proves very clearly that conjugate additions in the cycloheptenone series does take place with the same perfect diastereoselectivity as was observed for cyclopentenones and cyclohexenones. The same turned out to be true for enolate alkylations as can be judged from the *cis*-alkylation products **113a–d** which are formed as single diastereomers.

The monosubstituted intermediate **114** on heating gave rise to the enantiopure cycloheptenone **115**, which according to reports in the literature may serve as a key intermediate for an enantioselective total synthesis of (–)-octalactin A:[58] Epoxidation and subsequent rearrangement of vinylsilane **113d**, which resulted from enolate alkylation, generated methylketone **118** which also underwent the thermal retro-process and very efficiently (91% yield) provided enantiopure (+)-clavularin A (see Scheme 22).[57]

VI. KINETIC RESOLUTION

For all the cycloadditions described so far one notices absolute face-selectivity accompanied by *endo*-selectivity and, as demonstrated by the just described experiments, also very reliable regioselectivity. This nicely fulfills all the requirements that were demanded in the introduction for an optical-pure diene eligible for chiral discrimination (see Scheme 1). Taking all this for granted, a transition state like the one portrayed in Scheme 23 should govern cycloaddition reactions with cyclic chiral dienophiles, e.g. substituted butenolides (see **119**), and one can safely draw the conclusion that only the (*S*)-enantiomer will be selected for the Diels–Alder process, while the (*R*)-enantiomer will not be accepted. This was shown to be true for butenolides as well as cyclohexenones (**122/123**) and cyclopentenones (**124/125**). As excellent discrimination was routinely noted (e.g. 99.6% ee for the butenolides),[59] we decided to use this type of kinetic resolution for the preparation of enantiopure intermediates for natural product synthesis.

Cyclopentenone adduct **124**, for instance, looked like an ideal candidate for an enantioselective synthesis of methyl cucurbate **132**, as well as the corresponding jasmonate **131b** and its epi-compound **131a**.[60] Owing to the cage-like structure of ketone **124**, enolate alkylation was expected to show very high β-selectivity and this was nicely born out by experiments which yielded the *cis*-disubstituted ketone **127** exclusively on alkylation with *cis*-pentenyl bromide. Selective Pd-catalyzed allyl–ester splitting followed by decarboxylation gave rise to ketoester **129**, which according to expectations underwent a very simple borohydride reduction again with exclusive β-attack of the hydride reagent to provide the single alcohol **128**. Thermal retro-Diels–Alder splitting was achieved in quantitative

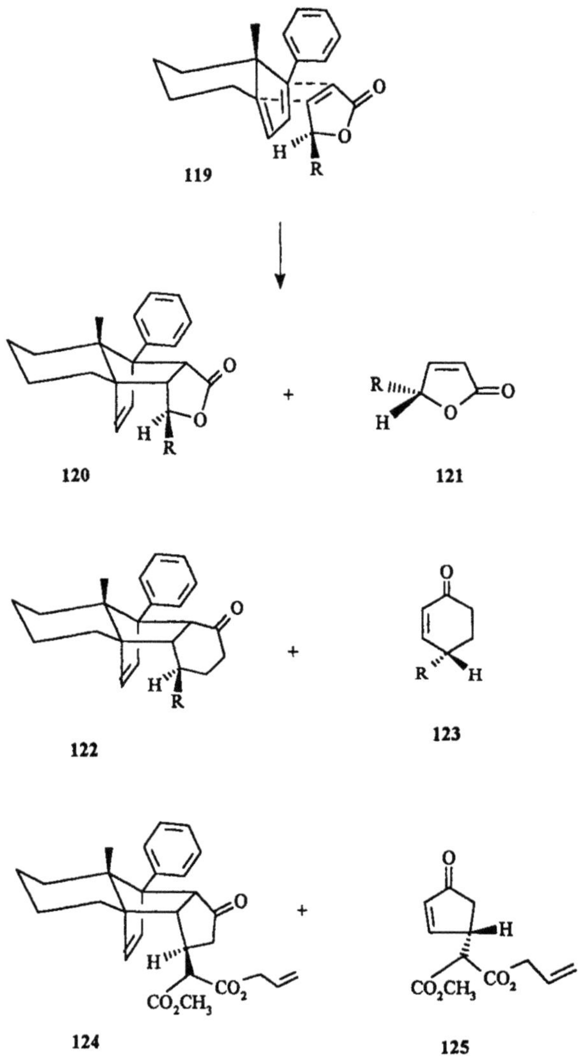

119
H
R
O
O
120
H
R
O
O
R
H
O
O
121
122
H
R
O
R
H
O
123
124
H
O
CO2
CO2CH3
125
O
H
CO2
CO2CH3

Scheme 23.

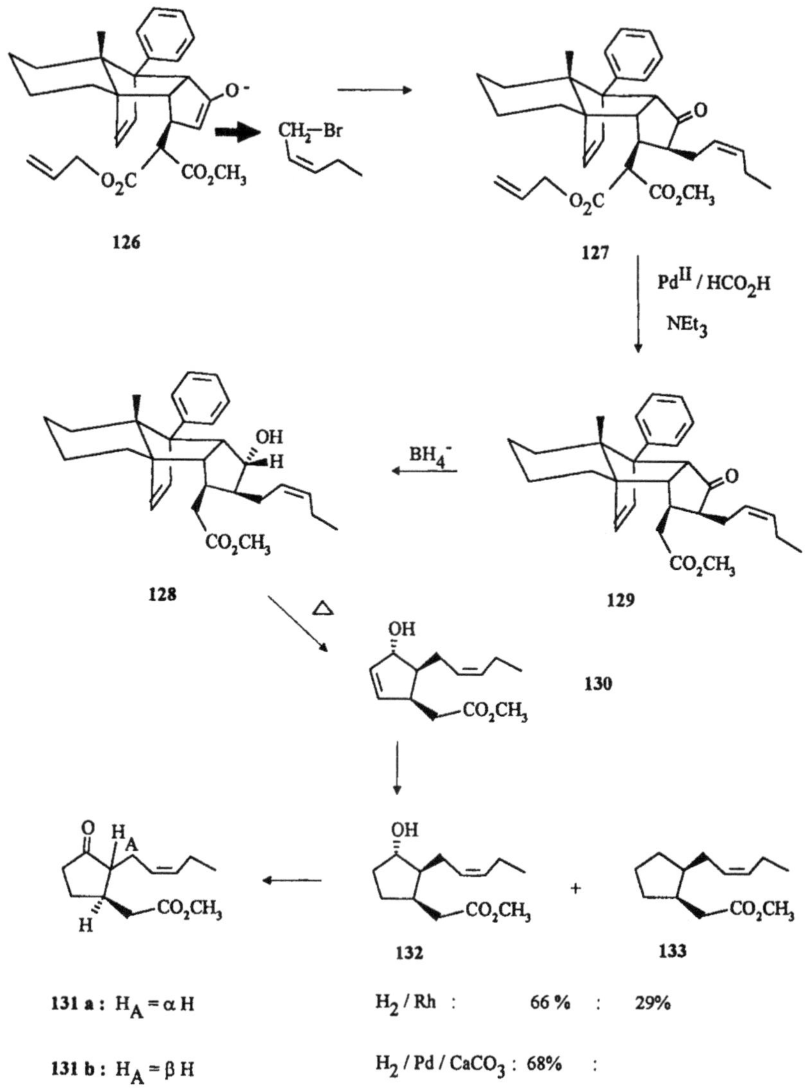
O⁻
CH2–Br
O_2C
CO_2CH_3
126
O
127
Pd^{II} / HCO_2H
NEt_3
OH
H
BH_4^-
128
129
△
OH
130
O
H_A
H
OH
+
132
133
131 a : H_A = α H
131 b : H_A = β H
H_2 / Rh : 66 % : 29%
H_2 / Pd / $CaCO_3$: 68% :

Scheme 24.

yield and gave rise to the enantiomerically pure allylic alcohol **130**, which has to be considered a very general chiral building block for various compounds in this series.

To prepare methyl cucurbate one has to solve a problem which is of rather general importance in connection with this type of synthesis. In order to arrive at this set of configurations and the corresponding substitution pattern one has to alkylate and reduce the ketone prior to the retro-process; hence, the product of thermolysis is bound to be an allylic alcohol. If, however, the corresponding cyclopentanol with an unsaturated side chain is needed, as is the case with methyl cucurbate (see **132**), one has to find a way to reduce or hydrogenate the double bond neighboring the hydroxy group selectively. This has been achieved in various ways which very generally took advantage of the hydroxy group as directing complexation of soluble catalysts or for orienting hydrogen bond interactions with polar surfaces in heterogenous catalysis.[61,62] In our hands complexation of rhodium catalysts, which needs basic conditions for the formation of alcoholates, met with failure, as alcohol **130** decomposed on addition of the deprotonating species. When we tried to run the reaction in the absence of base, the desired cucurbate **132** was indeed obtained, but since it was accompanied by the deoxygenated compound **133** in a 2:1 ratio, we considered this unsatisfactory and shifted to heterogenous catalysis employing palladium on calcium carbonate in the hope that the hydroxy group would strongly interact with the basic carrier surface. This turned out to be the case and particularly if the hydrogenation was run under dilute conditions, a good yield of methyl cucurbate **132** was obtained. To obtain jasmonate and epi-jasmonate a mild and selective oxidation is needed and since Knochel[63] recently proved the Dess–Martin reagent to give rise to epi-jasmoante without epimerization, we focused on the oxidation (PCC/NaOAc) epimerization sequence that provided a 95/5 ratio of methyl jamonate **131b** and its epi-compound **131a** in very high yield.

Having been successful with this comparatively simple cyclopentenone we shifted to the more complicated ketone **138** which in the racemic series had proven to be a useful intermediate for the preparation of the cyclic acetal **145** (see Scheme 26). This compound represents an important substructure of the azadirachtins and at least part of the biological activity of this very potent class of natural products is linked to this tricyclic system. The preparation of our first cyclopentenone target **138** does not pose any problems if one starts from racemic 4-*tert*-butoxycyclopentenone **134** which undergoes a very clean Michael addition with the anion of lactone **135** to provide the addition product **136** (Scheme 25). Treatment with peracids transformed this thioether into the corresponding sulfoxide and subsequent heating in dichloromethane gave rise to butenolide **137** which in a "flash-hydroxylation" afforded the corresponding *cis*-diol **139** in 93% yield.

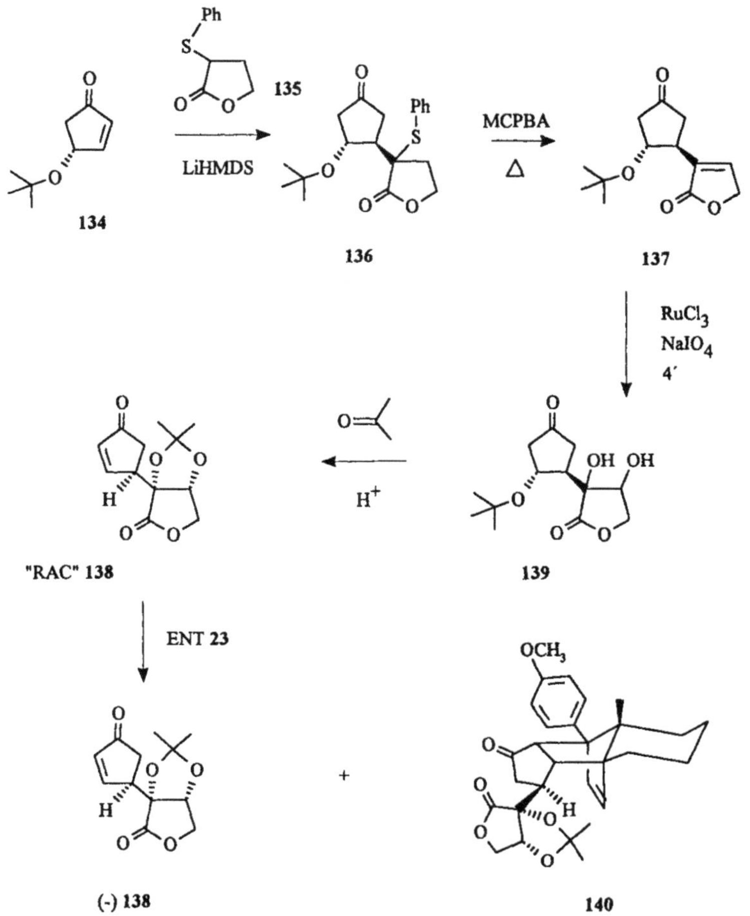

Scheme 25.

Acid catalyzed formation of the acetonide went along with an elimination process and gave rise to cyclopentenone **138**. Further treatment with the chiral cyclopentadiene **23** at 6.5 kbar resulted in the clear cut formation of Diels–Alder adduct **140**, while the pure enantiomer (–)**138** was left behind (Scheme 25).[64,65]

Enantiomer (–)**138** served very well for the preparation of our target-molecule **145** (Scheme 26). To protect the cyclopentenone double bond and discourage allylic rearrangements, thiophenol was added again. Directed by the lactone moiety the DIBAH reduction of the keto group in the five-membered ring occurred

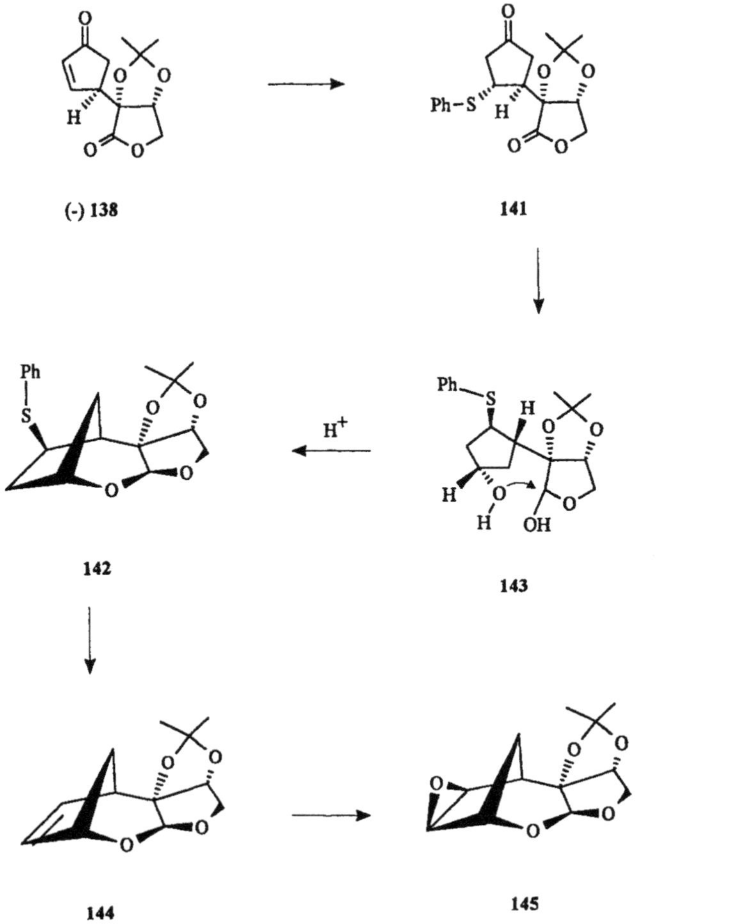

Scheme 26.

with high diastereoselectivity, and since the lactone itself was reduced to the lactol epimers (see **143**) one was not surprised to observe acetal formation on subsequent acid treatment (see **142**). After sulfoxide formation the cyclopentene double bond was reinstated in a thermal process and uneventfully transformed into the epoxide **145**.

In the final example describing the preparation of an intermediate for an enantioselective total synthesis by kinetic resolution, we were again dealing with a cyclopentenone (see **146**), but this time there were two new and important aspects to be considered. On the one hand the stereogenic center this time is not located next to the double bond, as in the two cases discussed above, but next to the carbonyl group, while on the other hand the compound does represent a spiroketal which means that in the transition state (see **147**), whatever the face-selectivity will be, one oxygen atom has to be placed in the "inside"-position. The question arose: Could this block the whole cycloaddition process? The answer was: No! We noticed, however, that the Diels–Alder reaction could not be run under the "normal" high-pressure conditions of 6–7 kbar, but needed 7 days at 12 kbar to go to completion. This indicates a lower reaction rate, but the cycloadduct isolated in nearly 50% yield proved to be a 96:4 mixture of the two diastereomers which could easily be separated by chromatography. The unreacted enantiomer (*S*)-**146** was already very pure (>98% ee) and the adduct **148** provided optically pure (*R*)-**146** on pyrolysis (Scheme 27). This proves the structure of the cycloadduct **148** to be as indicated, which is in line with the NMR data obtained from this compound and demonstrates very clearly that again only the one enantiomer that places the small hydrogen atom into the "inside" position (see **148**) is accepted for the cycloaddition process. The remaining enantiomer (*S*)-**146** was picked for an enantioselective total synthesis of the nonnatural antipode (**151b**) of the didemnenones.[66]

Highly diastereoselective nucleophilic attack at the carbonyl group followed by aldol addition with acrolein afforded hydroxylactone **149**, which after mesylation and base-catalyzed elimination, under the influence of base was focused into the one diene–isomer **150** as the thermodynamically stable stereoisomer. DIBAH reduction followed by acid treatment then gave rise to the protected didemnenone **151a** which was deprotected with "CAN" to generate the nonnatural configuration of didemnenone glycoside **151b**. It should be mentioned at this stage that this series showed biological activity very similar to the data reported for the natural enantiomer, indicating a low dependence of these properties on the absolute configuration.

The most important detail to remember at this stage is the fact that obviously the oxygen atom of the ketal group may be placed in the so-called "inside position" (see **148**). This led to lower reaction rates, but as the cycloaddition process remained possible in principle, we addressed *rac*-4-*tert*-butoxycyclopentenone (**152**) to investigate the possibilities with this dienophile for kinetic resolution (Scheme 28).[67] According to the expectations based on the results from Scheme 27, the difference between the reaction rates turned out to be very small. If the cycloaddition was run at 9 kbar for 3 weeks the two cycloadducts **153** and **154** were finally formed in a 48:41 ratio, respectively. This nicely confirms the suspicion that arose from the formation of **148** and proves very clearly that it is much easier to place an oxygen atom in the "inside position" than an alkyl residue.[43]

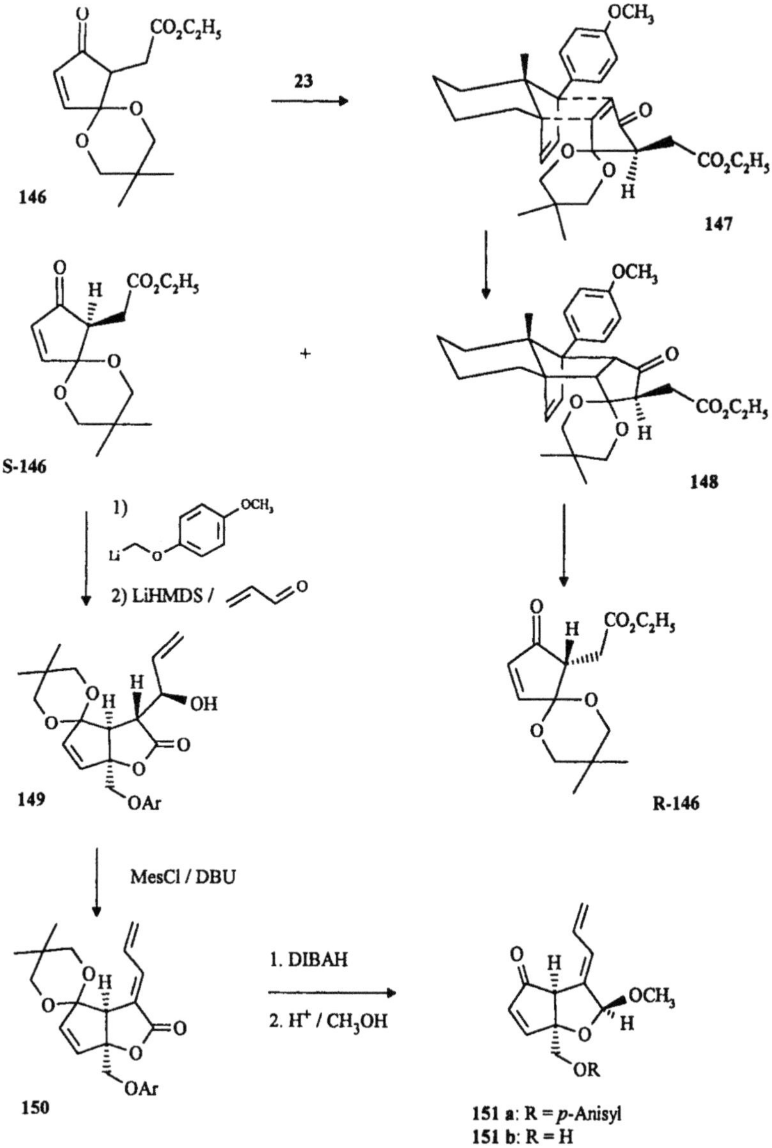

146
23
147
S-146
+
148
1)
2) LiHMDS /
R-146
149
MesCl / DBU
150
1. DIBAH
2. H+ / CH3OH
151 a: R = p-Anisyl
151 b: R = H

Scheme 27.

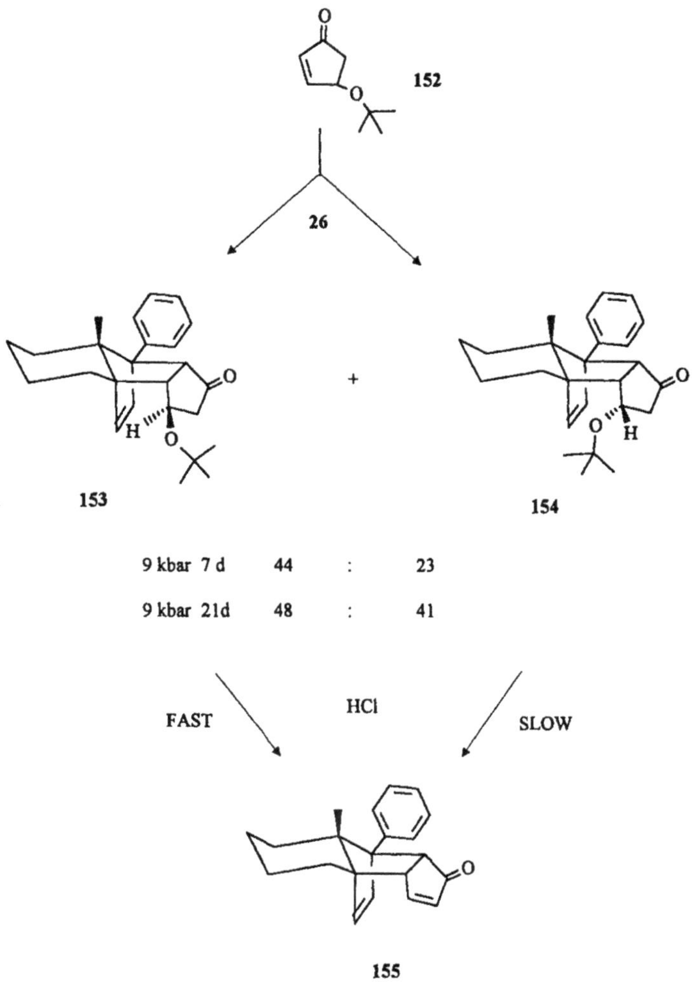

Scheme 28.

VII. DIFFERENTIATION OF ENANTIOTOPIC GROUPS

The results described above triggered a series of experiments aiming at the differentiation of enantiotopic groups. The general idea is most easily explained with a simple example of the cyclohexadienone **156** (Scheme 29). Since this lactone is a prochiral compound, face-selective attack on one of the two enantiotopic double

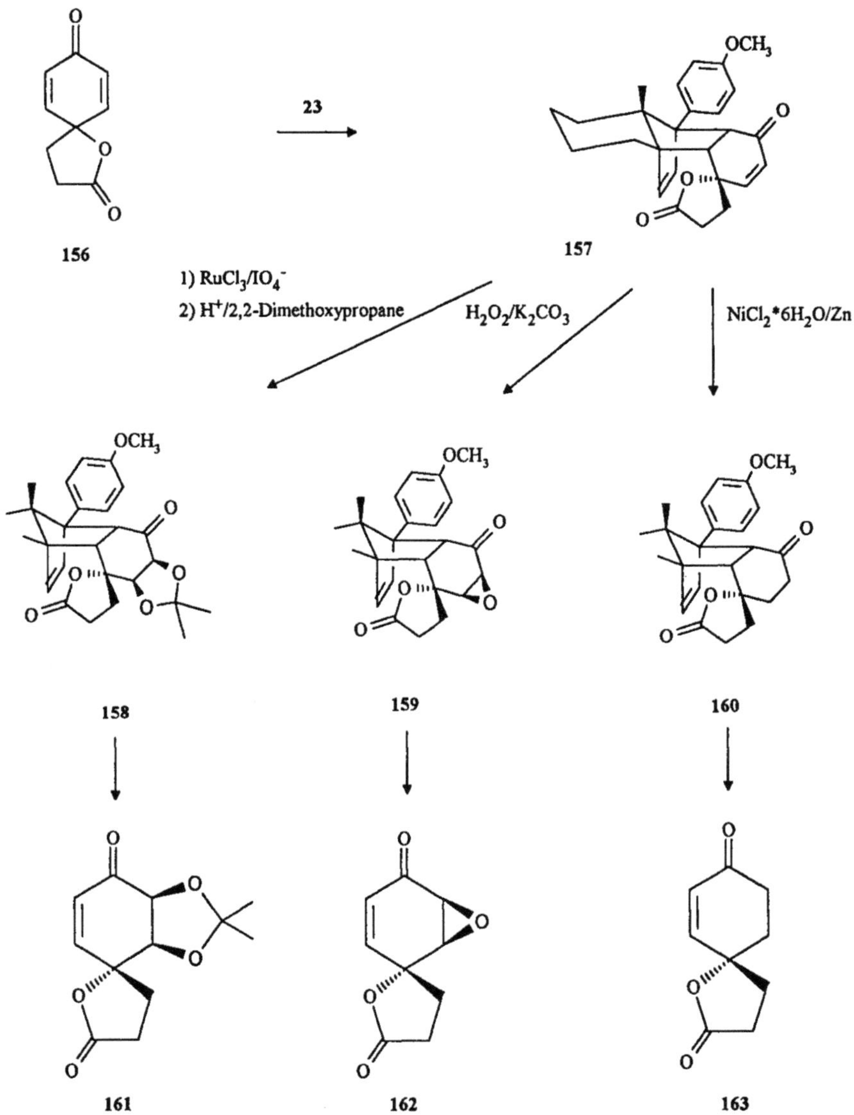
156
23
157
1) $RuCl_3/IO_4^-$
2) H^+/2,2-Dimethoxypropane
H_2O_2/K_2CO_3
$NiCl_2 \cdot 6H_2O/Zn$
158
159
160
161
162
163

Scheme 29.

bonds should lead to enantiomerically pure cycloadducts. In this particular case and in contrast to kinetic resolution the whole prochiral material is converted into one single enantiomer exclusively. On the other hand since both antipodes of diene **23** are easily available one can approach both enantiomers of **157** in a predictable way. To make sure that there would be no repulsive interactions between the π-cloud of the double bond and the free electrons on the oxygen atoms we chose lactone **156** first with the hope that the carbonyl group should reduce the electron density on this oxygen atom, and were pleased to note that one single cycloadduct was indeed formed in excellent yield under high-pressure conditions. The assumed structure **157** could not be proven beyond any doubt by NMR spectroscopy and so we retreated to X-ray structure determination to prove this particular configuration.

The various functional groups present in cycloadduct **157** invite a broad range of structural variations prior to the thermolytic generation of the configurationally well-defined cyclohexenones. At the beginning we addressed the double bond exclusively which was transformed into the *cis*-diol by flash hydroxylation, and this was followed by the easy formation of the corresponding acetonide (see **158**). Epoxidation and chemical reduction produced **159** and **160** respectively. On heating, these three adducts generated the highly functionalized cyclohexenones **161–163** in high yield as pure enantiomers (Scheme 29).

As may be expected the *cis*-hydroxylation–ketalization sequence (see **164**) could also be exercised with α-hydroxylactone **165** which was easily obtained in a Luche reduction (Scheme 30). Again the retroreaction provided an enantiomerically pure spirolactone (see **167** >98% ee), but in this case, with the carbonyl group lacking the yield was somewhat lower (68%). Although the epoxidation of hydroxylactone **165** could also be achieved in 69% yield, this adduct (**168**) turned out to be extremely sensitive to acids and this may be the reason that all our attempts to do the retro-reaction met with failure. On heating, the hydroxy epoxide decomposed completely. Since traces of acid may be responsible for this outcome, the formation of **169** may still become possible from highly purified **168**. From our experiments aiming at Mitsunobu inversions of the α-hydroxy group employing phthalimide as the nucleophile, we secured a 68% yield of the Gabriel intermediate **166**, which from the constitutional as well as from the configurational point of view widens the synthetic potential of spirolactone **157**.

Since later experiments proved spirocyclic cyclohexadienones to be highly superior to the corresponding ring open derivatives in the cycloaddition reaction (see below), we took the opportunity to ring open the spirolactone moiety at a later stage, thus circumventing the necessity to add ring open cyclohexadienones to the 4π-system **23**. This turned out to be much less straightforward than expected, but we were finally successful with the lithium hydroxide technique

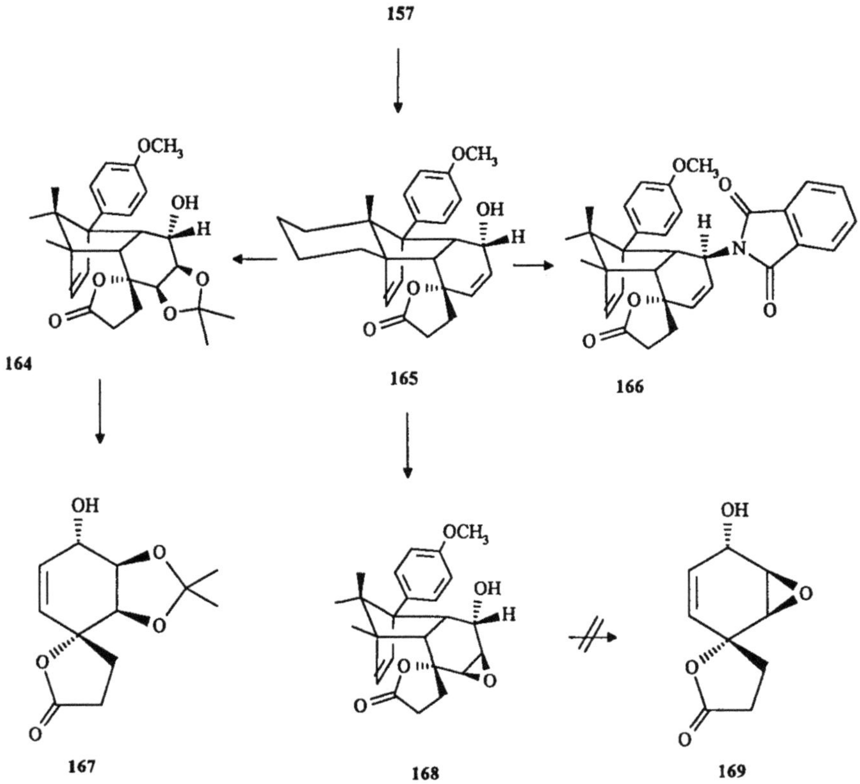

Scheme 30.

which provided hydroxyesters **170** and **171** from lactones **157** and **160** leading, however, to a much higher yield with the dihydro derivative **171** (Scheme 31). Additionally lactol **173** which results from a standard DIBAH reduction yields nicely to a Wittig olefination to give rise to the ring opened diol **172**.[68]

Because spirolactone **174** is easily accessible from tyrosin, we checked this dienone and were not surprised to get a single cycloadduct **175** (see Scheme 32). According to expectations the additional substituent did not only lead to lower reaction rates but did also shield the remaining double bond so effienctly that neither peracids nor dimethyl dioxirane gave rise to an epoxide. Only the well-established reduction techniques did work and afforded the corresponding dihydro compound. In order to gain information on the ring size–cycloaddition relationship, we examined δ-lactone **176** as the next dienophile, but realized very

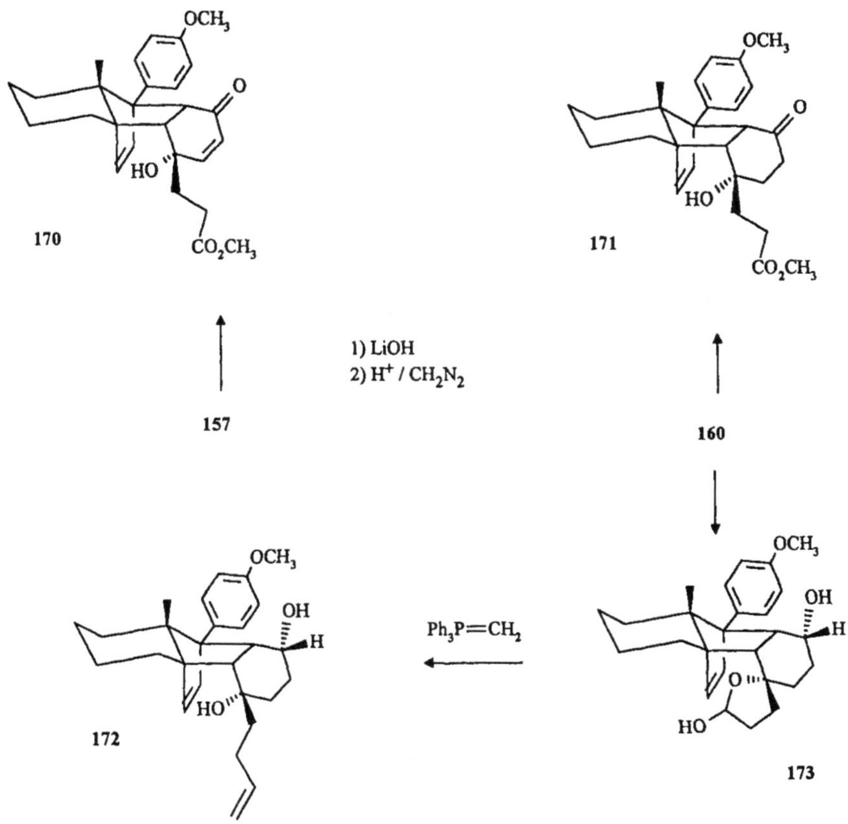

Scheme 31.

quickly that the diacetoxy–iodobenzene reagent that had served very satisfactorily with the spiro-type γ-lactones provided only very low yields with six-membered lactones. We never exceeded 18% yield with this material. Further improvements are warranted because the cycloaddition reactions with this compound, although they need quite forcing conditions (12 kbar 7d), provided again one single compound in 70% yield. The configuration **177** as given in Scheme 32 was assigned in analogy to the configuration of the corresponding γ-lactone, which had been proven by X-ray data. On the assumption that most of the transformations observed for **157** should comparably operate with **177**, an improved route to the dienophile is highly desirable.[69]

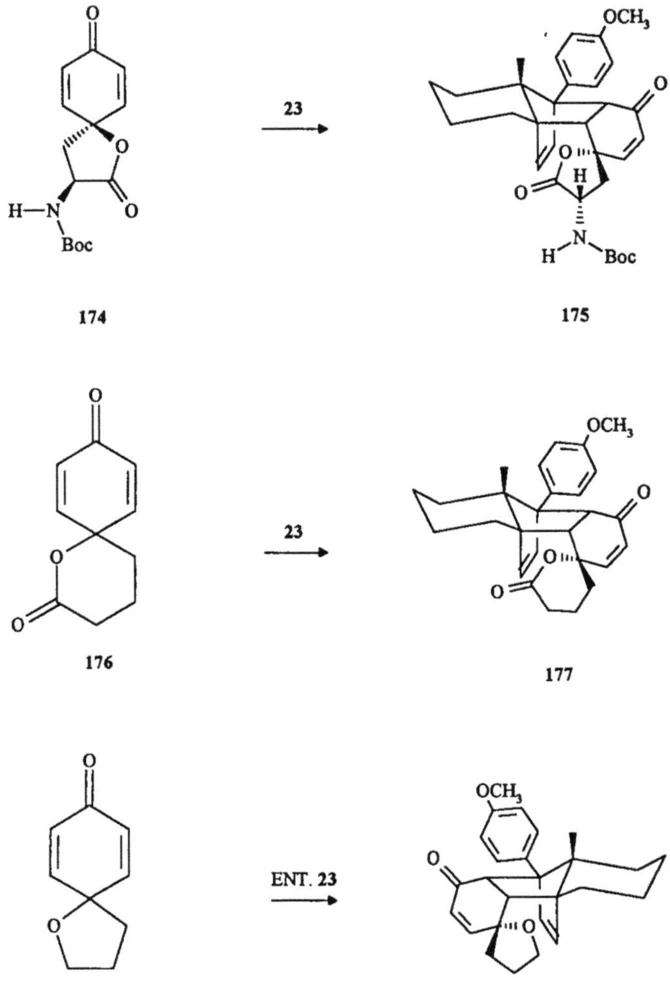

Scheme 32.

Having been successful with different spirolactones we next turned to spiroether **178** and were pleased to note smooth and highly selective cycloadditions in this case, too. It provided a high yield (86%) of adduct **179** (Scheme 32) and again the absolute configuration was secured in an X-ray structure determination.[67] According to expectations the reduction (see **181**) as well as the hydroxy-

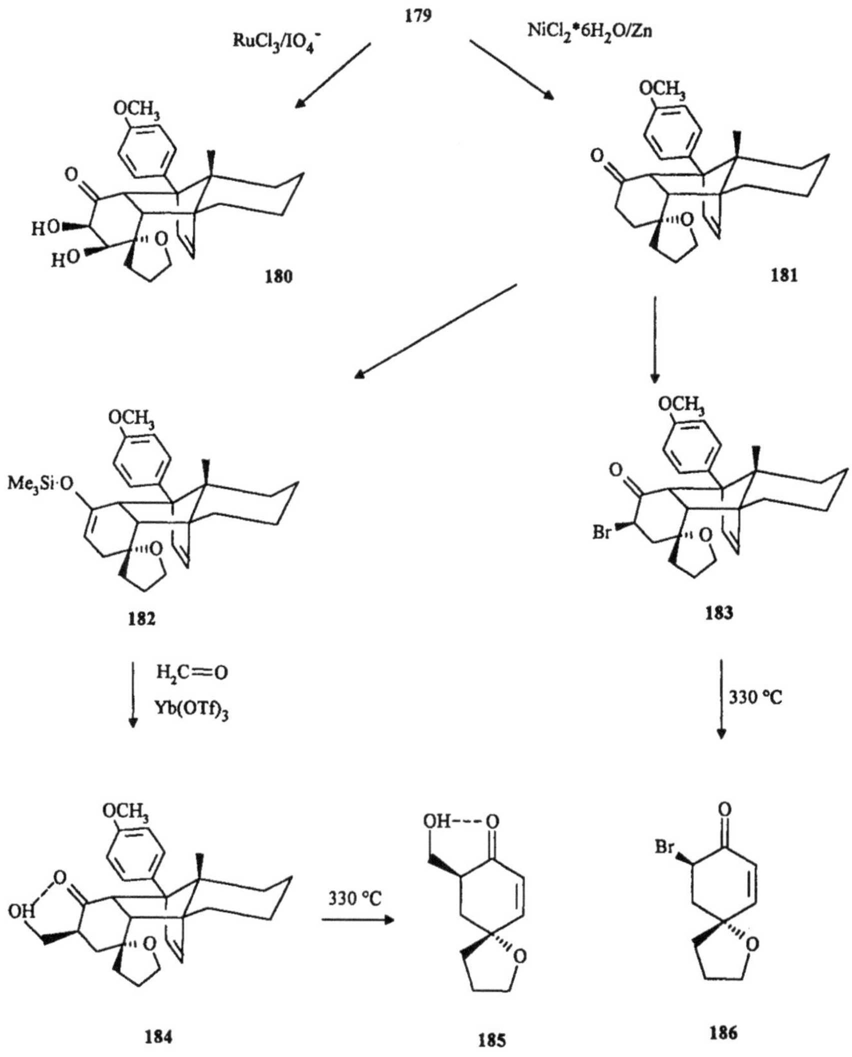

Scheme 33.

lation (see **180**) of the double bond also was easily achieved in this series, and since there is only one carbonyl group available we also studied some typical transformations of this moiety (Scheme 33). A highly diastereoselective α-bromination followed by pyrolysis provided the bromoketone **186** via **183** and

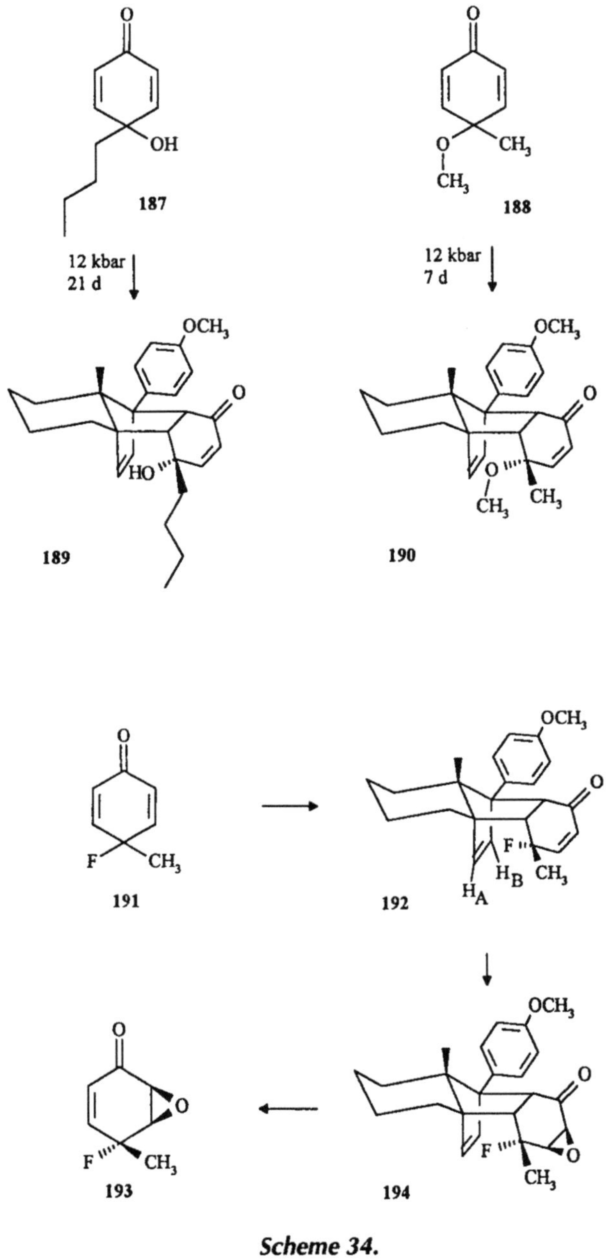

187
12 kbar
21 d
189
OH
OCH3
HO
188
12 kbar
7 d
190
191
192
193
194
F
CH3
HA
HB

Scheme 34.

treatment of the trimethylsilyl enol ether **182** with formaldehyde in the presence of ytterbium triflate gave rise to a 77% yield of aldol **184**, which on heating underwent the retro-reaction with high efficiency to generate hydroxyketone **185** as a pure enantiomer.[69]

As already mentioned above, the ring open compounds **187** and **188** were checked, too, but although they did form the Diels–Alder adducts **189** and **190** in principle,[69] the high pressure and the long reaction times needed for these comparatively sluggish transformations (see Scheme 34) made us prefer ring-splitting reactions of the spirocyclic adducts of type **157** (see above). Since we attributed the low reaction rates of **187** and **188** to the higher space demand of the conformationally more mobile substituents, we decided to also investigate cycloadditions to the corresponding fluoro compounds because the carbon–fluorine bond (1.39 Å) compares very nicely to the carbon–oxygen bond (1.43 Å), while the

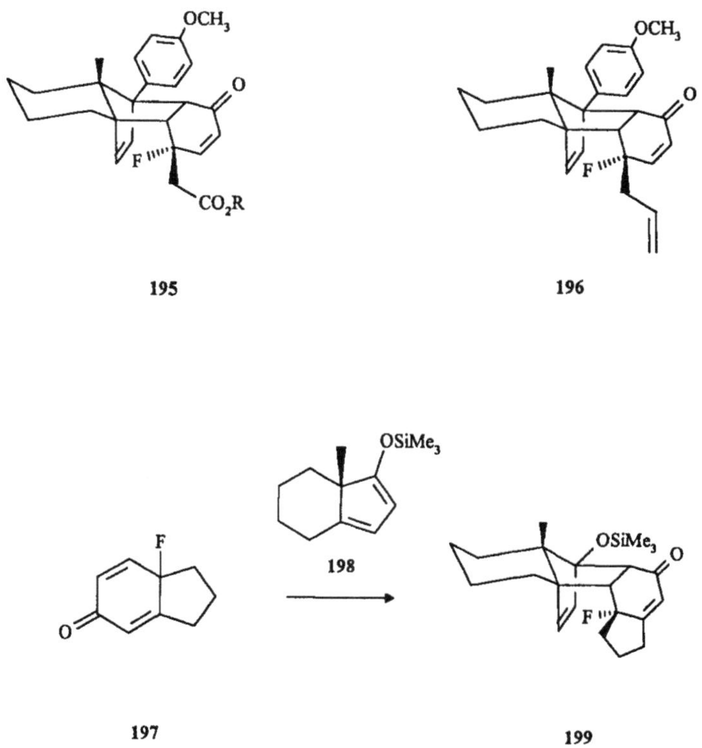

Scheme 35.

comparatively small fluorine van der Walls radius (1.35 Å) comes very close to that of hydrogen (1.20 Å). In essence there should be a comparatively small atom sitting in the position of the oxygen atom in the Diels–Alder transition state. As expected, a smooth cycloaddition was observed with **191** at 6.5 kbars and the absolute configuration of adduct **192** could be read quite convincingly from the very revealing 6 HZ through space-coupling of the fluoro atom with the proton H_A. This phenomen, called "direct coupling," is very common in fluorine compounds. Again, only one single stereoisomer was formed. This outcome is also quite interesting from the preparative point of view since intermediates like epoxide **194** or the corresponding diol on heating provided high yields of enantiopure fluoro compounds of type **193**.[70]

Considering the fact that the direct enantioselective introduction of fluorine atoms is still unsatisfactory, as far as enantiomeric excesses are concerned, this differentiation of enantiotopic groups provides easy access to enantiopure building blocks. In the meantime this face-selective cycloaddition was extended to higher functionalized analogs like **195** and **196** (Scheme 35), and we did also manage to achieve a kinetic resolution of the racemic dienone **197**. For acceptable reaction rates we had to retreat in this case to the very electron-rich cyclopentadiene **198**, which highlights again the importance of electron density in these cycloaddition reactions.[64,71]

VIII. CONFORMATIONALLY MOBILE DIENES

The highly selective additions reported so far had all been exercised with the very rigid diene systems listed in Scheme 5 and therefore we became interested in a conformationally more mobile 4π-system for a comparison to these compounds. Luckily an ideal diene (**200**) is easily available from the well-established butynone adduct **103**. With a few standard dienophiles an uneventful formation of the cycloadducts **201** through **203** was observed and, in the case of imide **203**, hydrolysis of the silylenol ether was shown to occur easily with camphorsulfonic acid (Scheme 36). The resulting ketone **204** yielded to the thermal retro-reaction at 350 °C and provided a 95% yield of the unsaturated ketone **205** without any shift of the double bond.[64]

Of particular interest is again the monoketal adduct **202** since it is formed with excellent α-*exo*-selectivity and regioselectivity according to NMR experiments with these adducts.[72] These results leave no doubt that even this more mobile 4π-system **200** may operate satisfactorily in kinetic resolution experiments; as for them this reliable selectivity was shown to be a prerequisite in the cycloadditions described above. The preferential α-addition is very probably due to the space

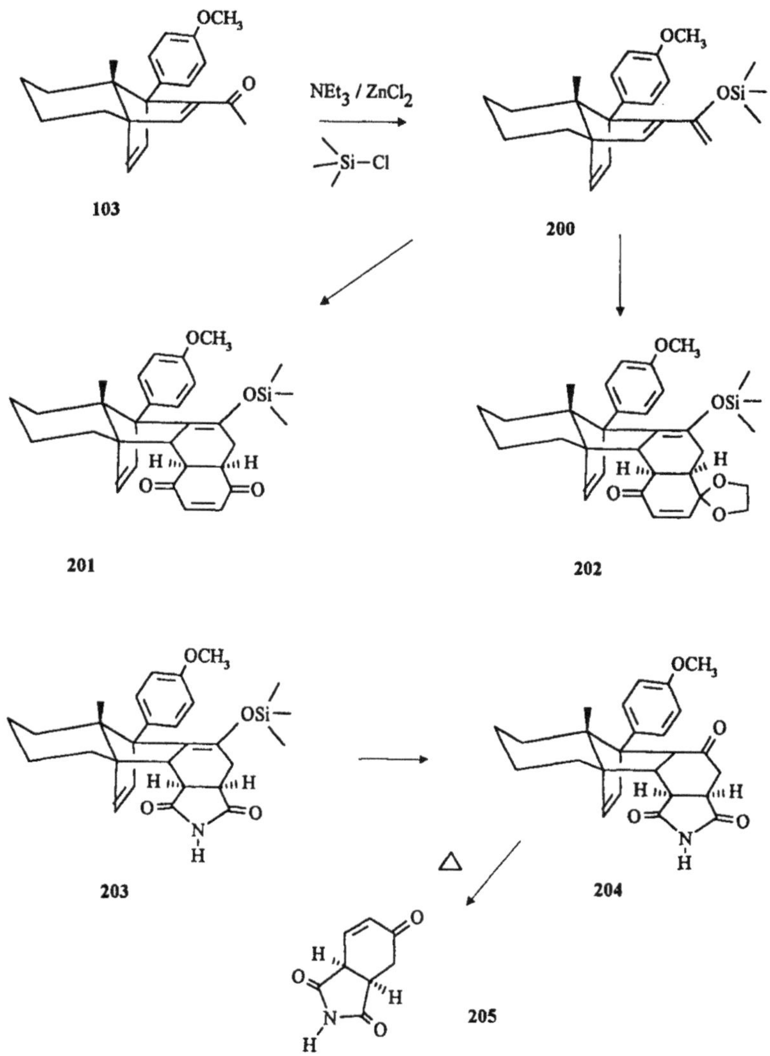

Scheme 36.

demand of the angular methyl group and the phenyl residue on the β-face of the diene system, while the ethylene bridge may be responsible for the *exo* approach of the dienophile. These observations encouraged us to investigate a few dienophiles that had performed very well in kinetic resolution and differentiation of enantiotopic groups with the rigid diene **23**.

For kinetic resolution we choose cyclopentenone **146** and were glad to observe a clean-cut resolution to provide crude adduct **206** in 51% yield, which could easily be purified by column chromatography (Scheme 37). The unreacted (−)**146** could be reisolated in 46% yield and proved to be enantiomerically pure. Since the absolute configuration of (−)**146** was known from our earlier experiments one could also be very sure about the configuration of **206**.[64] As had been demonstrated with imide **203**, the silyl enol ether **206** was hydrolyzed with camphorsulfonic acid and in a subsequent retro-reaction the interesting hydrindane derivative **208** was obtained in 68% yield.

To also investigate a much more simple case, we chose the cyclopentenone **209**, which had given a good kinetic resolution with diene **23** to form adduct **210** exclusively (Scheme 38).[60] However, when **209** was treated with the more mobile

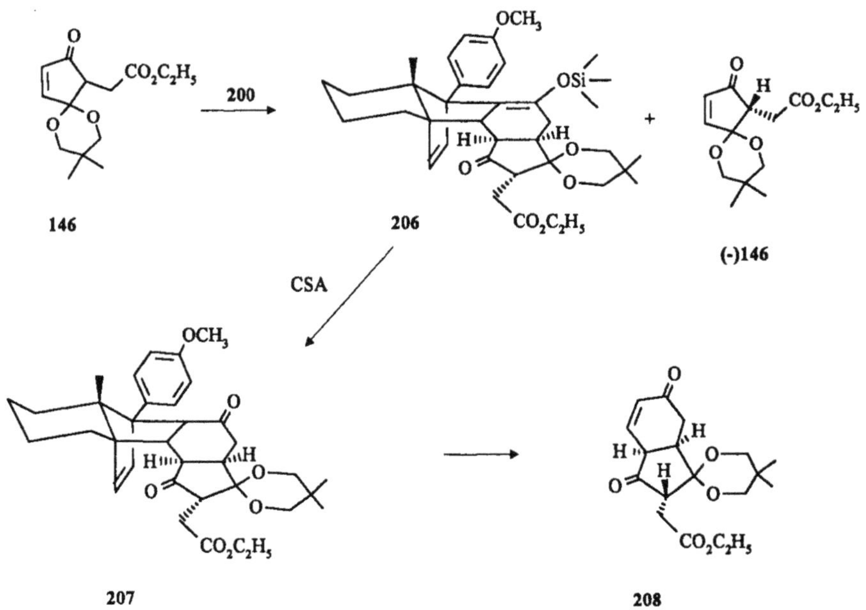

Scheme 37.

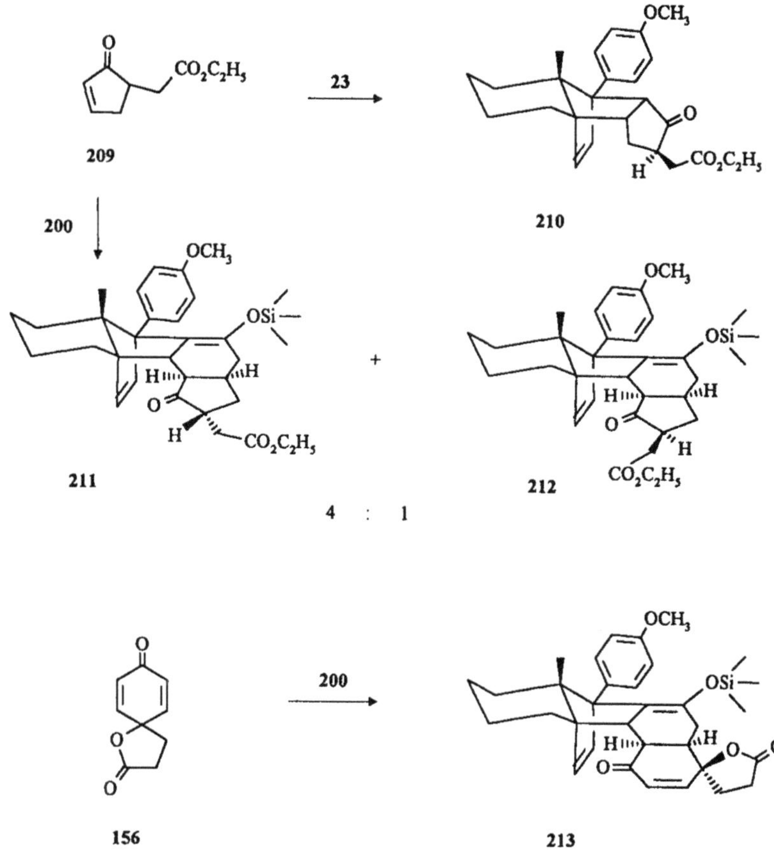

Scheme 38.

diene **200**, this simple ester resulted in a 4:1 mixture of diastereomers. In analogy to adduct **206** the main component was tentatively assigned configuration **211** but additional proof for this assumption is of course mandatory. Anyway this experiment clearly proves that as far as kinetic resolution goes the more flexible diene **200** is certainly inferior to the more rigid cyclopentadienes.[73] This raised the question of differentiation of enantiotopic groups and this was tested with the very reliable spirolactone **156**. Although quite long reaction times (2–3 weeks) were necessary at 6.5 kbars, the cycloadduct **213** formed under these conditions again proved to be one single diastereomer and the configuration given was secured by X-ray data.[73]

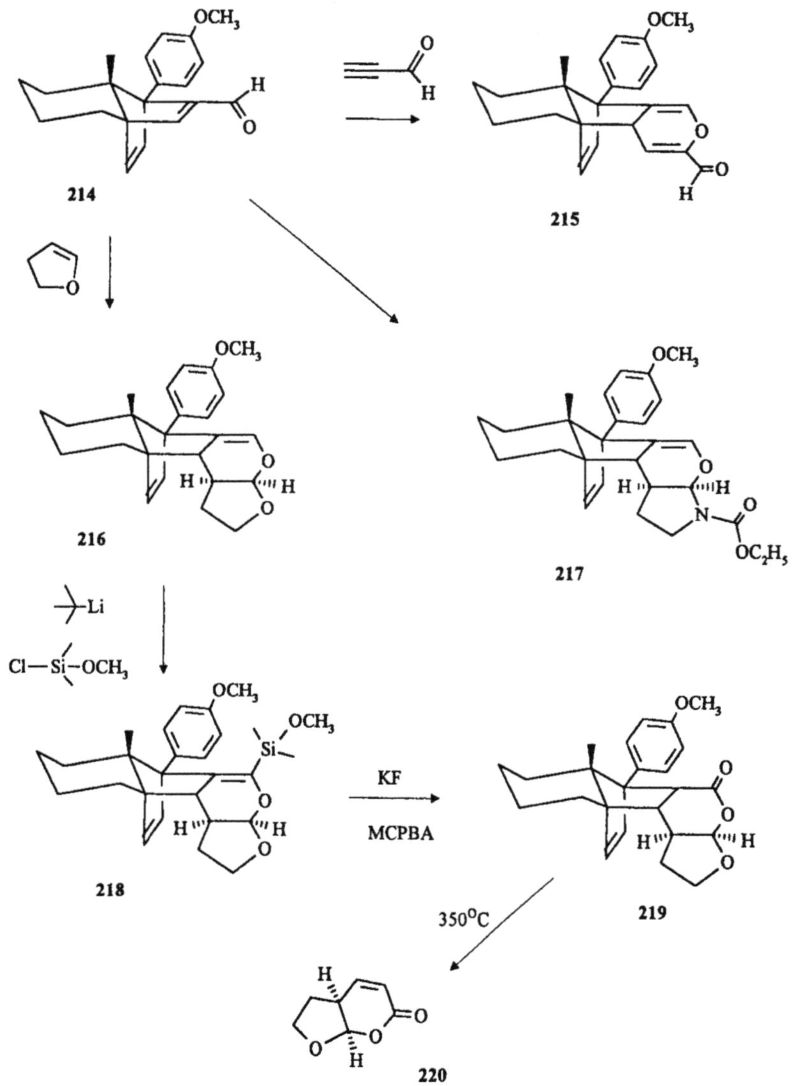
OCH3
H
O
214
O
H
215
O
H
216
OCH3
Li
Cl—Si—OCH3
217
N
OC2H5
Si
KF
MCPBA
218
219
350°C
220

Scheme 39.

To probe at least one electron-poor analog of the diene **200**, we conducted a few cycloadditions with the propargylaldehyde adduct **214** (Scheme 39).[8] Its high reactivity in cycloaddition processes had been noticed in its formation since an excess of the aldehyde easily led to the double addition product **215**.[72] Dealing with an electron-poor 4π-system we employed dihydrofuran and a dihydropyrrole very successfully in a noncatalyzed high-pressure cycloaddition. It should be mentioned at this stage that in the presence of even very mild Lewis acids the formation of by-products and decomposition products could not be avoided and led to very unattractive reaction mixtures. These transformations are therefore very good examples again for the power of the high-pressure technique, since under these conditions adducts **216** and **217** were exclusively formed in 60 and 75% yield, respectively (Scheme 39).[72] Although the enol ether moiety offers in principle various options to generate lactols or lactones, a few of the very obvious approaches, like proton-catalyzed addition of water or alcohols or various electrophiles, met with failure. The method of choice turned out to be a deprotonation–silylation sequence. If methoxy–dimethylsilylchloride was employed in the last step, the subsequent oxidation of **218** with *m*-chloroperbenzoic acid gave rise to an 88% yield of lactone **219**. The thermal retro-reaction of this material provided the unsaturated lactone **220** in high yield (90%) and with excellent optical purity (ee >98%).

IX. SUMMARY

As one can easily judge from the results reported here chiral dienes are very useful tools in many fields of highly selective transformations and in chiral discrimination including the differentiation of enantiotopic groups. This chapter was written with the intention to indicate the general scope and to show how many different enantiopure compounds can be approached with their help. The next goal will be to make also use of the many possibilities of attractive interactions for the recognition and orientation of diene and dienophile.

REFERENCES

1. Wassermann, A. *Diels–Alder Reactions*; Elsevier Publishing: Amsterdam, 1965.
2. Ciganek, E. *Organic Reactions*; New York, 1984, Vol. 32, p. 1.
3. Boger, D.L.; Weinreb, S.N. *Hetero-Diels-Alder Methodology in Organic Synthesis*; Organic Chemistry, Series of Monographs; Academic Press: New York, 1987, Vol. 47.
4. Taber, D.F. *Intramolecular Diels-Alder and Alder Ene Reactions*; Springer: Berlin, 1984.
5. Fringuelli, F.; Taticchi, A. *Dienes in the Diels-Alder Reaction*; J. Wiley and Sons: New York, 1990.

6. Paquette, L.A. In *Asymmetric Synthesis*; Morrison, J.D., Ed.; Academic Press: New York, 1984, Vol. 3, p. 455.
7. Kagan, H.B.; Riant, O. *Chem. Rev.* **1992**, *92*, 1007.
8. Winterfeldt, E. *Chem. Rev.* **1993**, *93*, 827.
9. Adams, H.; Jones, D.N.; Aversa, M.C.; Bonaccorsi, P.; Gianetto, P. *Tetrahedron Lett.* **1993**, *20*, 6481.
10. Grieco, P.A.; Beck, J.P. *Tetrahedron Lett.* **1993**, *20*, 7367.
11. Kerrigan, J.E.; McDougal, P.G.; van Derveer, D. *Tetrahedron Lett.* **1993**, *20*, 8055.
12. Larsen, D.S.; Trotter, N.S.; Stoodley, R.J. *Tetrahedron Lett.* **1993**, *20*, 8151.
13. Krohn, K. *Angew. Chem.* **1993**, *105*, 1651.
14. Aversa, M.C.; Bonaccorsi, P.; Gianetto, P.; Jones, D.N. *Tetrahedron Asymm.* **1994**, *5*, 805.
15. Beaudegnies, R.; Ghosez, L. *Tetrahedron Asymm.* **1994**, *5*, 557.
16. Schlessinger, R.H.; Pettus, T.R.R.; Springer, J.P.; Hogsteen, K. *J. Org. Chem.* **1994**, *59*, 3246.
17. Arce, E.; Carreno, M.C.; Cid, M.B.; Garcia Ruano, J.L. *J. Org. Chem.* **1994**, *59*, 3421.
18. Bloch, R.; Chaptal-Gradoz, N. *J. Org. Chem.* **1994**, *59*, 4162.
19. Schlessinger, R.H.; Wu, X.H.; Pettus, T.R.R. *Synlett* **1995**, 536.
20. Handy, S.T.; Grieco, P.A.; Mineur, C.; Ghosez, L. *Synlett* **1995**, 565.
21. Haltermann, R.L.; Tretyakov, A. *Tetrahedron* **1995**, *51*, 4371.
22. Hoffmann, R.; Mattay, J. *Liebigs Ann. Chem.* **1995**, 1455.
23. Krohn, K.; Khanbabaee, K.; Micheel, J. *Liebigs Ann. Chem.* **1995**, 1529.
24. Winterfeldt, E. In *Stereocontrolled Organic Synthesis*; Trost, B., Ed.; Blackwell Scientific Publications: 1994, 275.
25. Solo, A.J.; Singh, B.; Shefter, E.; Cooper, A. *Steroids* **1968**, *11*, 637.
26. Solo, A.J.; Singh, B.; Kapoor, J.N. *Tetrahedron* **1969**, *25*, 4579.
27. Solo, A.J.; Eng, S.; Singh, B. *J. Org. Chem.* **1972**, *37*, 3542.
28. Atwal, K.S.; Sahoo, S.P.; Tsai, T.Y.R.; Wiesner, K. *Heterocycles* **1982**, *19*, 641.
29. Bull, J.R.; Thomson, R.J. *J. Chem. Soc., Chem. Commun.* **1986**, 451.
30. Bull, J.R.; Steer, M.L.; Dillen, D.L.M. *Tetrahedron* **1990**, *46*, 5401.
31. Bull, J.R.; Thomson, R.J. *J. Chem. Soc., Perkin Trans. 1* **1990**, 241.
32. Bull, J.R.; Bischofberger, K. *J. Chem. Soc., Chem. Commun.* **1989**, 1405.
33. Bull, J.R.; Bischofberger, K. *J. Chem. Soc., Perkin Trans. 1* **1991**, 2859.
34. Bull, J.R.; Thomson, R.J.; Laurent, H.; Schroeder, H.; Wiechert, R.; Schering, P.A. AG, Ger. Offen. 362189P: 5pp. CL: C07J53/00, YR: 86 (20.08.1988).
35. Kirsch, G.; Neef, G.; Laurent, H.; Wiechert, R.; Beier, S.; Elger, W.; Bull, J.R.; Schering, P.A. AG, Ger. Offen. 3838779, P: 16pp. CL: C07J1/00, YR: 88 (11.11.1988).
36. Beckmann, M.; Hildebrandt, H.; Winterfeldt, E. *Tetrahedron Asymm.* **1990**, *1*, 335.
37. Eder, U.; Sauer, G.; Weichert, R. *Angew. Chem.* **1971**, *83*, 492.
38. Hajos, Z.G.; Parrish, D.R. *J. Org. Chem.* **1974**, *39*, 1612.
39. Beckmann, M.; Meyer, T.; Schulz, F.; Winterfeldt, E. *Chem. Ber.* **1994**, *127*, 2505.
40. Dockner, M.; Meyer, T.; Nemes, P.; Otten, M.G.; Winterfeldt, E. *Bull. Soc. Chim. Belg.* **1994**, *103*, 379.
41. Rivière, P.; Mauvais, A.; Winterfeldt, E. *Tetrahedron Asymm.* **1994**, *5*, 1831.
42. Rivière, P. Dissertation, Universität Hannover, 1995.
43. Wegener, B. Dissertation, Universität Hannover, 1994.
44. Meyers, A.I.; Lefker, B.A.; Sowin, T.J.; Westrum, L.J. *J. Org. Chem.* **1989**, *54*, 4243.
45. Davis, F.A.; Vishwakarma, L.C.; Billmers, J.M.; Finn, J. *J. Org. Chem.* **1984**, *49*, 3243.
46. Klunder, A.J.H.; Bos, W.; Verlaak, J.M.M.; Zwanenburg, B. *Tetrahedron Lett.* **1981**, *22*, 4553.
47. Barltrop, J.A.; Day, A.C.; Samuel, C.J. *J. Am. Chem.. Soc.* **1979**, *101*, 7521.

48. Brünjes, R.; Tilstam, U.; Winterfeldt, E. *Chem. Ber.* **1991**, *124*, 1677.
49. Mauvais, A.; Winterfeldt, E. *Tetrahedron* **1993**, *49*, 5817.
50. Hansen, M. Dissertation, Universität Hannover, 1995.
51. Szablikowski, I. Dissertation, Universität Hannover, 1996.
52. Kamenecka, Th.M.; Overman, L. *Tetrahedron Lett.* **1994**, *35*, 4279.
53. Schomburg, D.; Thielmann, M.; Winterfeldt, E. *Tetrahedron Lett.* **1985**, *26*, 1705.
54. Schomburg, D.; Thielmann, M.; Winterfeldt, E. *Tetrahedron Lett.* **1986**, *27*, 5833.
55. Winterfeldt, E. *Chimia* **1993**, *47*, 39.
56. Brünjes, R. Dissertation, Universität Hannover.
57. Weinmann, H.; Winterfeldt, E. *Synthesis* **1995**, 1097.
58. McWilliams, J.C.; Clardy, J. *J. Am. Chem. Soc.* **1984**, *116*, 8378.
59. Wegener, B.; Hansen, M.; Winterfeldt, E. *Tetrahedron Asymm.* **1993**, *4*, 345.
60. Borm, C.; Winterfeldt, E. *Liebigs Ann. Chem.* **1996**, 1209.
61. Brown, J.M. *Angew. Chem.* **1987**, *3*, 196.
62. Thompson, H.W.; McPherson, E. *J. Am. Chem. Soc.* **1974**, *96*, 6232.
63. Stadtmüller, H.; Knochel, P. *Synlett* **1995**, 463.
64. Borm, C.; Meibom, D.; Winterfeldt, E. *J. Chem. Soc., Chem. Commun.* **1996**, 887.
65. Schlesiger, H.; Winterfeldt, E. *Chirality*, in preparation.
66. Göres, M.; Winterfeldt, E. *J. Chem. Soc., Perkin Trans. 1* **1994**, 3525.
67. Jones, P.G.; Weinmann, H.; Winterfeldt, E. *Angew. Chem. Int. Ed. Engl.* **1995**, *34*, 448.
68. Nerenz, F. Dissertation, Universität Hannover, 1996.
69. Weinmann, H. Dissertation, Universität Hannover, 1995.
70. Weinmann, H.; Winterfeldt, E. *Synthesis* **1996**, 357.
71. Meibom, D. Dissertation, Universität Hannover, 1996.
72. Merten, E. Dissertation, Unversität Hannover, 1995.
73. Borm, C. Dissertation, Universität Hannover, 1996.

SYNTHESIS OF NONRACEMIC AMINES

Lendon N. Pridgen

I. INTRODUCTION

The synthesis of nonracemic amines is one of the most important problems of contemporary organic synthesis. Increasingly, pure enantiomers are highly sought after commodities as chiral building blocks, especially in the field of phar-

Advances in Asymmetric Synthesis
Volume 2, pages 55–117

ISBN 1-55938-797-1

maceutical products. The design of practical methods for effectively controlling product stereochemistry in nucleophilic additions remains an important challenge in synthetic chemistry. Although, the majority of the literature has focused on the addition of carbon nucleophiles to aldehydes and ketones,[1] synthetic and biological interest in highly functionalized acyclic and cyclic amines has contributed to the wealth of experimental methodology developed for the addition of carbanions to the carbon–nitrogen double bond of imines/imine derivatives (azomethines). While there exist certain inherent structural limitations in developing practical synthesis methods for chiral imines,[2] imine derivatives, by virtue of nitrogen substitution, add a synthetic dimension available to ketones only through derivatization. In addition, improved procedures for the preparation and activation of imine derivatives have increased the scope of the imine addition reaction.

During recent years, asymmetric synthesis has greatly contributed to progress in highly controlled formation of new chiral centers. The most economical approach has been to utilize naturally occurring precursors as the source of the inherent chirality.[3] This approach, which will be the focus of this chapter, will more specifically utilize naturally occurring α-amino acids as the source of chirality. α-Amino acids are considered to be the second most important natural source of chiral substrates, after monosaccharides, which are useful in stereocontrolled organic synthesis.[4]

In recent years, two strategies for controlling product stereochemistry in nucleophilic additions have emerged: (a) chelation control, in which Lewis acid reagents form intermediate chelates, rendering one face of the carbonyl moiety more accessible for nucleophilic addition (Cram's cyclic model **1a**, Figure 1); and (b) nonchelation control, in which the addition of reagents to substrates incapable of internal chelation is governed by steric and/or electronic factors. The bias for addition from a diastereotopic face in nonchelation-controlled additions was predicted empirically by Cram and is supported by the Cornforth and Felkin–Ahn models (Figure 1; **1b** and **1c**).[2a]

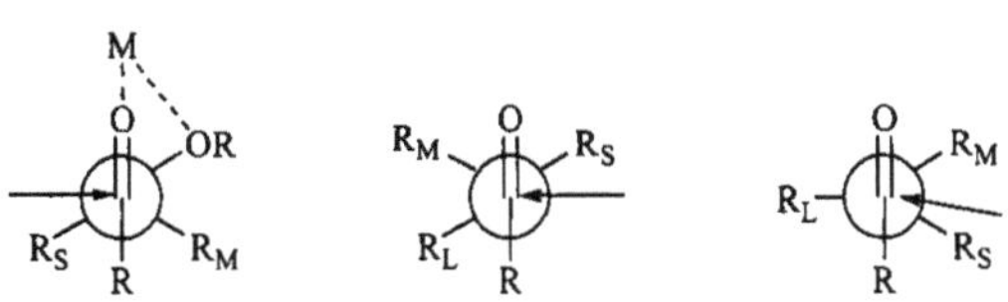

1a **1b** **1c**

Figure 1. Cram's cyclic model **1a**, Cornforth model **1b**, and Felkin–Ahn model **1c**.

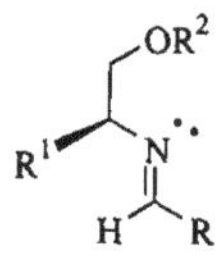

2

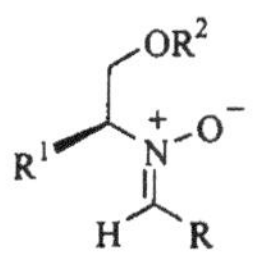

3

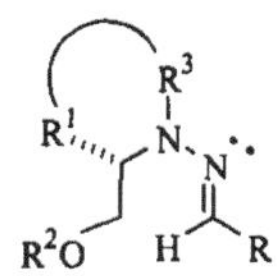

4

A major commitment has been placed on studying the environmental parameters which influence asymmetric induction resulting from the addition of organometallic reagents to aldimine/aldimine derivatives derived from chiral amines and/or hydrazines. A variety of chiral aldimines **2**,[5] nitrones **3**,[6] and hydrazones **4**[7] have been examined which contain as the chiral auxiliary alcohol, ether, or ester moieties, all of which are capable of chelation.

High diastereoselectivity can be achieved even though the exact mode of action is not known. However, it is suggested that the mechanism involves: (a) chelation control (Figure 2a); and (b) chelation-mediated delivery of the organometallic reagent (Figure 2b). The susceptibility to reductive or oxidative cleavage of the N–O, N–N, and N–C bonds within the addition products contributes to their value for the enantioselective synthesis of amines.[2a]

This chapter will present a comprehensive review of nonstabilized carbanion additions to the carbon–nitrogen double bond of chiral 1,3-oxazolidines, which may be considered chiral imine derivatives. These oxazolidines are derived from naturally occurring α-amino acids, their source of chirality, and may be considered to be masked chiral aldehydes.[2,8] The resulting chiral 1,3-oxazolidines may be employed as reactive electrophilic amino methine functionalities suitable for further transformations, wherein now the incipient chirality is transferred from the substrate imine/oxazolidine to a newly functionalized amino alcohol derivative. The pendant chiral auxiliary may be easily removed by either oxidative cleavage or hydrogenolysis, although both procedures destroy the auxiliary. Alternatively, the side chain may be incorporated into the formation of another

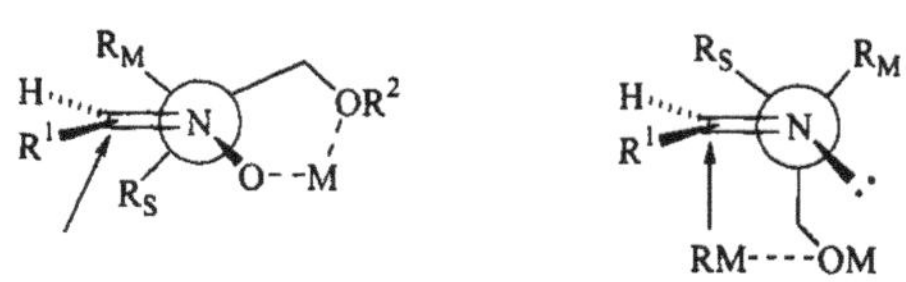

a **b**

Figure 2. (**a**) Chelation control; (**b**) organometallic delivery.

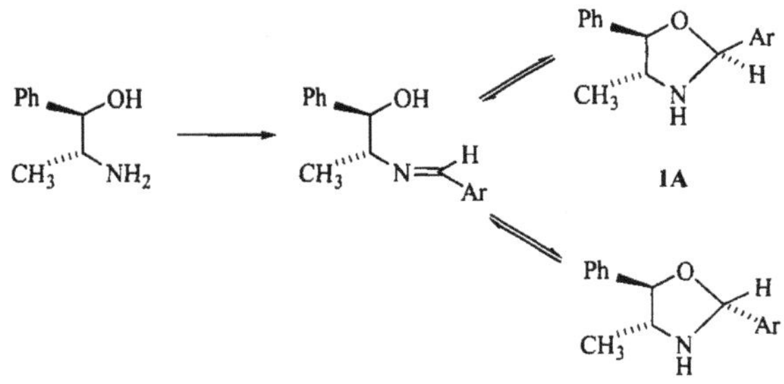

Figure 3. The ring–chain tautomerism in 1,3-oxazolidines.

heterocycle.[9] The above features, as well as the fact that both antipodal forms of amino alcohol chiral auxiliaries are commercially available, indirectly via the chiral pool, lends support to the use of chiral oxazolidines as extremely valuable chiral amine precursors.

The literature has already well documented the addition of Grignard reagents to imino ethers and oxazolidines,[10] which generally are known to exist in a tautomeric equilibrium with its imino form.[11–13] This ring closure is formally classified as a 5-*endo*-trig-type closure that is disallowed according to Baldwin's rules (Figure 3, **A/B**).[14] There are several reports of 5-*endo*-trigonal processes being more rapid that the 6-*endo*-trigonal process.[11–13] Pelleteir and Mody suggested that the immonium bond is much more accessible than the carbonyl bond and hence less subject to vectorial constraints.[15] Further, the tautomeric equilibria of oxazolidines and other 1,3-heterocycles have been the focus of several ^{1}H NMR studies.[16,17] In any event, organometallic addition appears to be independent of the extent of the ring–chain tautomerism, i.e. similar amino alcohol adducts are formed from either **A** or **B**.

The original concept to employ chiral 1,3-oxazolidines as surrogate imines, serving as substrates leading to the syntheses of chiral α-amines via organometallic addition, was first effectively demonstrated by Takahashi,[5a,7d] although the originally employed β-amino alcohol substrate was depicted by them as an imine.[5a] More recent studies have shown that the diastereoselective addition of various organometallic reagents to nonracemic imines and 1,3-oxazolidines derived from nonracemic amino alcohols proceeds in a highly efficient manner, ultimately providing a route for generating chiral amines in high chemical and optical yields (Scheme 1).[2]

Nonracemic 1,3-oxazolidines are most commonly prepared by condensation of nonracemic amino alcohols with aldehydes to yield, usually, a mixture of

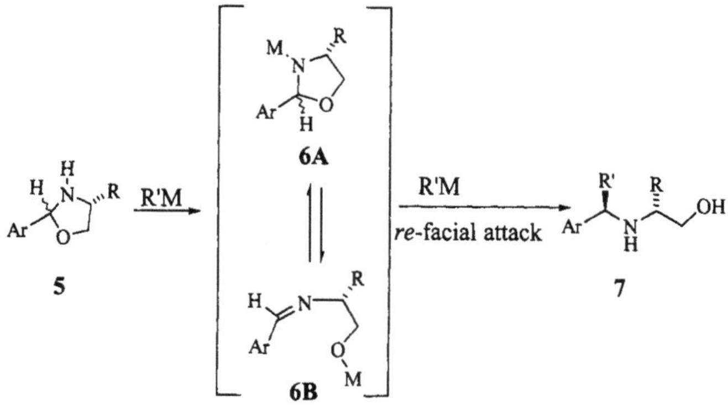

Scheme 1.

inseparable diastereoisomers.[8,18–26] However, it has been shown that the isomeric ratio of these 1,3-oxazolidine diastereomers is solvent-dependent. Consequently, care should be exercised when making configurational assignments and/or mechanistically interpreting results.[21] As an example, the configuration of the oxazolidine formed from L-ephedrine and *p*-bromobenzaldehyde was initially thought to be *trans* at the 2-position of **8**,[25] but was later revised to be the all *cis*-structure **9** based on X-ray data (Eq. 1).[21–26] In most cases, as alluded to earlier, in solution, the secondary oxazolidine exists as an isomeric mixture that does not affect the stereochemical outcome of the reaction.[8,27] Conversely, for tertiary oxazolidine, the stereochemical outcome of the addition reactions appears to be somewhat dependent on the stereochemical integrity of the substrate since the reaction is most probably S_N2-like in character, although this point has not been rigorously proven. In general, the latter is also less selective than the former, wherein chelation is highly important in stabilizing the reactive transition state.[8,28] A most recent approach was to employ nonracemic very sterically bulky perhydrobenzoxazines instead of 1,3-oxazolidines. For the former class of com-

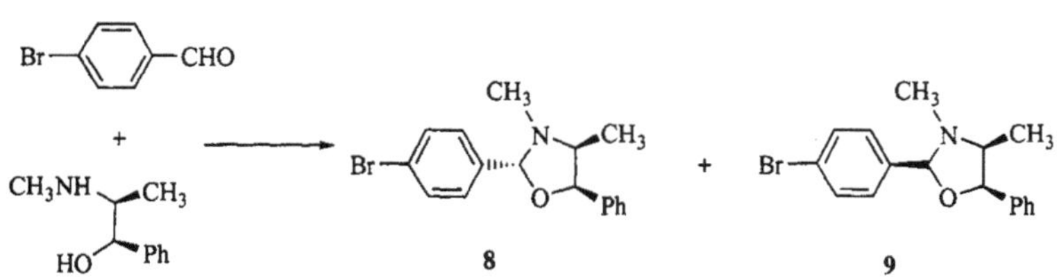

(1)

Table 1. Diastereoselective Reaction of *N*-Substituted 4-Phenyl-1,3-oxazolidines (**10a–g**) with Grignard Reagents

Substrate (**10**)	*Reagent (R)*	*Reaction Time (h)*	*Product* **11 & 12**	*Yield (%)*	*Ratio of (1S,1′R): 1R,1′R)*
a	CH_3MgBr	20	**a**	80.5	66:34
b	CH_3MgBr	20	**b**	86.2	66:34
c	CH_3MgBr	20	**c**	94.4	84:16
d	CH_3MgBr	72	**a**	89.6	97:3
e	C_6H_5MgBr	15	**c**	90.3	22:78
f	C_6H_5MgBr	15	**d**	92.1	23:77
g	C_6H_5MgBr	15	**d**	89.4	11:88

pounds, only one diastereomer was obtained on amino alcohol condensation with the aldehyde.[27e]

Grignard addition to oxazolidines **10a–g** as reported by Takahashi[29] yielded a mixture of diastereomers with the diastereomeric ratios of the products **11a–g** and **12a–g** as shown in Table 1 and Eq. 2.

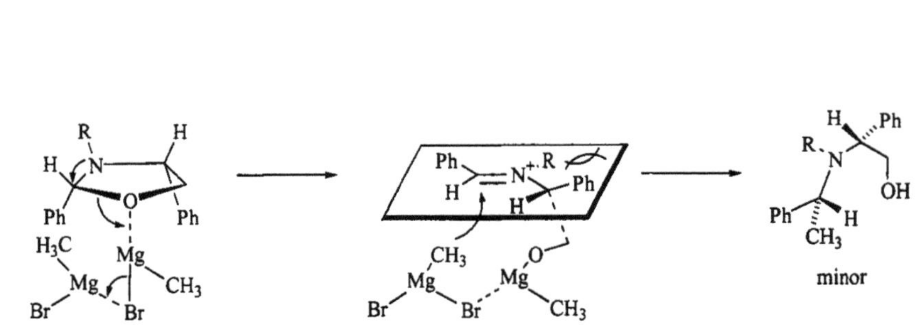

Scheme 2.

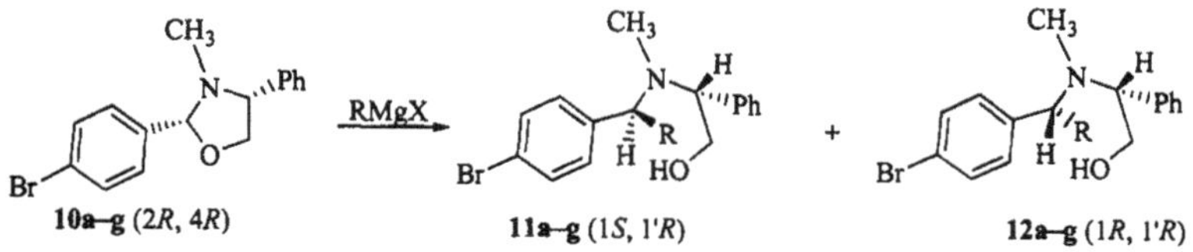

(2)

The selectivity of this reaction was suggested to increase with increasing bulkiness of the N-substituent, where methyl < benzyl < isopropyl. Consequently, a possible mechanism was proposed in which the Grignard reagent approaches the oxygen atom of the 1,3-oxazolidine ring to give a favorable intermediate immonium salt, and nucleophilic attack occurs from the *si*-face of the carbon–nitrogen double bond as shown in Scheme 2.

II. GRIGNARD ADDITION TO 1,3-OXAZOLIDINES TO YIELD NONRACEMIC AMINES

Substituted nonracemic 1-phenylethylamines are of interest as enzyme inhibitors,[30] as derivatizing agents for the analysis and/or resolution of enantiomeric mixtures,[31] and as potentially improved reagents or catalysts in asymmetric synthesis.[32] Moreover, especially when bearing oxygen functions on the aromatic ring, 1-phenylethylamines are partial structures, and thus useful synthetic precursors of important natural products. These are represented, not only by the simple 1-methyl-tetrahydroisoquinoline alkaloids,[33] but also by more complex substances, such as naphthylisoquinoline alkaloids,[34] tetrahydroprotoberberines,[35] and certain antibiotics.[36]

(*R*)-and (*S*)-N-Alkyl-1-cyclohexyl-2-phenylethylamines have been shown to exhibit analgesic activity more potent than that of (−)pentazocine hydrochloride and are not antagonized by (−)naloxone hydrochloride.[27b–d] Thus, an asymmetric synthesis of these phenylethyl amines was initiated by Takahashi based on Grignard addition to 2-cyclohexyl-4-phenyl-1,3-oxazolidine. Another requirement for this general class of amines was needed in the area of asymmetric resolution. α-Alkyl phenylmethylamines have been used for some time as resolving agents and are normally readily commercially available for small-scale use. However, in those cases where large quantities of these amines are required, supply difficulties could arise. Therefore, an asymmetric synthesis that would permit facile access to these amines in high optical purity in a manner that would be amenable to kilogram quantity development was highly desirable.

Many of the published routes to α-alkyl phenylmethylamines require a tedious resolution of the corresponding racemate.[31] The few reported asymmetric approaches are unacceptable for a number of reasons, such as high cost, multiple steps, low chemical yields, or low diastereoselectivity.[37] Based on Takahashi's asymmetric synthesis of *optically pure* *N*-alkyl-1-cyclohexyl-2-phenylmethylamines by stereoselective addition of benzylmagnesium chloride to (4*R*)-2-cyclohexyl-4-phenyl-1,3-oxazolidine,[27b] the analogous 2-aryl-4-phenyl-1,3-oxazolidine (**13**) was explored for use as a general substrate for organometallic additions (Scheme 3).[8] Prior to the latter report, only one example of Grignard addition to a secondary 2,4-disubstituted oxazolidine was known.[27b–d] The results of a study to explore the generality of organometallic addition to secondary 1,3-oxazolidines, with the ultimate intention of employing the aminoalcohol adduct as a source of nonracemic α-substituted phenylmethylamines, is depicted in Scheme 3 and Table 2.

Table 2 outlines results of the stereoselective organometallic nucleophilic addition to **13** with its subsequent oxidative cleavage to α-substituted phenylmethylamines.[8] Most reactions were performed utilizing the readily available (*R*)- or (*S*)-phenylglycinol, prepared in bulk via a recently reported BH_3/DME amino acid reduction procedure.[38] Grignard addition to oxazolidine **13** occurred quite readily under THF reflux (4–24 h) usually with very high diastereoselection for **14** (averaging ~96% de) as determined by 400 MHz ^{1}H NMR spectroscopy. Typically, 2.5–3.0 equiv of Grignard reagent was required to force the reaction to completion *and* achieve high diastereoselectivity. In fact Grignard addition was reported not to occur cleanly until at least ~1.5 equiv of the organometallic had been added.

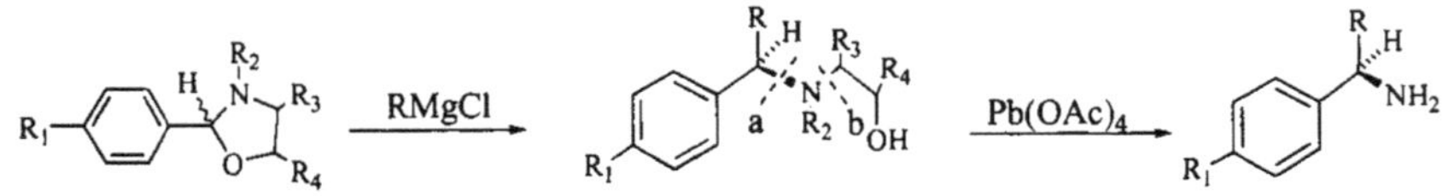

	R_1	R_2	R_3	R_4
a,	H	H	Ph	H
b,	Br	H	Ph	H
c,	OMe	H	Ph	H
d,	Br	H	Me	Ph
e,	Br	Me	Me	Ph

	R	R_1	R_2	R_3	R_4
14 a,	C_2H_5	H	H	Ph	H
b,	C_4H_9	H	H	Ph	H
c,	CH_3	H	H	Ph	H
d,	CH_2Ph	H	H	Ph	H
e,	CH_3	Br	H	Ph	H
f,	CH_3	MeO	H	Ph	H
g,	CH_3	Br	H	CH_3	Ph
h,	CH_3	Br	Br	CH_3	Ph

	R	R_1
15a,	C_4H_9	H
b,	CH_3	H
c,	C_2H_5	H
d,	CH_2Ph	H
e,	CH_3	Br
f,	CH_3	MeO

Scheme 3.

Table 2. Yields and Selectivities for Grignard Additions to Oxazolidines **13** and Cleavage Yields of **15**

Entry	Substrate	RMgCl	Yield %	Diastereo-selectivity **13**[a]		Cleavage Yield, %	**15**	$[\alpha]_D^{25}$ (c, $CHCl_3$)	% ee
1	**a**	Et	62	98:2	**a**	—		—	—
2	**a**	Bu	47	98:2	**b**	52	**a**	+11.7(1.0)	92
3	**a**	Me	56	95:5	**c**	50	**b**	+28.8(1.0)	94
4	**a**	Et (3 equiv, $CeCl_3$)	85	>99:	**a**	46	**c**	+35.2(1.2)	95
5	**a**	Me [$MgBr_2 \cdot O(Et)_2$]	31	96:4	**c**	—		—	—
6	**a**	Benzyl	87	94:6	**d**	—		—	—
7	**a**	Benzyl (3 equiv, $CeCl_3$)	78	98:2	**d**	59	**d**	−10.9(1.6)	>99
8	**b**	Me	60	95:5	**e**	65	**e**	+24.1(1.6)	98
9	**c**	Me	45	97:3	**f**	46	**f**	+24.6(1.0)	96
10	**d**	Me	51	94:6	**g**	—			—
11	**e**	Me	77	60:40	**h**	—			—
12	**a**	MeLi (−78 °C)	48	95:5	**c**	—			—
13	**a**	MeLi (25 °C)	49	77:23	**c**	—			—
14	[b]	Me	50	87:13	[c]	—			—

Notes: [a]Diastereomeric isomer ratios were determined by using 400 MHz ^1H NMR.
[b]The imine in this case was the methyl ether of the imine **6B**.
[c]This product is the methyl ether analog of **14e** [−9.7 (c 1.0, $CHCl_3$)].

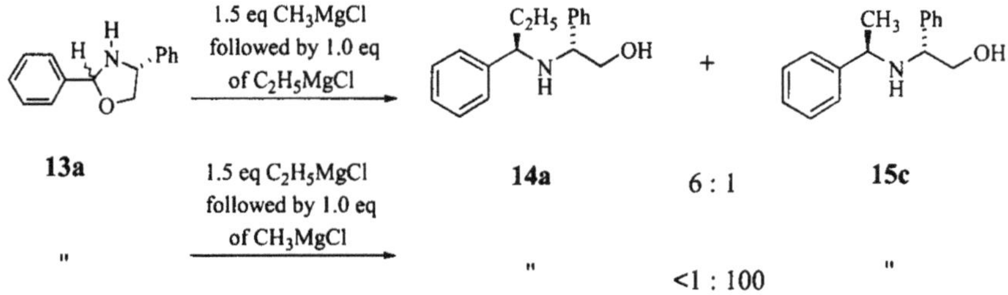

Scheme 4.

In an interesting experiment, 1.5 equiv of methyl Grignard was added to **13a** in THF at room temperature followed shortly by 1.0 equiv of ethyl Grignard. The reaction mixture was then heated to THF reflux. The predominate resulting product was that formed via addition of the ethyl Grignard rather than methyl (6:1 ratio). Reversing the order of addition led to a reversal of product selectivity (Scheme 4), but with an even greater disparity in the ratio of products (>100:1). This unprecedented high level of asymmetric induction for Grignard addition to the normally tautomeric oxazolidine/imino functionality may be attributed to a highly ordered transition state resulting from significant chelation of the alkoxy substituent and imino nitrogen to at least half of a magnesium cation. The Grignard reagent then attacks the *re*-face of C-2 of either **A** or **B** from the less hindered side, in an S_N1-like fashion distal to the stereogenic center of the amino alcohol moiety (Scheme 5).[27b]

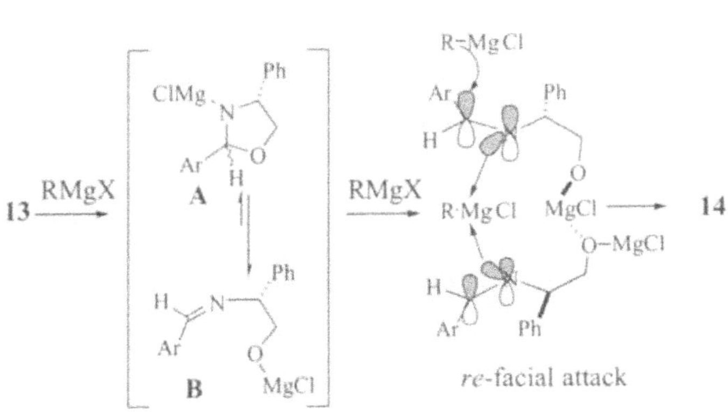

Scheme 5.

The data reported to date does not allow one to discern whether or not Grignard addition is preceded by ring opening since the stereochemical result would be equivalent. However, the ^{1}H NMR ($CDCl_3$) solution spectrum of **13a** indicates that the oxazolinyl/imino tautomeric equilibrium lies predominately toward the imine.[27b,29] This observation is in agreement with the postulate of Hauser[39] who suggested that amino ethers form a 2:1 strongly coordinated nitrogen/oxygen to magnesium complex, which in our case forms after deprotonation with the first equivalent of Grignard. Consequently, 1.5 equiv of Grignard is unavailable for addition to carbon. This mode of addition is not dissimilar to the one invoked by Takahashi[7d] and Koga[4b] in their reported chelation-controlled nucleophilic addition to nonracemic valinol-derived hydrazones and *t*-leucine-derived α,β-unsaturated aldimines, respectively.

Lewis acid chelators were used in an attempt to further exploit this "chelation handle" and also possibly decrease the quantity of Grignard reagent required. For example, when either $ZnCl_2$, $TiCl_4$, $BF_3{\cdot}O(Et)_2$, CuI, or $CuBr_2{\cdot}S(CH_3)_2$ were used with methyl Grignard, at best a 77:23 ratio of isomeric products was obtained. However, with a 1:1 ratio of cerium chloride to Grignard reagent (3 equiv) under the standard reaction conditions, essentially one stereoisomer was obtained. Thus, the strong oxophilicity of cerium has enhanced the selectivity of the organometallic addition to the extent that a single diastereomer may be produced.[40] Unlike other organometallic reagents explored, the cerium organometallic is highly regioselective for C-2 of **13** and is the reagent of choice in a nucleophilic addition to **13**. The methoxy analog of imine **B** gave a diminished ratio of isomers (~87:13) as did the *p*-bromophenyl oxazolidine analog of ephedrine (Table 2, entries 10, 11, and 14). Takahashi[41] and Davidsen[42] obtained comparable or worse results with nonracemic N-substituted valinol derived oxazolidines. Thus, at least one of the nitrogen or oxygen–magnesium bonds in Scheme 5 should be covalent in order to produce the higher selectivity.

Additional support for a very strong chelated transition state complex, of the type possibly similar to the one represented in Scheme 5, was found on examination of the X-ray structure of aminoalcohol hydrochloride **16** (Figure 4). The molecule crystallizes in two different conformations and is shown in Figure 4 with hydrogen bonding of the two independent molecules with the Cl anions. The absolute configuration of the chiral centers C-1 and C-1′ are *R* and *S*, respectively. The independent molecules differ mainly in the dihedral angles of the ethyl ester group. The phenyl rings are roughly planar where the rings stack above one another with the dihedral angle between phenyl rings being close to 30° in both molecules. The average distance between atoms in each ring is within 4 Å. The most interesting feature concerning Figure 4 is the strong cationic bonding to the chloride anions. The average N–H to Cl interaction is 3.2 Å, while the

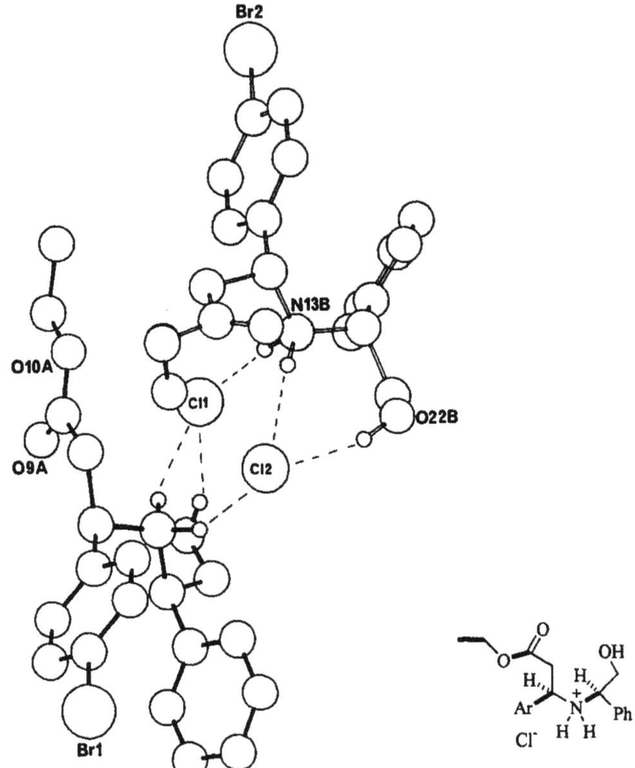

16

Figure 4. (1′, 3*R*)-Ethyl 3-[(2′-hydroxy-1′-phenyl)ethylamino-3-(4-bromophenyl)-propionate (**16**).

average O–H to Cl interaction is 2.3 Å. The structure exists as a dimeric cell where one could easily envision metallic cations replacing the chloride anions within the spherical cavity, resulting in very strong nitrogen/oxygen to magnesium interactions necessary to stabilize the diastereomeric transition state.[43]

A report by Corey[44] is a recent example where a metal centered 1,2-amino alcohol complex played a dominant role in determining which diastereomeric transition state undergoes reaction with an attacking nucleophilic organometallic. Enantiomerically enriched (2*S*, 5*R*)-(–)-2-(hydroxydiphenylmethyl)-5-[((dimethylamino)methyl)]pyrrolidine was used as a chiral controller ligand in a catalyzed reaction between benzaldehyde and diethylzinc (Scheme 6). No reaction was reported to occur between the catalyst **17** (10 mol%) and benzaldehyde (0.05 M each) in C_6H_6 at 23 °C over 19 h. Only after adding additional diethylzinc (~1.2 equiv), in several experiments, either at 0 °C for 16 h or at 32 °C for 4 h, did ethyl addition to benzaldehyde occur to yield (*S*)-(−)-1-phenylpropanol as the sole product, in 94–100% yield and 93–95% ee. When dibutylzinc (1.0 equiv) was substituted for diethylzinc (Scheme 6) employing 0.1 equiv of catalyst **17**, a 10:1 mixture of (*S*)-(−)-1-phenylpentanol and (*S*)-(−)-1-phenylpropanol was obtained.

This suggests the catalytic mechanism shown in Scheme 6. The initial step—coordination of the aldehyde with the zinc atom of complex **17** to form **18**—is accompanied by replacement of the dimethylamino ligand to maintain the tetracoordinate status of zinc. Subsequent complexation of diethylzinc (a weak Lewis acid) with the nucleophilic oxygen of **18** leads to the termolecular assembly **19** from which transfer of an ethyl group from zinc to the nearby aldehyde carbonyl function can occur via a six-membered cyclic structure to form the complexed carbonyl adduct **20**. Elimination of zinc alkoxide from **20** produces the product **21** with regeneration of the catalyst **17**. In assembly **19**, the aldehyde is coordinated to zinc through the sterically more accessible lone pair *anti* to the phenyl and is arranged spatially to minimize steric repulsion with neighboring atoms. This preferred three-dimensional arrangement exposes the *si*-face of the formyl carbon to the attacking ethyl group and leads to the *S*-configuration of the adduct

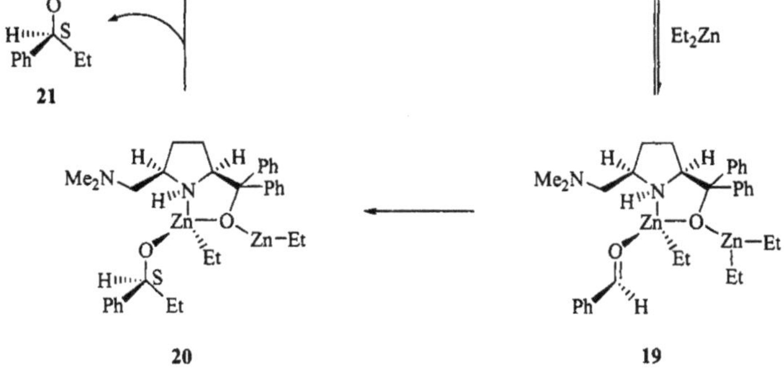

Scheme 6.

20, in accord with experimental results. The alternative three-dimensional arrangement of assembly **19**, which leads to the carbonyl adduct of *R* absolute configuration, is disfavored sterically because of repulsion between the aldehyde and the (dimethylamino)methyl substituent.

Evidence that the conversion of **19** to **20** is the rate-limiting step was provided by the experimental observation of the relative rates of reaction of benzaldehyde and its *p*-trifluoromethyl and *p*-methoxy derivatives, the relative reactivities being *p*-$CF_3C_6H_4CHO > C_6H_5CHO >$ *p*-$CH_3OC_6H_4CHO$. The reaction of *p*-(trifluoromethyl)benzaldehyde with diethylzinc and **17** (10 mol%) under the conditions described above for benzaldehyde was complete in 3 h at 0 °C and gave cleanly (*S*)-1-(*p*-trifluoromethylphenyl)propanol of 95.5% ee. The corresponding reaction of *p*-methoxybenzaldehyde required 18 h at 23 °C for 90% conversion to product, which was obtained in only 83% ee, mainly because of a competing uncatalyzed reaction of the aldehyde and diethylzinc as demonstrated by a control experiment (*p*-$MeOC_6H_4CHO$, Et_2Zn, toluene, 23 °C, with gas chromatographic analysis of rate).

Takahashi[27b] and Davidsen[42] both demonstrated the applicability of *aliphatic* oxazolidines as substrates. Similarly, (4*R*)-2-^{n}butyl-4-phenyloxazolidine was employed as a substrate in a reaction with phenyl Grignard and gave a 9:91 ratio of **14b**. As expected, the minor isomer in Table 2 (entry 2) is now major. Thus, with the exception of the unstable 2-methyl-4-phenyloxazolidine, *both* aliphatic and aryl-2-oxazolidines may be employed in this reaction. Therefore, in addition to being able to employ either enantiomer of the chiral auxiliary, one has an alternative entry to the different diastereoisomers. In related work, diastereoselective organometallic addition to nitrones bearing stereogenic N-substituents have been reported.[6]

In an approach to prepare pseudodipeptides, Davidsen[42] added Grignard reagents to amino-acid-derived oxazolidines as depicted in Scheme 7. Condensation of crude aldehyde **22** or **23** with the appropriate *N*-benzyl valinol **24** gave the oxazolidine **25** or **26** in 75% yield after recrystallization from methanol. The stereochemical assignment of **25** and **26** was corroborated by NOE studies; irradiation of the C-2 methine in both **25** and **26** caused an enhancement of the signal corresponding to the C-4 methine, thus supporting a *cis* relationship across the oxazolidine ring.[45] In addition, X-ray crystallography analysis of **26** confirmed the stereochemical assignment at C-2. Addition of oxazolidine **25** to an ethereal solution of ethyl Grignard at 0 °C gave a 2.4:1 ratio of readily separable diamino alcohols **27** and **29** in 81% yield. The ratio of products was essentially the same in the case of **26**, which gave **28** and **30** in a 2.5:1 ratio in 83% yield after chromatographic separation.

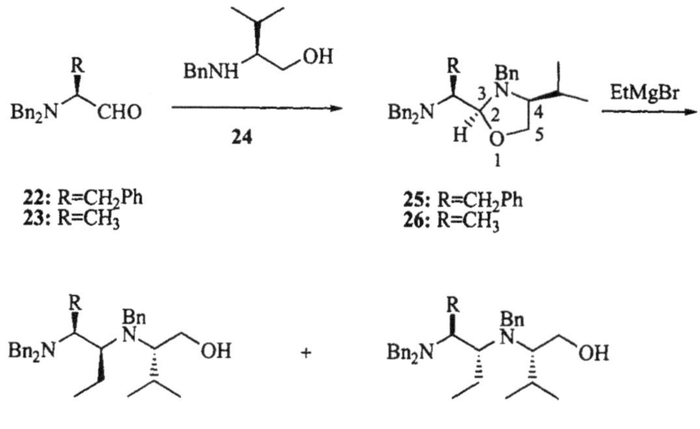

22: R=CH$_2$Ph
23: R=CH$_3$

25: R=CH$_2$Ph
26: R=CH$_3$

27: R=CH$_2$Ph
28: R=CH$_3$

29: R=CH$_2$Ph
30: R=CH$_3$

Scheme 7.

Subsequent reaction of the enantiomer of **22** with **24** yielded a 10:1 inseparable mixture of oxazolidines, the stereochemisty of which could not be determined by NOE irradiation. Addition of ethyl Grignard in this case surprisingly gave a single diamino alcohol by ^{1}H NMR, ^{13}C NMR, and silica gel TLC analyses. The fact that this latter adduct is a different isomer than either **27** or **29** establishes that the asymmetric center from the original benzyl-protected amino aldehyde does not racemize during oxazolidine formation and subsequent Grignard addition.

A report by Jones[46] described the use of the equilibrium mixture of 2-pyridyl-4-substituted-1,3-oxazolidine (**31a**) and the corresponding imine **31b** in a reaction with the Grignard reagent which had been prepared from 1-bromomethylcyclohexane (Scheme 8). They found the solution equilibrium ratio of **31a/31b** to be solvent-dependent. In THF, the imino form was the major tautomer for Y = ^{i}Pr by 1:3. While in $CHCl_3$, the oxazolidinyl form predominates in a 80:20 ratio for Y = ^{i}Pr. Upon reaction of the mixture with the Grignard **32**, an 87:13 mixture of isomeric alcohols **33** was obtained, as determined by analysis of the crude reaction mixture by ^{1}H NMR. The resulting amino alcohol was oxidatively cleaved with $NaIO_4$ to yield amine **33a** which was shown by chiral HPLC analysis to have a 75% ee. Higher selectivity (as high as 96% ee) was eventually found by employing the alternative pathway of reducing the nonracemic 1,3-oxazolidine prepared from (2-pyridyl)-cyclohexylmethyl ketone.

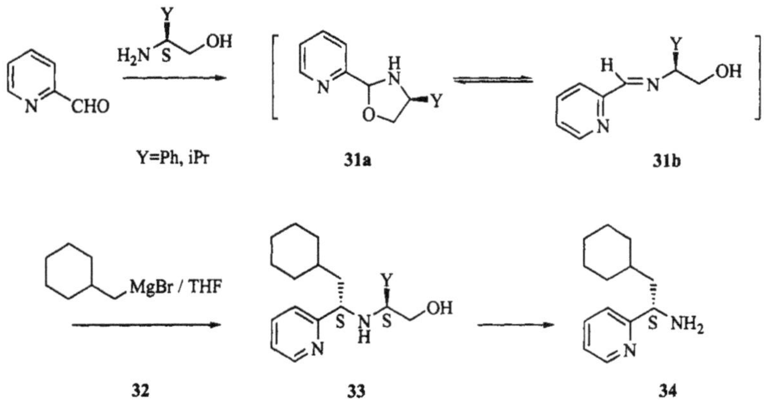

Scheme 8.

Higashiyama added Grignard reagents to 1,3-oxazolidines as the central theme in a total synthesis of the naturally occurring alkaloids (−)-coniine and (−)-dihydropinidine.[47a] The application of this methodology resulted in an enantioselective synthesis of these piperidine alkaloids with good to moderate selectivity.

Two approaches were employed by Higashiyama to prepare (−)-coniine and (−)-dihydropinidine. Both procedures began by preparing the nonracemic tertiary 1,3-oxazolidine **34** as normal by condensing *N*-benzylphenylglycinol with butyraldehyde in CH_2Cl_2 in the presence of $MgSO_4$ (Scheme 9). The 1,3-oxazolidine was obtained as a 84:16 mixture of C-2 epimeric isomers which was used without purification. Reaction of 1,3-oxazolidine (**34**) with the Grignard reagent prepared from 2-(chloropropyl)-1,3-dioxolane in THF afforded pairs of the diastereomeric adducts in a quantitative yield. The isomer ratio determined before purification was 93:7 but the major isomer was isolated in 84% yield after chromatography. The alcohol **35a** was hydrogenated under acidic conditions to afford a primary amino group and to liberate the aldehyde, which immediately formed an imine, that was reduced to the desired (*R*)-(−)-coniine in 90% yield, all in one step.

The second approach, employing the same methodology, began by treating the **34** with 4-pentenylmagnesium bromide in THF at −15 °C to furnish the alcohol **36** in 98% yield as an inseparable diastereomeric mixture, again in a 86:14 ratio of isomers. The amino alcohol adducts were subjected to the Wacker procedure to form a mixture of the methyl ketones **37a,b** in quantitative yield. The major ketone was isolated by chromatography in 85% yield and ring closed as described above to provide exclusively (2*R*, 6*S*)-(−)-dihydropinidine with the

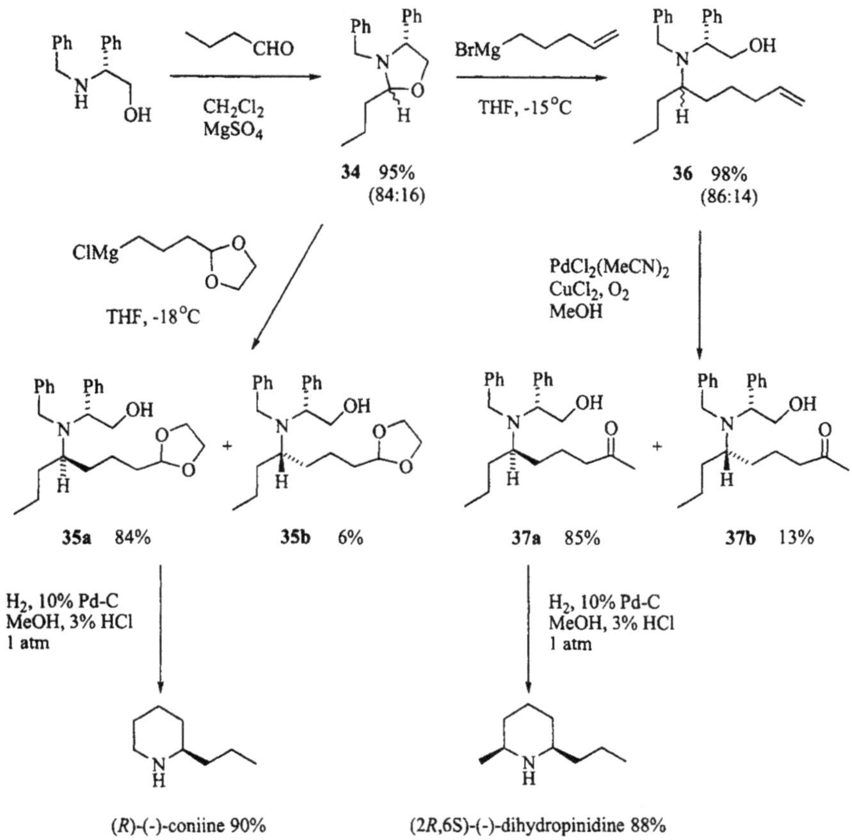

Scheme 9.

unnatural configuration. Thus, these latter two examples demonstrate the potential of this Grignard addition to 1,3-oxazolidines to selectively form many varied nonracemic amines. The range of functionality that this procedure will tolerate has yet to be fully explored, especially considering that a wide range of nucleophilic agents other than magnesium reagents could conceivably be employed.

A third very concise approach to coniine was reported by Amat[47b] where oxazolopiperidines, similar to 2-cyano-6-oxazolopiperidine (**38**), were subjected to titanium tetrachloride catalyzed addition of allyltrimethylsilane. A 9:1 ratio of diastereomers was obtained.

In a series of papers beginning in 1983, Husson described the preparation and chemistry of **38**, a new type of chiral synthon which shows considerable promise

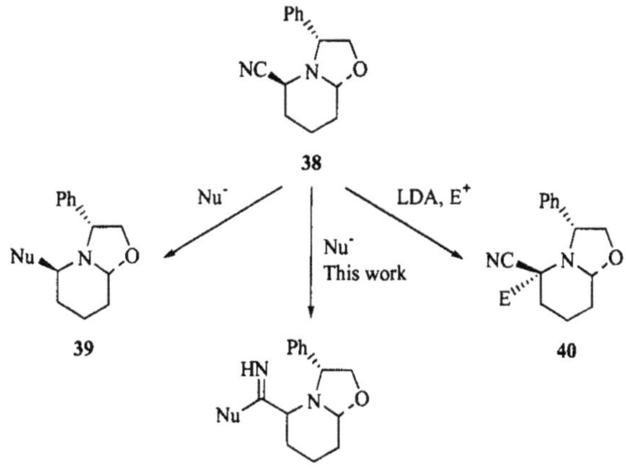

Scheme 10.

for the asymmetric synthesis of heterocyclics in general and alkaloids in particular.[48] The primary synthetic potential of **38** lies in the fact that its amino–nitrile functionality may be regarded as a potential iminium ion reacting with nucleophiles to yield **39**, or as a C-6 proton activator leading to an anion ready for electrophilic substitution, yielding **40** (Scheme 10).[49]

Previous literature reports indicated that lithium reagents would react specifically on the α-aminonitrile without disrupting the oxazolidine function.[50] It was actually observed that the butyl cuprate as well as butyllithium reacted as expected to furnish **41**. The stereoselectivity was explained by a preferred attack on the *re*-face of the more stable conformation in which the imine group is oriented above the oxazolidine ring to minimize steric hindrance (Figure 5).[48b]

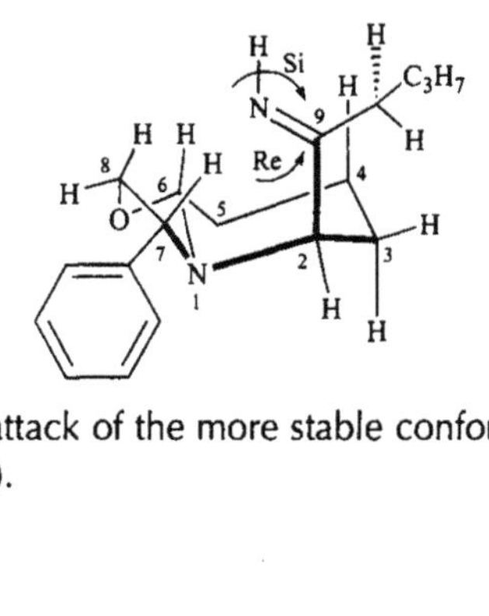

Figure 5. The *re*-facial attack of the more stable conformation of 2-cyano-6-phenyloxazolopiperidine (**38**).

An earlier alternative strategy was employed by the same laboratory using the same substrate for the asymmetric synthesis of the indolizidine alkaloids (–)-monomorine,[51a] (–)-gephyrotoxin-223 AB (**47**),[51a] (+) and (–)-coniine,[49] dihydropinidine,[49] and (+)-solenopsin A,[51b] wherein after the alkylation of the anion of **38** the cyano group was selectively cleaved by complexation with $AgBF_4$ and treatment with $Zn(BH_4)$ at low temperature. The synthesis of (–)-gephyrotoxin-223 AB (**47**) is shown in Scheme 11.

Compound **43** was obtained as a 3:2 mixture of C-6 epimeric oxazolidines. The reaction of this epimeric mix with propylmagnesium bromide gave the desired *cis*-compound **44** accompanied by small amounts of its *trans*-epimer (*cis/trans* : 85/15). Hydrogenolysis of **44** led to **45**, which was treated in a biphasic

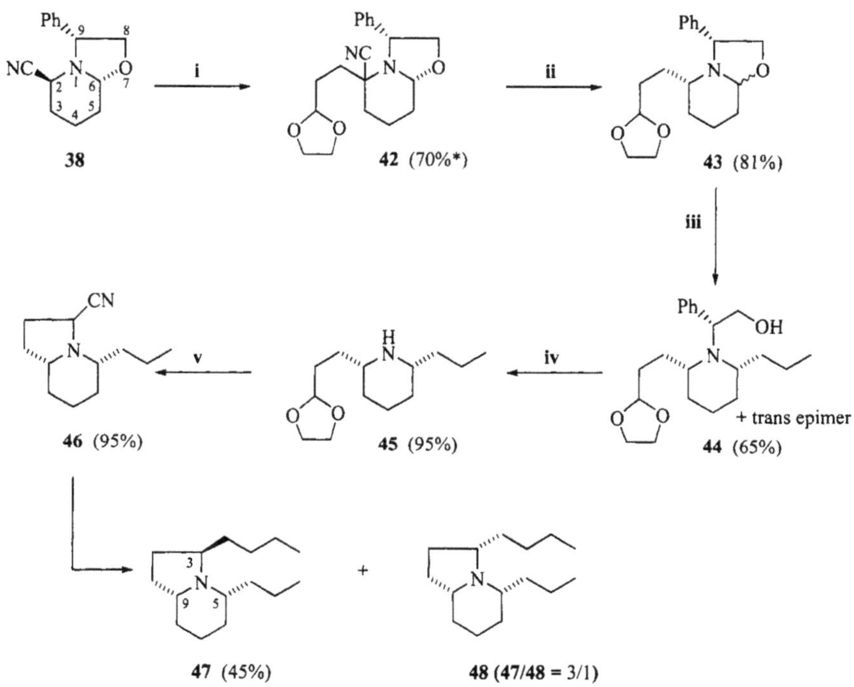

38 **42** (70%*) **43** (81%) **44** (65%) **45** (95%) **46** (95%)

47 (45%)
(-)-Gephrotoxin 223 AB

48 (**47**/**48** = 3/1)

Reagents: i) LDA, THF, -78°C, $BrCH_2CH_2CH\text{-}(OCH_2)_2$, 3h. ; ii) $AgBF_4$, THF, 5 min., r.t. $Zn(BH_4)_2$, -50°C, 1h. ; iii) PrMgBr, ether, -50°C, 1h. ; iv) H_2, Pd/C 10%, MeOH, 30 min. ; v) CH_2Cl_2-HCl 1N, KCN ; vi) nBuMgBr, ether, 0°C.

Scheme 11.

medium (CH_2Cl_2–H_2O) with HCl 1N and KCN (pH 3–4) to furnish the aminonitrile **46** as a single isomer. Finally, the butyl side chain was introduced without incident stereoselectively at C-3 of **46** with butyl magnesium bromide. This reaction generated preferentially the desired (*R*)-epimer (*R/S* : 3/1) (–)-gephyrotoxin-223 AB (**47**) in 45% yield from **46**.

Husson has recently extended this methodology to include the stereoselective synthesis of *cis-* or *trans-*1,3-disubstituted tetrahydroisoquinolines from the same oxazolidine precursor.[51c]

An asymmetric synthesis of another natural product, (*S*)-(–)- and (*R*)-(+)-salsolidine, was achieved via a Grignard addition to a 1,3-oxazolidine appended tetrahydroisoquinoline.[52] Yamato synthesized a 19/1 isomeric mixture of the tetrahydroisoquinoline **49** from which the major isomer was isolated by recrystallization from ethanol in 95% yield and in 90% de. Asymmetric methylation of the nonracemic **49** to yield **50** was achieved by reaction with methylmagnesium iodide at –78 °C for 1 h. Subsequent hydrogenation on Pd–carbon in acidic ethanol yielded optically pure (*S*)-(–)-salsolidine **51** (Scheme 12). Another approach to functionalized piperidines from *N*-Boc-oxazolidines was reported by Agami.[9f]

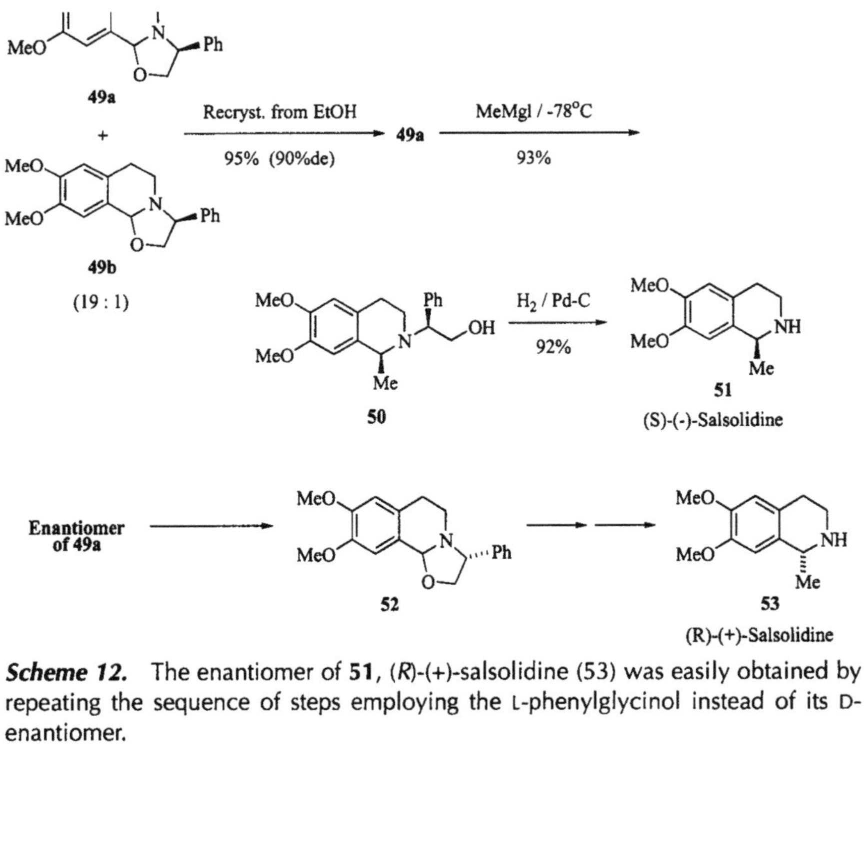

Scheme 12. The enantiomer of **51**, (*R*)-(+)-salsolidine (53) was easily obtained by repeating the sequence of steps employing the L-phenylglycinol instead of its D-enantiomer.

III. NONRACEMIC PYRROLIDINES

Nonracemic 2-arylpyrrolidines and 2,5-bis(aryl)pyrrolidines may be synthesized by Grignard additions to nonracemic 1,3-oxazolidines. Enantiomerically pure pyrrolidines have been efficiently employed as chiral auxiliaries[53] and intermediates[54] for the synthesis of a variety of chiral, nonracemic compounds. A highly selective and efficient synthesis of nonracemic pyrrolidines was reported by Higashiyama,[55] according to Scheme 13.

The reaction of the nonracemic imino/1,3-oxazolidine equilibrium mixture **54a–c** with excess Grignard reagent **55**, which was prepared from 3-bromopropionaldehyde dimethyl acetal in THF, afforded the desired products **56a–c** in 58–64% yields with very high diastereomeric excesses. The high degree of stereocontrol of this reaction is again attributed to the highly ordered transition state resulting from significant chelation of the alkoxy substitutent and imino nitrogen to the magnesium atom and then delivery of the other Grignard reagent from the least hindered side of the carbon–nitrogen double bond.

Treatment of **58a–c** with triethylphosphine and diphenyl disulfide yielded the thiophenyl derivatives, which, without isolation were then treated with the lithium radical anion of 4,4′-di-*tert*-butylbiphenyl to afford the desired (*R*)-2-arylpyrrolidines (**59a–c**). Bicyclo compounds **57a,b** reacted smoothly with the phenylmagnesium bromide or 4-methoxyphenylmagnesium bromide in THF to

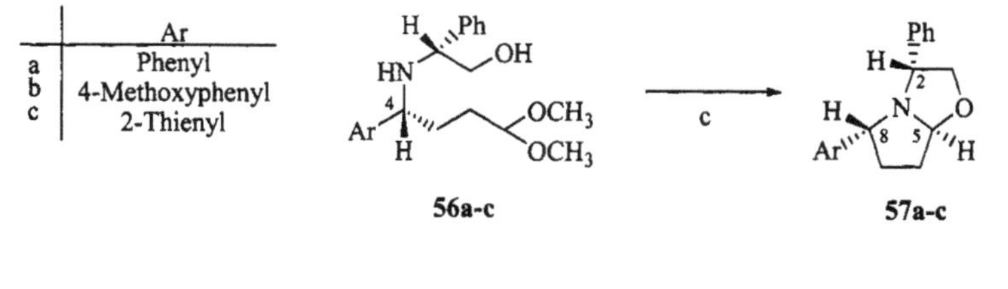

Reagents: (a) aromatic aldehyde (1.0 eq.), C_6H_6, reflux, 4h, **54a:** 94%, **54b:** 71%, **54c:** 98%; (b) Grignard reagent (3.0 eq.), THF, 0°C, 30min, then rt, 15h, **56a:** 63%, **56b:** 64%, **56c:** 58%, (c) MeOH-6N.HCl (pH4), 2 days, **57a:** 53%, **57b:** 49%, **57c:** 63%.

Scheme 13.

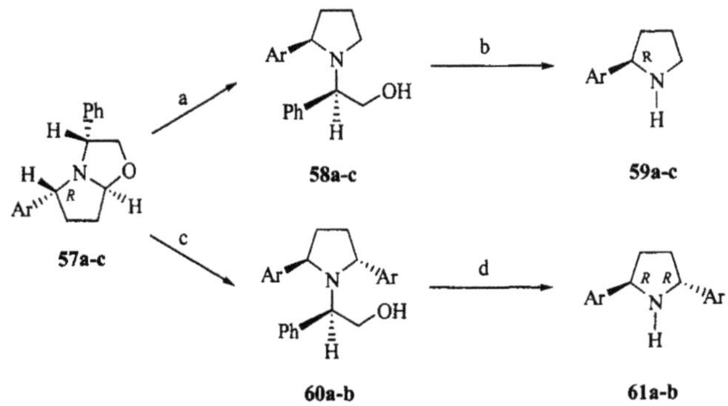

Reagents: (a) $NaBH_4$ (3.0 eq.), MeOH, rt, 2h, **58a**:87%, **58b**:81%, **58c**:76%; (b) 1)PhSSPh(5.0 eq.), Et_3P(5.0 eq.), THF, rt, 16h. 2)LiDBB(3.6 eq.), THF, -60°C, 15 min., then 0°C, 16h, **59a**:62%, **59b**:30%, **59c**:13% (c) Grignard Reagent(3.0 eq.), THF, rt, 2.5h, **60a**:82%, **60b**:80%; (d) In the same manner as (b) **61a**:60%, **61b**:42%

Scheme 14.

give essentially single adducts **60a,b** in 82 and 80% yields, respectively, as evidenced by the ^1H NMR (270 MHz) spectra of the crude product mixture (Scheme 14). The stereochemistry of the newly formed asymmetric center in the adducts **60a,b** was assumed to be *R*-based on previous results[28] and unambiguously determined by their conversion into 2,5-bis(aryl)pyrrolidines (**61a,b**) in the same manner as described for the preparation of **59a–c**.[55]

The observed diastereoselectivity for the Grignard reaction of **57a,b** to yield **60a,b** can be rationalized by assuming that the Grignard reagent approaches the oxygen atom of the 1,3-oxazolidine ring to give a favorable intermediary imin-

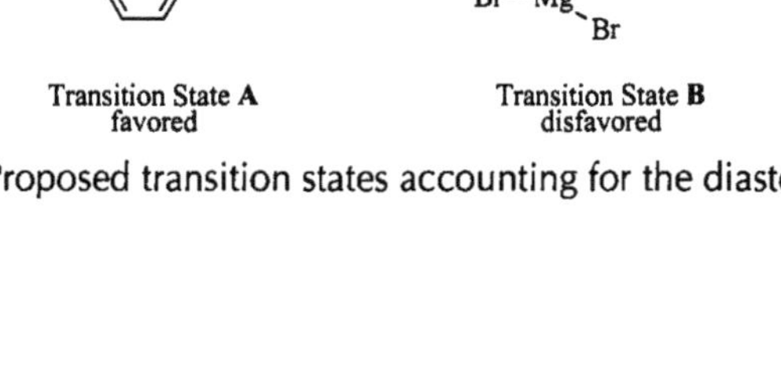

Figure 6. Proposed transition states accounting for the diastereoselectivity

ium salt (**A**). Nucleophilic attack is favored from the *si*-face of the carbon–nitrogen double bond of the intermediate as shown in Figure 6.

The laboratories of Husson[53,56] have continued to report on their exploration of the versatility of structures similar to 2-cyano-6-oxazolopiperidine (**38**). In these latter examples, they extended their research to cover the synthesis of pyrrolizidine alkaloids, by employing the analogous 2-cyano-6-oxazolopyrrolizidine (**62**). Alkylation of the anion of **62** (Scheme 15) with 1-bromo-2,3-(ethylenedioxy)butane produced a mixture of stereoisomers **63**, isolated in 65% yield after flash chromatography on silica gel. A stereospecific decyanation of the diastereomeric mixture **63** was achieved using Li/NH_3 reduction. Compound **64** was obtained in 63% yield as a single isomer. The *S* absolute configuration at the new asymmetric center was ascertained unambiguously on the basis of X-ray analysis performed on an analog of **64** and by the use of 1D difference NOE and NOESY

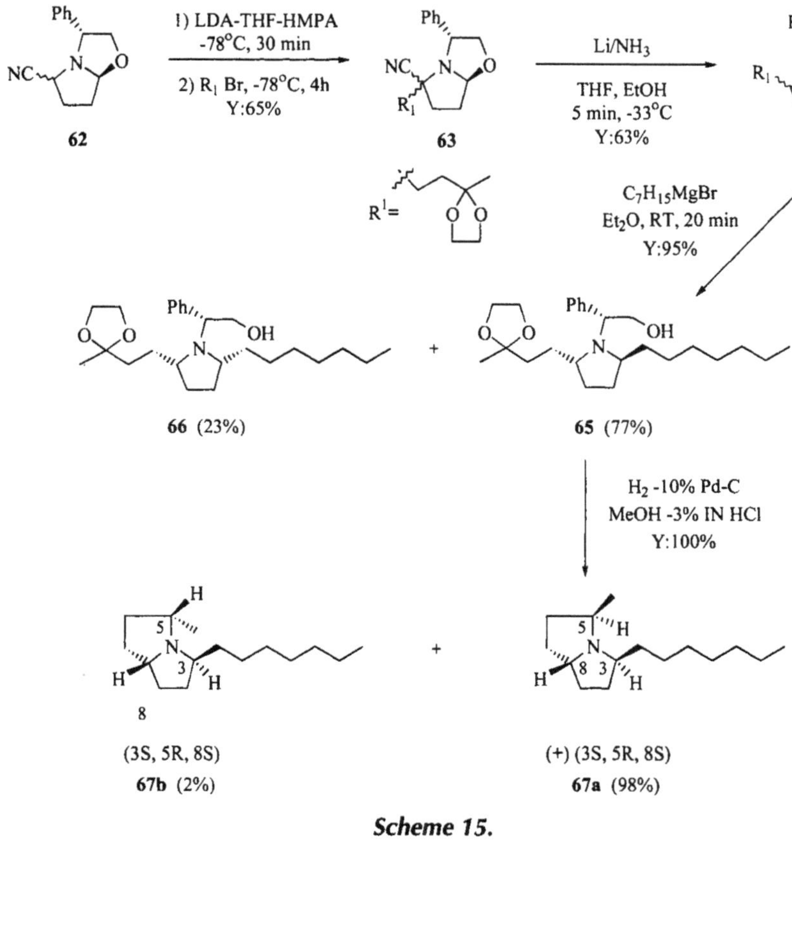

Scheme 15.

NMR techniques. The heptyl chain was introduced at C-5 (pyrrolidine ring numbering) with $C_7H_{15}MgBr$ in 95% yield providing predominantly the 2,5-*trans*-dialkylpyrrolidine (**65**) accompanied by the *cis*-isomer **66** (**65/66** : 77/23).

After separation of the two isomers by silica gel column chromatography, compound **65** was catalytically hydrogenated. Thus the *trans*-isomer **65** was converted in 98% yield to **67a**.

The synthesis of nonracemic pyrrolidines employing a novel bicyclic 1,3-oxazolidine lactam derivative as starting material has been recently reported by Meyers.[57] The seminal reports in this area employed organocuprate addition to the vinylogous amide **68** to furnish **69** which was eventually carried on to nonracemic 2- or 2,3-disubstituted pyrrolidines (**70**) (Eq. 3).[57a,b]

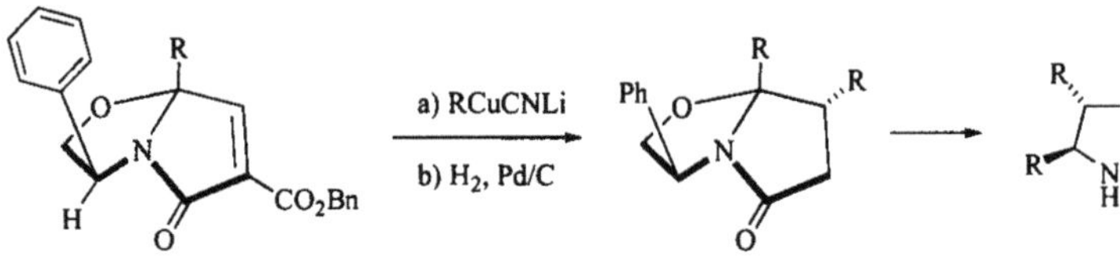

(3)

68 (R-H) **69** (*endo* entry) **70**

It was observed that the selectivity of the conjugate addition was determined primarily by whether or not an angular substituent was present, predominately *endo*-addition occurs when the substituent is present. Alternatively, *exo*-addition occurs when the substituent is absent (Eq. 4). It should be noted that under these reaction conditions the oxazolidine ring remains intact.

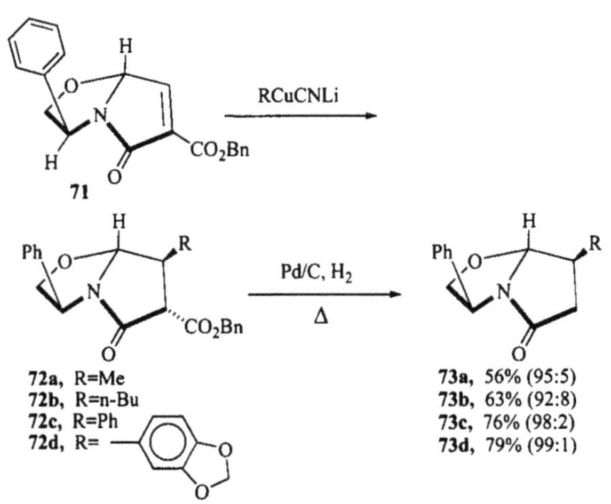

(4)

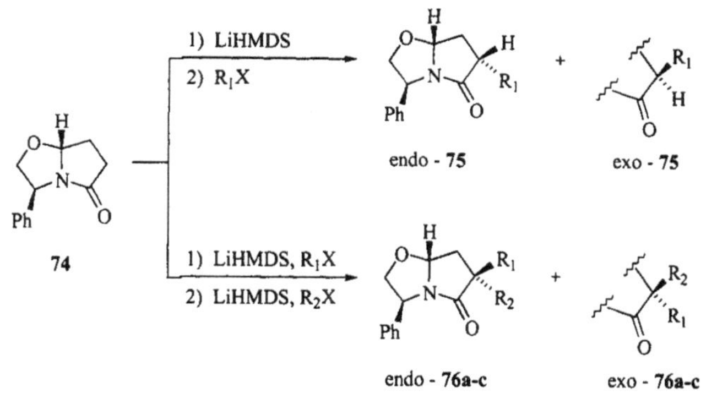

Scheme 16.

Another variation reported by the same laboratory used 1,3-oxazolidinyl-pyrrolidinone (**74**) to synthesize enantiomerically pure 3-mono- and 3,3-disubstituted pyrrolidines via stereoselective enolate alkylation of **74** (Scheme 16 and Eq. 5).[58]

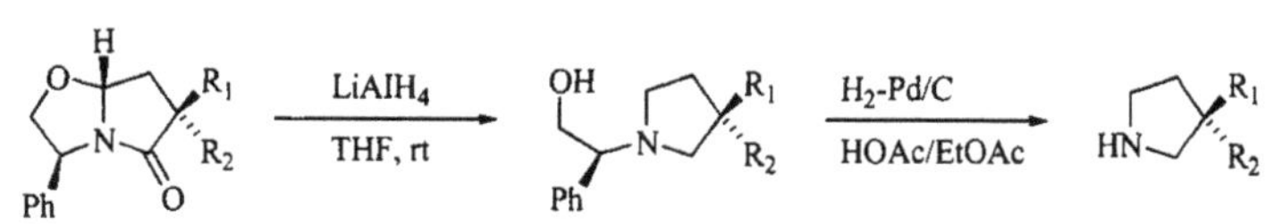

75a,b
76a-c
77
78

(5)

Diastereoselectivity from the first alkylation was usually modest (Table 3, entry 1, 46% de) with the slightly improved exception of methylation (entry 2, 66% de). Alkylation with LiHMDS/allyl bromide gave an *endo/exo* product ratio

Table 3. Mono- and Disubstituted Bicyclic Lactams **75** and **76**

Entry	R_1X	R_2X	*endo : exo*	*Yield %*[a]	*endo/exo Products*
1	$PhCH_2Br$	—	73 : 27	95	**75a**
2	CH_3I	—	83 : 17	97	**75b**
3	CH_2=$CHCH_2Br$	CH_3I	84 : 16	98[b]	**76a**
4	$PhCH_2Br$	CH_3I	82 : 18	95[b]	**76b**
5	CH_3I	4-(CH_3O)-$PhCH_2Br$	82 : 18	99[b]	**76c**

Notes: [a]Combined *endo* and *exo* products.
[b]Yield of second alkylation.

of 60 : 40 (20% de) and an isolated yield of 76%. The disubstituting alkylations typically resulted in better selectivity (entry 3–5), probably due to higher steric congestion around the tertiary enolate.

IV. α- AND β-AMINO ESTERS

Nonracemic β-amino acids are highly important intermediates in certain very attractive syntheses of β-lactam derivatives.[59] As a consequence, several varied approaches have been reported for their synthesis.[60–62] One of these approaches, employed principally by Kunz,[63] is based on a diastereoselective addition of allylsilane to carbohydrate derived Schiff bases followed by oxidative cleavage of the *protected* amino olefin (Eq. 6).

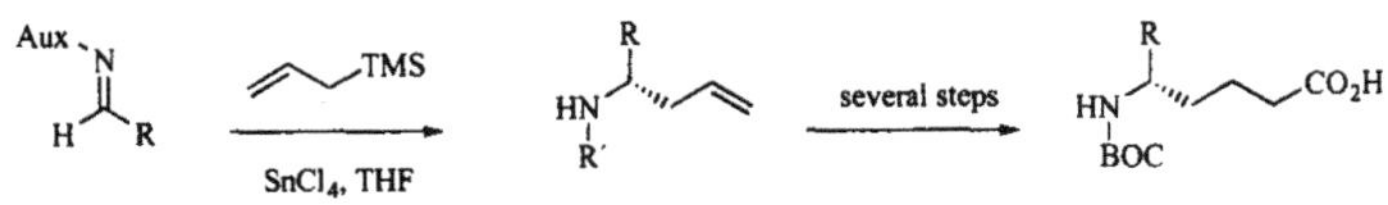

79 80 R′ = Aux 81 (6)

Another entry into this area relied on the more efficiently prepared nonracemic 2-aryl-1,3-oxazolidine (**82**) as a substrate in an allyl organocerium stereoselective addition reaction (~88% de).[64] Allylcerium chloride, prepared in situ by stirring allylmagnesium chloride with cerium chloride,[65] was added to various nonracemic oxazolidines derived from (*R*)-phenylglycinol, yielding amines **83** in excellent chemical yield with remarkable stereocontrol (Table 4 and Eq. 7). The amino alcohol adduct **83** was oxidized upon treatment with lead tetraacetate to yield homoallylamine (**84**), which was subsequently transformed to **81**. The configuration of the amine adduct **83** is dependent only on the configuration of the nonracemic precursor phenylglycinol as previously determined.[8]

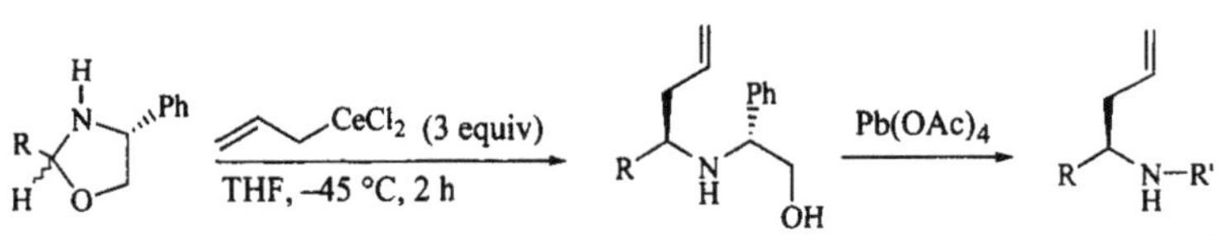

82 **83** **84** R' = H (7)

Other approaches have also been sought to further increase the efficiency of β-amino ester synthesis *without using protecting groups*. After several unsuccessful attempts at employing the zinc derivative of ethyl bromoacetate (Reformatsky

Table 4. Allyl Organocerium Addition to 2-Aryl-1,3-(R)-phenyloxazolidine (**82**)

R	*Crude Yield (83) (R,R), %*	*Diastereoselectivity (Isomeric Ratio)*	*Yield (84)(R)%*	*Enantio-selectivity, %*	$[\alpha]_D^{25}$	*Mp (C) (HCl)*
Phenyl	94	(93:7)	87	86	+36.2 (1.4)	233–235
3-Pyridyl	52	(90:10)	69	92	+48.7 (1.5)	212–214
4-Bromophenyl	98	(94:6)	82	88	+29.8 (1.3)	255–259
1-Naphthyl	99	(94:6)	75	89	+82.3 (1.3)	226–230
4-Tolyl	94	(95:5)	78	78	+25.9 (0.9)	217–219

reagent) and ethyl trimethylsilylacetate as nucleophilic acetates, it was anticipated that the highly stable ethyl tributylstannylacetate (**85**) would serve as a nucleophile under appropriate Lewis acid catalyzed conditions. It was subsequently shown that this addition indeed occurs and with remarkable stereocontrol, despite the prolonged reaction times (Table 5 and Eq. 8).[66]

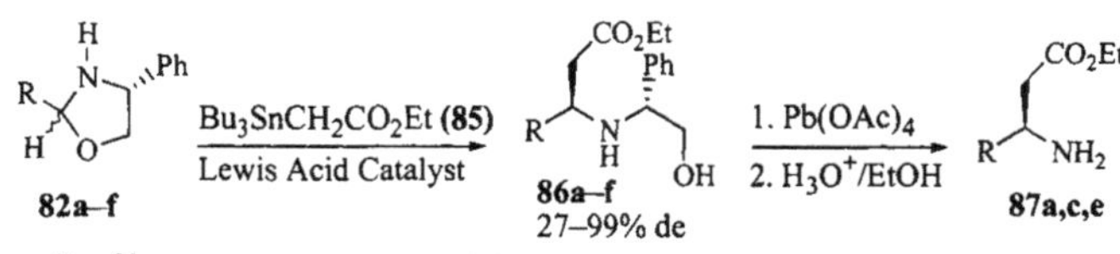

(8)

a, R = Ph **d**, R = 1-naphthyl
b, R = *p*-$CH_3C_6H_4$ **e**, R = cyclohexyl
c, R = *p*-BrC_6H_4 **f**, R = *i*-propyl

Table 5 shows the results of the addition **85** to oxazolidines **82a–f** yielding amino alcohols **86a–f**. Reaction of **82a** with **85** in the presence of $ZnCl_2$ in refluxing THF resulted in only a 39% crude yield of **86a** after a 100-h reflux, but at the same time providing a 98:2 ratio of diastereomeric amino alcohols. Using boron trifluoride etherate as a catalyst on the same substrate accelerated the rate of addition, albeit at the expense of stereoselectivity (28% de).

A variety of other Lewis acid catalysts were also employed (Table 5), but none were as effective as either $ZnCl_2$ or $Et_2O{\cdot}BF_3$. Reasoning that $ZnCl_2$ was acting primarily as a chelator that would restrict flexibility in a transition state similar to (A) and $Et_2O{\cdot}BF_3$ was most probably activating **85** by complexation with the ester carbonyl, combinations of these two catalysts were used. One-half equivalent of each of the Lewis acids, $ZnCl_2$, and $Et_2O{\cdot}BF_3$, represents the best found combination to obtain both good chemical yields and high diastereofacial selectivity (Table 5). The activation of **85** through the use of a fluoride source (e.g. KF + 18-Crown-6) was ineffective.

Table 5. Lewis Acid Catalyzed Addition of Ethyl Tributylstannylacetate (**85**) to 2-Substituted-4-phyenyl-1,3-oxazolidines (**82a–f**) and Conversion of **86a,c,e** to **87a,c,e**

Substrate	*Lewis Acid*	*Rxn Time (h)*	*Yield **86** (%)*	*de, %*	*Yield **87** (%)*	*ee, %*
82a	$ZnCl_2$	100	39	96		
82a	$Et_2O{\cdot}BF_3$	6	82	28		
82a	$Ti(^{i}OPr)_3Cl$	48	43	—		
82a	$TiCl_4$	15	43	40		
82a	$Sn(OTf)_2$	72	53	70		
82c	$ZnCl_2$	72	21	96		
82c	$Et_2O{\cdot}BF_3$	72	68	27		
82a	$ZnCl_2/Et_2O{\cdot}BF_3$	66	71	92	86	91
82b	$ZnCl_2/Et_2O{\cdot}BF_3$	192	58	92		
82c	$ZnCl_2/Et_2O{\cdot}BF_3$	96	67	91	87	95
82d	$ZnCl_2/Et_2O{\cdot}BF_3$	144	60	99		
82e	$ZnCl_2/Et_2O{\cdot}BF_3$	72	33	94	73	94
82f	$ZnCl_2/Et_2O{\cdot}BF_3$	264	43	96		

Equation 9 with its transition state (A) appears to best depict experimental observations which are similar to earlier reports. Namely, the nucleophilic acetate adds across the *re*-face of the imino/oxazolidine moiety, opposite the (*R*)-phenyl of the auxiliary. This highly selective and efficient synthesis of β-amino esters is applicable for synthesizing both antipodal isomers of *aliphatic and aromatic* derivatives.

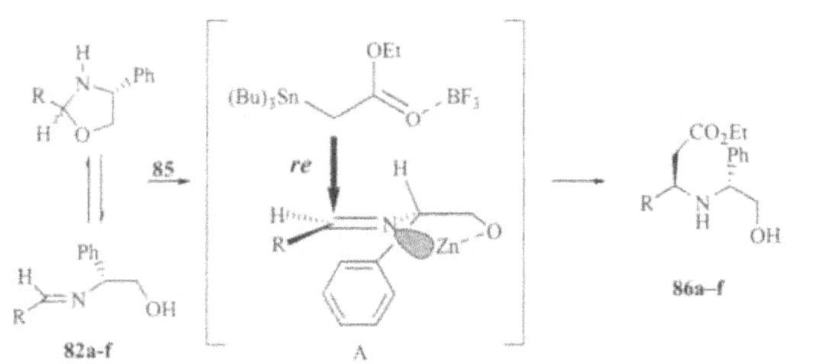

(9)

A Possible Transition State

A recent report by Pedrosa[67] describes the moderately selective addition of the moisture sensitive Reformatsky reagent to a tertiary analog of **82**, i.e. **89a–f** (Eq.

10). The 1,3-oxazolidines **89a–c** readily react with the zinc derivative of ethyl bromoacetate under mild conditions (dimethyl ether, or CH_2Cl_2, 0 °C) affording β-amino esters **90a–c** with moderate stereoselectivity (entries 1–6 in Table 6). Substrates with larger substituents **89d–f** lead to **90d–f** in very good diastereomeric excess (entries 7–10 in Table 6). In addition, the change in solvent of the reaction has very little effect on both the chemical yields and de's (compare entries 4, 6, and 8 vs. 3, 5, and 7 in Table 6).

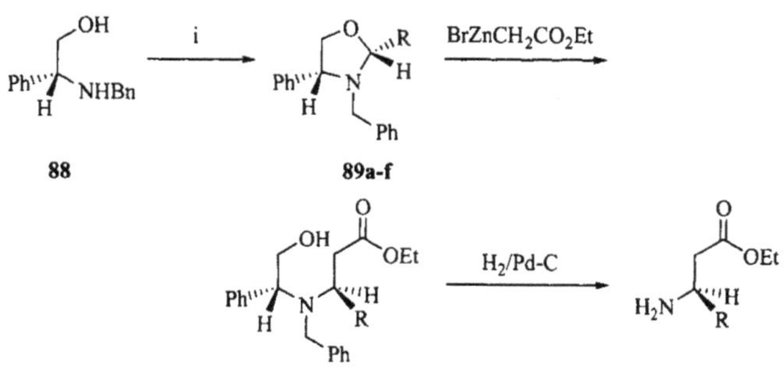

(10)

An exception to this behavior occurs with the 2-isopropyl analog **89g** (entry 11). This compound does not react with the Reformatsky reagent at 0 °C, and it is necessary to heat the reaction mixture to reflux for 3 h, leading to a 69:31 mixture of epimers. This suggests that other sterically bulky 2-substituents may suffer the same fate. This approach was only attempted for the synthesis of aliphatic β-amino esters, presumably, because the use of Pd/C to remove the auxiliary would also cleave the desired amino benzylic bond.

As part of extensive research devoted to the preparation of optically active amines by the organometallic addition to imines carrying a homochiral substituent on the nitrogen, Savoia and Umani-Ronchi[68] reported results of a Barbier-type allylation on imine/oxazolidine **92** (Scheme 17).

A clean reaction took place with the imine/oxazolidine **92a**, affording essentially (*S, S*)-**93a** in > 99% yield and with > 99% stereocontrol. However, it was noted that the diastereoisomeric ratio decreased with increasing reaction time, even after the reaction was complete, unless the reaction mixture was quenched in a timely manner with water. Furthermore, the treatment of an isolated sample of > 99% pure **93** with a catalytic amount of zinc chloride in THF for 3 h led to a 78:22 mixture of diastereoisomers.

By examining the effect of several metal salts in the Barbier reaction with zinc, it was observed that the epimerization was suppressed in the presence of a

Table 6. Enantimoselective Synthesis of Chiral Ethyl β-Amino Carboxylates by Diastereoselective Nucleophilic Ring Opening of 1,3-oxazolidines by Reformatsky Reagents

Entry	*Substrate* (**89**)	*R*	*Reaction Conditions* *Solvent*	*T(°C)*	*t(min)*	*Product* (%)[a](**90**)	*Diast. Ratio*[b]	*β-Amino Esters (%)*[a] (**91**)	*ee, %* (*Conf.*)[c,d]
1	**a**	Me	Et_2O	0	15	**a** (65)	80:20	**a** (71)	60 (*R*)
2	**ent-a**		Et_2O	0	15	**a** (68)	19:81	**ent-a** (73)	62 (*S*)
3	**b**	Et	Et_2O	0	20	**b** (80)	86:14	**b** (69)	72 (*R*)
4	**b**		CH_2Cl_2	0	20	**c** (80)	87:13	**b** (70)	74 (*R*)
5	**c**	*n*-Pr	Et_2O	0	20	**c** (77)	89:11	**c** (70)	78 (*R*)
6	**c**		CH_2Cl_2	0	20	**c** (72)	89:11	**c** (70)	78 (*R*)
7	**d**	Bu	Et_2O	0	30	**d** (74)	96:4	**d** (73)	92 (*R*)
8	**d**		CH_2Cl_2	0	30	**d** (75)	95:5	**d** (72)	90 (*R*)
9	**e**	*i*-Bu	Et_2O	0	60	**e** (63)	93:7	**e** (75)	86 (*R*)
10	**f**	CH_2CH_2 C_6H_5	Et_2O	0	60	**f** (74)	96:4	**f** (76)	92 (*R*)
11	**g**	*i*-Pr	Et_2O	35	180	**g** (67)	69:31	**g** (55)	38 (*R*)

Notes: [a]Yields refer to pure compounds, after column chromatography.
[b]Diastereomeric ratios were measured by ^{13}C NMR.
[c]Enantiomeric excesses were determined by GC or HPLC as indicated in the text.
[d]The given configurations correspond to the major enantiomer in the mixture.

catalytic amount of cerium bichloride heptahydrate since (*S, S*)-**93a** was formed quantitatively with almost complete diastereoselectivity (> 99% de). It was proposed by the authors that the cerium salt acts as a strong Lewis acid activating both the allyl bromide towards zinc and the imine toward allylzinc bromide. The

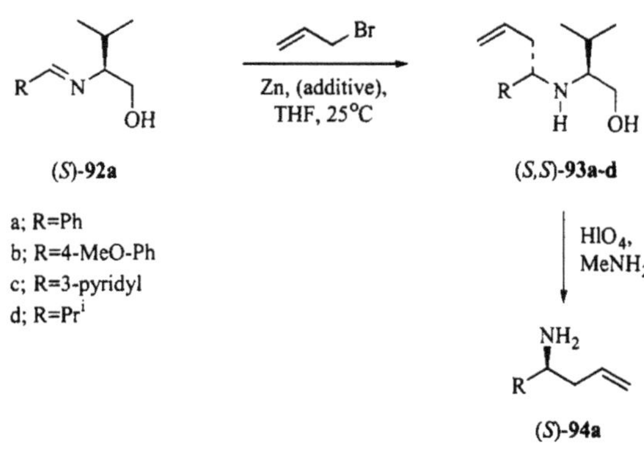

Scheme 17.

purpose of the water is to act simply as a proton source for the organometallic–imine adduct to afford in situ the homoallylic amine **93** and a less acidic zinc hydroxide species. Treatment of (*S*)-**93** with periodic acid in the presence of aqueous methylamine[6b] allowed the optically pure (*S*)-1-phenyl-2-butenamine **94a** to be obtained directly in 88% yield.

The availability of α-phenylglycinol in both enantiomeric forms serves as an excellent chiral auxiliary for the diastereoselective Strecker synthesis of optically pure (L- and D-isomers) as well as nonprotegenic D-amino acids. In addition, the facile removal of the chiral auxiliary makes it an ideal choice for the synthesis of α-aryl glycines, which are among the most difficult amino acids to obtain in optically pure form. Thus, imine/oxazolidine **95** was treated with trimethylsilyl cyanide to yield **96** in moderate diastereoselectivity (~85:15) (Eq. 11). With phenylacetaldehyde, essentially no selectivity was observed.[69]

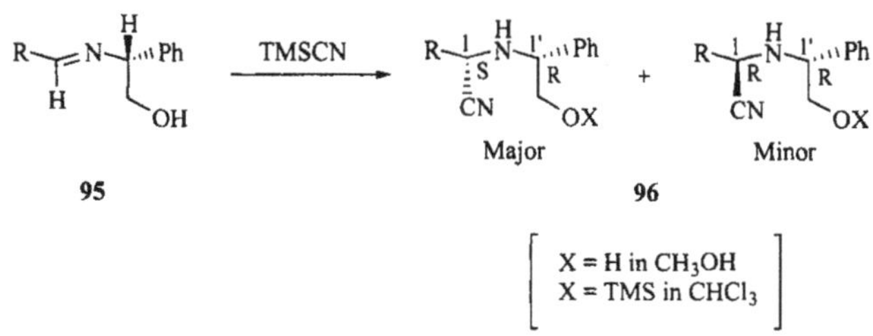

(11)

The Ahn–Eisenstein hypothesis[70] was invoked to explain the formation of (1*S*, 1′*R*)-**96a** as the major diastereomer. Accordingly, nucleophilic attack on the imine should take place antiperiplanar to the α-phenyl group. This group, by having the lowest allylic σ*-orbital energy, occupies, in the reactive conformer, a position perpendicular to the imine π-plane. This enables maximum overlap with the π*-orbital, thereby lowering its energy and minimizing the free energy of activation for reaction. The two possible transition states are conformations **A** and **B** (Eq. 12) where the phenyl group is perpendicular to the imine π-plane. Of these, conformer **A**, in which the attack takes place from the *re*-face of the imine giving rise to major (1*S*, 1′*R*)-**96a** diastereomer, is more favorable not only for steric reasons, but also for the extra stabilization through intramolecular 5-membered hydrogen bonding. This explains the enhanced diastereoselectivity in comparison with α-methylbenzylamine as chiral auxiliary where such stabilization is not possible.

Reactions conducted in THF, which acts as a strong H-bond acceptor, resulted in loss of diastereoselectivity because of loss of intramolecular H-bonding through solvation of the hydroxyl group. Addition of magnesium chloride

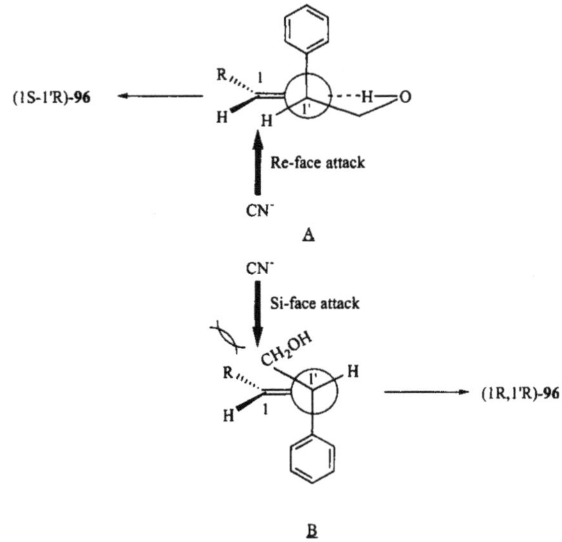

(12)

restored the five-membered transition state through chelation and diastereoselectivity was regained. The best results were obtained in $CHCl_3$ (dielectric constant 4.8).[69]

Chakraborty applied the above discussed TMSCN reaction conditions in $CHCl_3$ toward the total synthesis of the antibiotic teicoplanin, a glycopeptide structurally related to vancomycin and ristocetine.[71a] To obtain the required *S* configuration of an amino acid functionality within the teicoplanin structure, it was necessary to employ the (*R*)-stereoisomer of α-phenylglycinol as chiral auxiliary to form **97** (Scheme 18). Sequential addition of methanol to the reaction mixture (to prevent the formation of O-silylated product) followed by TMSCN gave a mixture of diastereomers in a ratio of 73:27. The major product **98** in ethanol was treated with ethereal HCl to convert the nitrile to the ester **99**. The chiral auxiliary was removed by oxidative cleavage using lead tetraacetate to furnish aldimine (**100**) which was hydrolyzed with dilute HCl to amino ester **101**. A similar protocol was used to carry out a second Strecker reaction in the same synthesis, but using the (*S*)-isomer. In this case, with the bulkier substrate, a 89:11 ratio of isomers was obtained. Zhu employed a similar strategy in his synthesis of the C-O-D rings of vancomycin.[71b,c]

Addition of diethylaluminum cyanide[72] to the *N*-benzyl analog of **95**, (2*R*, 4*R*)-3-benzyl-2-methyl-4-phenyl oxazolidine (**102**), which is readily prepared

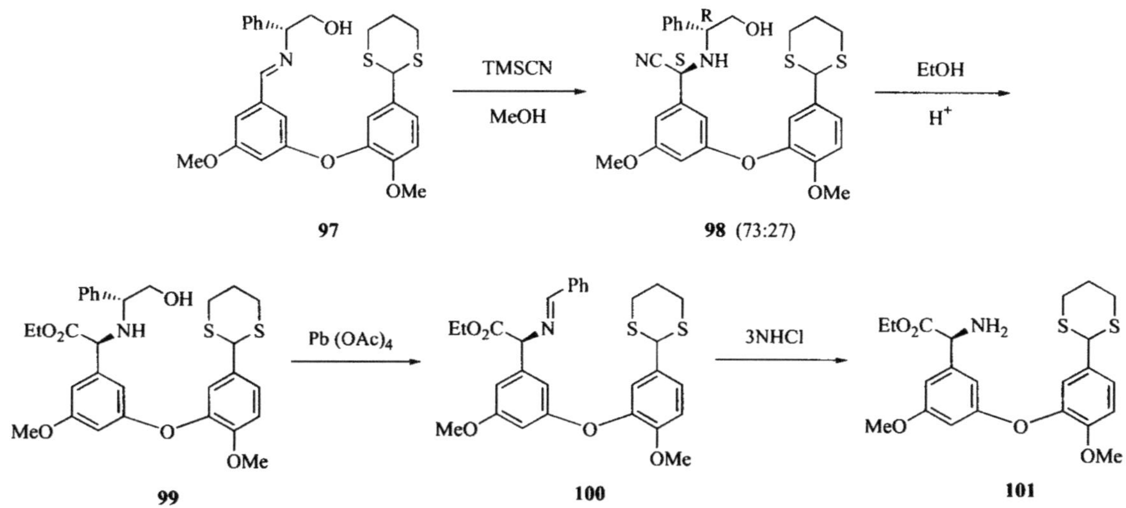

Ph
OH
N
S
S
MeO
O
OMe
97
TMSCN
MeOH
Ph
R
OH
NC
S
NH
S
S
MeO
O
OMe
98 (73:27)
EtOH
H+
Ph
OH
EtO2C
NH
S
S
MeO
O
OMe
99
Pb (OAc)4
Ph
EtO2C
N
S
S
MeO
O
OMe
100
3NHCl
EtO2C
NH2
S
S
MeO
O
OMe
101

Scheme 18.

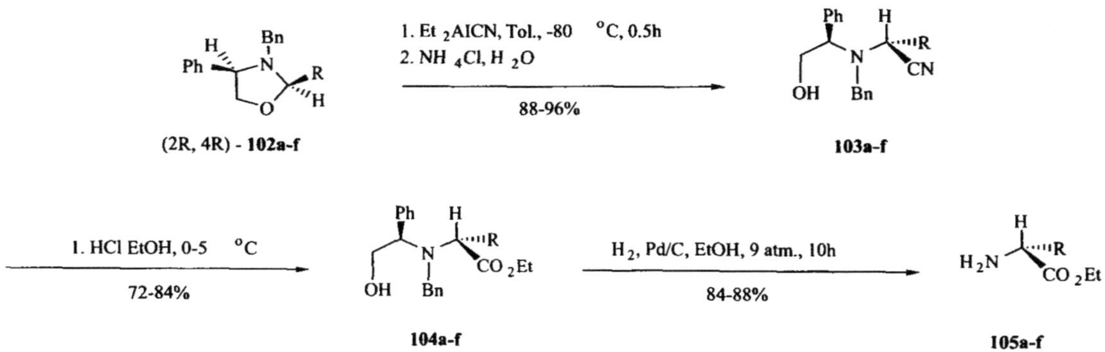
Bn
H
Ph
N
R
H
O
(2R, 4R) - 102a-f
1. Et 2AlCN, Tol., -80 °C, 0.5h
2. NH 4Cl, H 2O
88-96%
Ph
H
R
N
CN
OH
Bn
103a-f
1. HCl EtOH, 0-5 °C
72-84%
Ph
H
R
N
CO 2Et
OH
Bn
104a-f
H 2, Pd/C, EtOH, 9 atm., 10h
84-88%
H
R
H 2N
CO 2Et
105a-f

Scheme 19.

from (–)-(*R*)-*N*-benzyl-2-phenylglycinol (**88**), has also been shown to occur in very good chemical yields with moderate to good enantiomeric excess (Scheme 19). The isomeric ratios ranged from 79:21 to 93:7. All attempts to debenzylate the nitrile **103** failed, leading to recovered starting materials. Instead, α-amino nitriles **103** were easily transformed, without loss of the stereochemical integrity, into esters **104** by treatment with saturated ethanolic HCl. Debenzylation then led to (*R*)-α-amino esters **105** in excellent chemical yields with good to moderate diastereoselectivities.

The stereoselectivity in the nucleophilic ring opening of nonracemic oxazolidines by Et_2AlCN is reported to proceed much better than with a TMSCN/$ZnBr_2$ combination in the key step in a synthesis of (+)-Solenopsin A,[51b] whereas the reaction of nonracemic acetals with trimethylaluminum is reported to be not selective.[73] This stereochemical result may be explained as a preferential complexation of the diethylaluminum cyanide to the oxygen in the heterocycle, followed by an "invertive"[74] substitution of the departing oxygen atom by the entering cyanide group.

Pedrosa has recently reported a novel synthesis of α-aminocarboxylates **105** starting with 2-ethoxycarbonyl-1,3-oxazolidine (**106**), thereby beginning with the carboxyl functionality already in place and using organozinc derivatives as nucleophilic agents (Scheme 20).[75a]

The condensation of (–)-(*R*)-*N*-benzyl-2-phenylglycinol (**88**) with the methyl hemiacetal of ethyl glyoxylate in refluxing CH_2Cl_2 for 2 h in the presence of molecular sieves gave the usual epimeric mixture (83:17) of 2-ethoxycarbonyl-

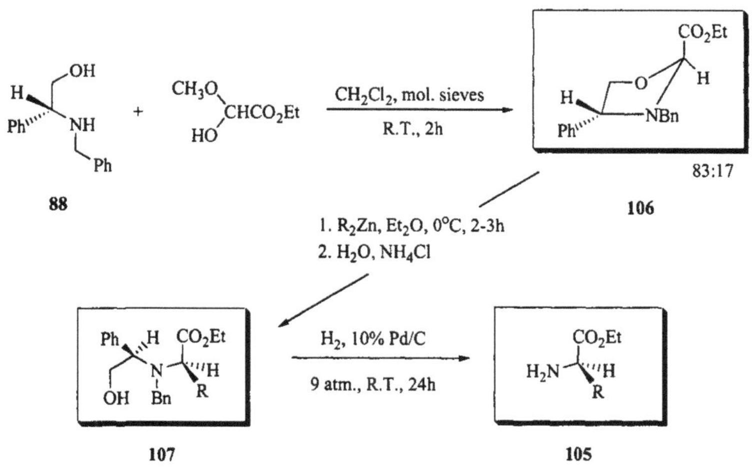

Scheme 20.

1,3-oxazolidines (**106**). A pure isomer was isolated and allowed to react with several dialkylzincs to furnish the ring opened compounds **107** with good to excellent diastereoselectivity. Compound **107** was debenzylated by hydrogenolysis while maintaining the stereochemical integrity to yield (*R*)-ethyl α-aminocarboxylate (**105**). The sense of asymmetric induction in the ring opening of **106** was explained by analogy to the behavior of nonracemic acetals.[75b,c]

V. 1,2-ADDITION VERSUS CONJUGATE ADDITION

A logical extension of 1,2-addition to 1,3-oxazolidines would be to examine conjugate addition pathways that are available with appropriately substituted substrates. Nonracemic oxazolidines derived from ephedrine have been used as 1,3-asymmetric inductors in cuprate additions to unactivated double bonds for the stereoselective synthesis of β-alkyl aldehydes with fairly good results.[76] As part of a study regarding π-face differentiation induced by a norephedrine derived oxazolidine, Scolastico[45,77] examined conjugate addition to unsaturated 1,3-oxazolidinyl esters **109** (Scheme 21).

Scolastico[45,77] and co-workers have reported that 2-(alk-1-enyl)oxazolidines bearing an electron-withdrawing group at the 3-position (e.g. **109**), which are available by treating the corresponding acetals with the appropriate norephedrine derivative in the presence of pyridinium tosylate, are perfectly stable to water and to chromatography. They report that cuprate reagents add to **109** with excellent

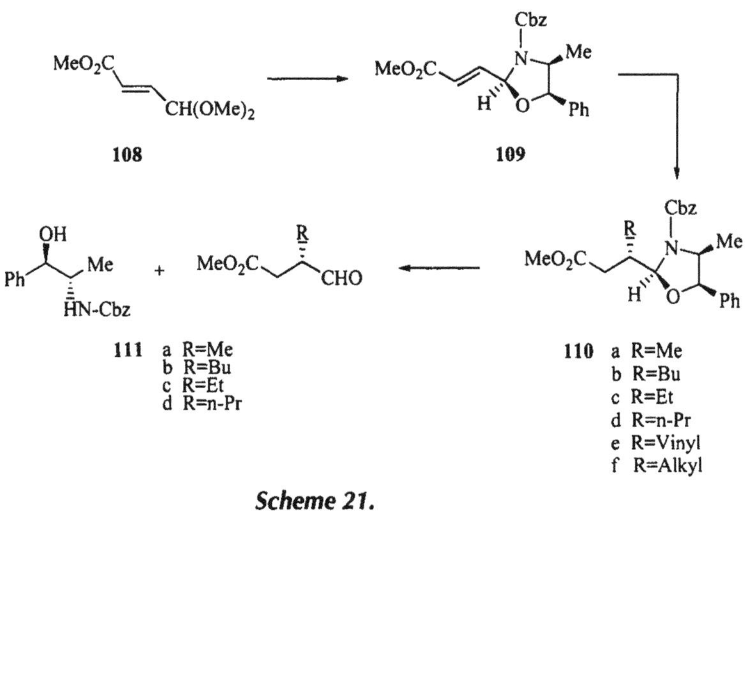

Scheme 21.

Table 7. Cuprate Addition to **109**

Entry	*Cuprate Reagent*	*Temp.*	*Product* (**110**)	*Diast. Ratio*	*Config.*	*Yield %*
1	Me_2CuLi	-25 °C	**a**	≥95:5	*S*	70
2	Me_2CuLi/Me_3SiCl	-78 °C	**a**	≥95:5	*S*	72
3	Bu_2CuLi	-25 °C	**b**	≥95:5	*S*	70
4	Et_2CuLi	-25 °C	**c**	≥95:5	*S*	70
5	$(H_2C{=}CH)_2CuLi$	-50 → -25 °C	**e**	≥95:5	*R*	75
6	$(H_2C{=}CHCH_2)_2CuLi$	-78 °C → RT	**f**	89:11	*S*	54

diastereoselectivity (> 95:5) to give, in good yield, esters **110a–f** (Table 7). Adducts **110** were transformed into aldehydes **111a–d** of known configuration by a two-step conversion that allows for the recovery of the intact chiral auxiliary. Cuprate addition occurs from the *si*-face, even in the case of the allylic cuprate (Table 7, entry 6), which usually exhibits diastereofacial selection opposite to that of aliphatic and vinylic reagents.[78] Similar selectivity is observed when the reaction is conducted in the presence of trimethylsilyl chloride (entry 2).[77]

Although the mechanism of cuprate reagents in conjugate addition is still open to question, the authors[77,79] propose reversible formation of a cuprate–substrate (d–π*) complex and a Cu^{III} β-adduct in the initial step to explain their selectivity. The stereochemical outcome of nucleophilic additions to an α,β-unsaturated carbonyl system has been interpreted in terms of Felkin–Ahn models **A** and **B** (Figure 7). On the basis of MO considerations, these models are believed to be the most likely conformations involved. For most favorable mixing of the π*-orbital with the lowest energy σ*-orbital, the most electronegative substituent must occupy a perpendicular position,[80] but conformation **A** would be favored on steric grounds.

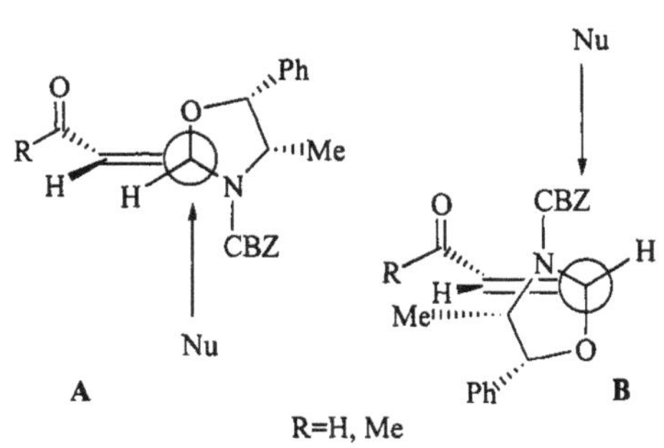

Figure 7. Transition state models for conjugate addition.

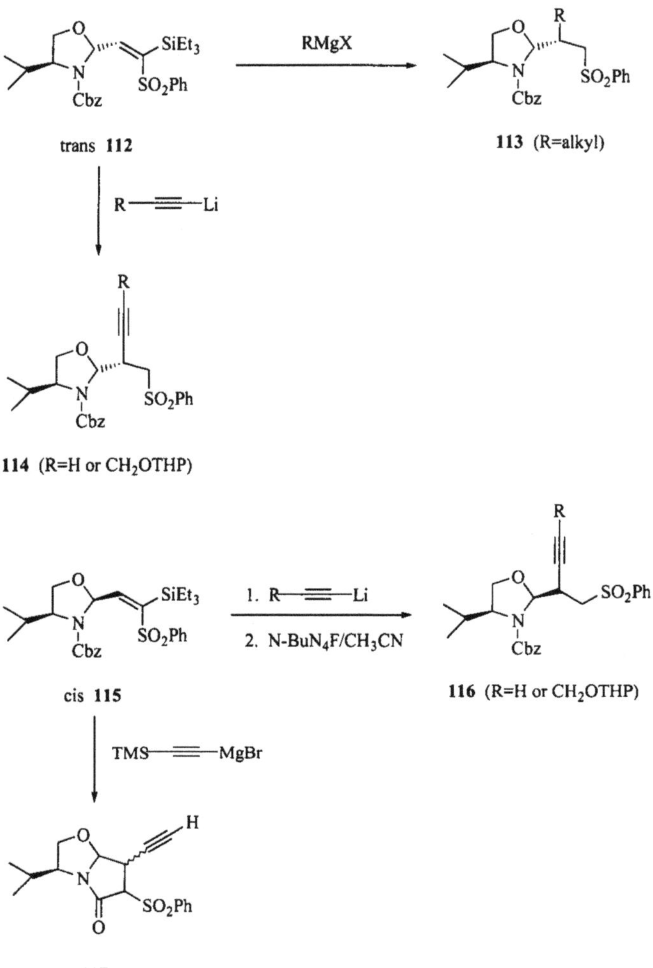

Scheme 22.

Isobe found that the valinol derived oxazolidine **112** reacts with organometallics with interesting regio- as well as stereochemical selectivity. Addition of alkyl Grignard reagents such as MeMgBr, EtMgBr, *n*-BuMgBr to the *trans*-oxazolidine **112** in THF at 0 °C, followed by treatment with *n*-BuN_4F in CH_3CN, afforded only α-product **113** in 72–75% yield (Scheme 22).[81a] Conjugate cyanide addition to 2-alkenyl-4,5-1,3-oxazolidines using diethyl aluminum cyanide has also

Table 8. 1,2-Addition of Organometallic Reagents to (*R*)-2-(1-Naphthyl)-4-phenyl-1,3-oxazolidine (**118**) and Oxidative Cleavage Results

Entry	*Cpd.*	*RM*	*Yield, %* **119**	*de, %*	*Yield, %* **120**	*ee, %*
1	**a**	CH_3Li (−78 °C)	46	95		
2	**a**	CH_3CeCl_2 (−45 °C)	75	96	61	>99
3	**a**	$CH_3Cu{\cdot}BF_3$(−78 °C)	71	52	80	54
4	**b**	$C_2H_5CeCl_2$ (−45 °C)	75	>99	57	>99
5	**c**	BuLi (−78 °C)	81	26		
6	**d**	PhLi (−78 °C)	48	>99		

been reported.[81b] Alkynyl Grignard reagents, under similar conditions, reacted to give a 1:1 mix of diastereoisomers. The analogous alkynyllithiums added, on the other hand, to **112** to yield the desilylated products **114** in an average of 60% yield. The stereochemistry of the acetylenic adducts **114** was found to be the opposite that of the alkyl α-adducts, i.e. β-addition. These results were proven through conversion of the products to common compounds.

Addition of carbon nucleophiles to the *cis*-oxazolidine **115** gave slightly different results than did the *trans*-isomer. Employment of alkyl metals (MeLi, EtMgBr, etc.) did not yield any addition product, but two alkynyllithiums reacted to produce β-isomers such as **116** in 46% and 61% yield, respectively, after removing the silyl group. When TMS ethynyl Grignard was added to **115**, a new γ-lactim **117** was isolated, after *n*-Bu_4NF treatment, as the only product in 59% yield. The stereochemistry of acetylenic addition was shown to be β through a different independent synthesis.

On expansion of a study of organometallic additions to 1,3-oxazolidines to include naphthalenes, Pridgen[82] observed that 2-(1-naphthyl) derivatives **118** underwent exclusively *1,4-addition with Grignard reagents and 1,2-addition with lithium, cerium, and copper organometallic reagents* (Tables 8 and 9). Such a result, particularly for a cuprate, was highly surprising in light of its propensity for predominantly conjugate-type addition. Two of the amino alcohol products resulting from the 1,2-addition were ultimately converted via oxidative cleavage to enantiomerically enriched α-(1-naphthyl)alkylamines analogous to the aryl

Table 9. 1,4-Addition of Grignard Reagents to (*R*)-2-(1-Naphthyl)-4-phenyl1,3-oxazolidine (**118**)

Entry	*Cpd.*	*R*	*Yield, %* **122**	*ee %*
1	**a**	Ph	50	93 (94)
2	**b**	Bu	83	93
3	**c**	Et	65	96

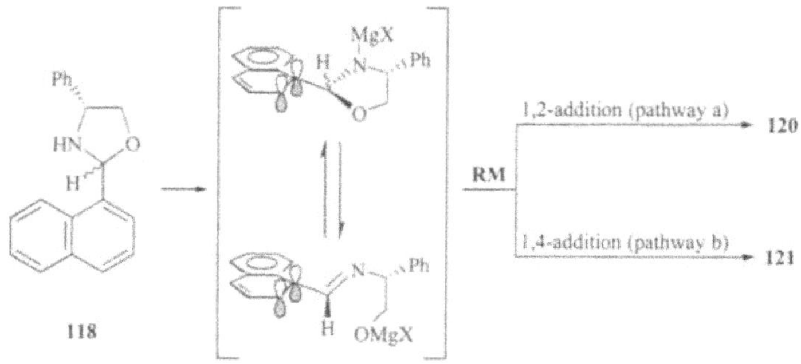

Scheme 23.

examples reported above. The 1,4-conjugate addition products, which are enantiomerically enriched disubstituted 1,2-dihydronaphthalenes, have been previously reported by Meyers and demonstrated by him to be valuable intermediates in syntheses of the natural products (–)-podophyllotoxin,[83] (+)-phyltetralin,[84] and the AB-ring of aklavinone.[85] Tomioka has also formed these same dihydronaphthalenes by organolithium addition to the cyclohexylimine of 1-naphthalenecarboxaldehyde in the presence of an enantiomerically enriched diether cocatalyst.[86] The majority of the previous work on organometallic addition to optically active α,β-unsaturated imines and/or oxazolines originated from the laboratories of Tomioka[87] and Meyers.[88,89]

Organometallic addition to the naphthalene nucleus has been limited so far to lithium reagents. In Tomioka's report of organolithium addition to naphthalene carboxylates, predominantly *cis*-addition products were obtained, although epimerization to *trans*-derivatives was easily accomplished with sodium methylate in THF. The report by Pridgen[82] that organomagnesium reagents add to the naphthalene ring in THF yielding *trans*-disubstituted 1,2-dihydronaphthalenes, apparently was the first report of organomagnesium reagents being capable of adding to the naphthalene nucleus.

Oxazolines,[90] imines,[86,91] carboxylates,[87] and now apparently oxazolidines, activate, through induction, the *ortho*-position of the naphthyl ring toward nucleophilic attack. Thus, the proposed mode of addition accounting for the subsequently obtained absolute stereochemistry of the resulting product, is analogous to that depicted previously in Scheme 23. The strong Lewis base character of both the imino nitrogen and amino alcohol oxygen results in very strong complexation to at least one-half equivalent of Grignard reagent, thus limiting the flexibility of the transition state. Consequently, nucleophilic attack occurs predominantly from the less sterically demanding face, namely that opposite the 4-phenyl. Attack by lithium, cerium, and copper reagents occur at the hard aminal/

imino carbon (pathway a) leading to 1,2-addition products (Table 8). It should be noted that all the organolithium derivatives gave exclusively 1,2-addition products, with butyllithium being almost nonselective. The resulting disubstituted amino alcohols were oxidatively cleaved to enantiomerically enriched α-(1-naphthyl)alkylamines. Pathway b requires attack at the softer electrophilic γ-carbon and is the exclusive pathway for Grignard addition in this ring system (Table 9).

Lithium, cerium, and magnesium organometallic reagents exhibited addition behaviors similar to that presented above for **118** when they were reacted with oxazolidine **123**, the cinnamyl derivative of **118**. Scheme 24 shows the results where lithium and cerium derivatives added in a 1,2-fashion, while Grignard reagents afforded 1,4-addition products. Reaction of **123** with ethyl and butyl organomagnesium reagents yielded **124** (40% yield, 96% ee) and **125** (42% yield, 98% ee), respectively. Conversely, reaction of **123** with butyllithium or butylcerium chloride gave **126** in 40% yield (64% de) and 63% yield (96% de), respectively.

Thus, the choice of metal becomes important in adding organometallics to 2-(l-naphthyl)-1,3-oxazolidines or to 2-cinnamyl-1,3-oxazolidines. Either conjugate addition products or products resulting from addition at the aminal/imino carbon are obtained in high stereoselectivity. The stereoselectivities reported herein for the Grignard 1,4-additions are comparable to those reported by Meyers[90] and Tomioka.[87] Thus, this work as well as the previously reported

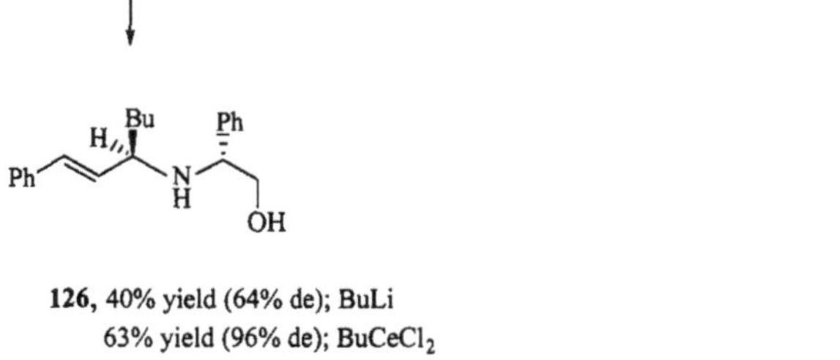

Scheme 24.

work,[8] establishes organoceriums as the organometallics of choice for nucleophilic 1,2-addition to 2-substituted 1,3-oxazolidines.

In 1984, Meyers demonstrated a tandem organolithium/electrophile addition procedure on nonracemic 1-naphthyloxazolines yielding stereoselectively nonracemic 1,1,2-trisubstituted 1,2-dihydronaphthalenes.[92] The intermediate oxazolinyl 1,2-dihydronaphthalene derivative was subjected to nitrogen alkylation, sodium borohydride reduction and/or a second reduction which yielded either nonracemic trisubstituted naphthyl aldehydes or carbinols in > 95% ee.[84] The absolute configuration of the adducts was confirmed by X-ray crystallography and was thought to be dependent on the configuration of the C-4 substituent on the oxazolinyl ring.

Pridgen et. al.,[93] therefore, sought to determine if a Grignard tandem addition procedure could be applied to the oxazolidinyl/imino system with its potential multisites of alkylation (Eq. 13).

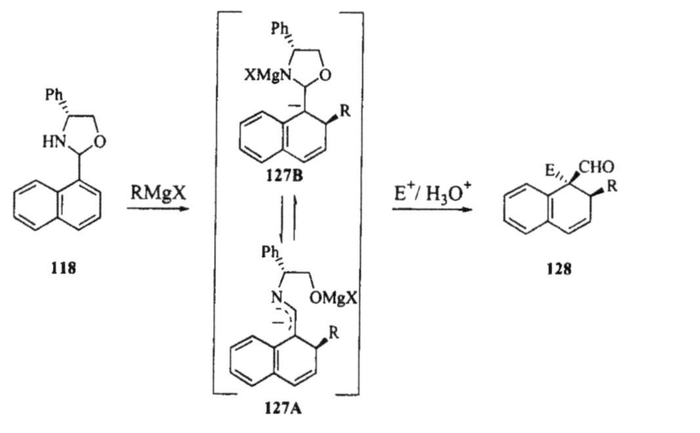

(13)

There are several advantages for choosing an oxazolidinyl system over the oxazolines for the synthesis of nonracemic 1,2-dihydronaphthalenes **128**:

1. In most cases, the more easily prepared organomagnesium reagents are sufficiently nucleophilic to effect the initial diastereomeric carbon–carbon bond formation. This avoids the necessity of having to form, through stannyl transmetallation, the lithium reagent which is required in the oxazolinyl series.
2. The C-2 oxazolidinyl carbon is in the appropriate oxidation level for simple hydrolysis to an aldehyde. Consequently, the alkylation/BH_4-

reduction sequence used to lower the oxidation level at C-2 of oxazolines is no longer a requirement. Facile acidic hydrolysis of the reaction mixture after sequential addition of the Grignard reagent and electrophile as specified in Table 10 is all that is required to transform achiral 1-naphthalenecarboxaldehyde to nonracemic trisubstituted **128** in the same flask.

3. The low temperature restraints typically required to achieve highest stereoselectivity for oxazolines are not required for the oxazolidines. A temperature range of –45 °C to ambient was found to be optimum.

As evident from the results in Table 10, the selectivity of the tandem additions are both exceptionally high. Furthermore, the isolated yields are remarkably high for this combined two-step reaction. It was found that adherence to the temperature constraints as described in Table 10 to be crucial since it was observed that the diastereoselectivity for the Grignard addition eroded at higher temperatures. At much lower temperatures Grignard addition did not occur. In all examples, the diastereofacial selectivity for the Grignard addition as well as electrophilic trap-

Table 10. Organometallic Additions to 2-(1-Naphthyl)-4-(4*R*)-phenyl-1,3-oxazolidines (**118**)

Entry	*R*	*E*$^+$	*Reaction Conditions, THF*	**128**,[a] *(Yield %)*	**128**,[b] *(% ee)*	*[a]D (CHCl*$_3$*)*
1	Et	Me	–45 °C (2 h) → 0 °C (18 h); MeI (0 °C) → rt (18 h)	86	97	–80.2 (c 1.0)
2	Et	allyl	–45 °C (2 h) → 0 °C (18 h); allyl iodide (–45 °C) → rt (18 h)	82	98[c]	+19.1 (c 0.8)
3	Et	CO_2Et	–45 °C (2 h) → 0 °C (18 h); $ClCO_2Et$ (–45 °C) → rt (18 h)	89	98	+167.3 (c 1.3)
4	Me	Me	–45 °C (2 h) → rt (18 h); MeI (–60 °C) → rt (18 h)	31	98	—[d]
5	Bu	Me	–45 °C (2 h) → 0 °C (4 h); MeI (0 °C) → rt (18 h)	87	98	–36.5 (c 1.0)
6	Ph	Me	–45 °C (2 h) → rt (24 h); MeI (0 °C) → rt (18 h)	84	99	+213.0 (c 0.8)
7	Vinyl	Me	–45 °C (2 h) → 0 °C (18 h); MeI (0 °C) → rt (18 h)	70	74	+60.8 (c 1.1)[c,d]

Notes: [a]Flash chromatographed yields. All new compounds gave satisfactory combustion analyses and/or spectral data.

[b]The enantiomeric purity was determined in most cases by HPLC on a CHIRALCEL OD or OJ column. The opposite enantiomer was prepared for each example to be certain of proper peak assignments.

[c]The enantiomeric purity was determined by ^1H NMR (400 MHz) using [Eu(hfc)$_3$] shift reagent on the acetate of the $NaBH_4$ reduced aldehyde.

[d]Contained isomeric material.

ping appears to be almost totally selective. This method of forming nonracemic trisubstituted 1,2-dihydronaphthalenes appears to be quite general and highly efficient. It was, however, observed that *t*-butyl and *s*-butyl Grignard reagent gave complex mixtures while a methyl Grignard reagent gave lower than expected yields. For the latter, about 5% of an unidentified diastereomeric isomer was detected, although enantioselectivity still remained high (entry 4).

It is assumed, based on the similarity of the stereochemical outcome, that the stereoselectivity of the Grignard addition as well as the electrophilic trapping of the azaenolate react as previously reported by Meyers.[84,92,94,95] Namely, the overall result is net *trans*-addition beginning with the Grignard reagent adding on the face remote from the stereogenic center on the magnesium complexed chiral auxiliary.

VI. SILYL ENOL ETHER ADDITION

The use of heteroatom dienophiles in the construction of nitrogen heterocycles through the use of imino–dienophiles in the Diels–Alder reaction has been the focus of several research efforts.[96] The diastereoselectivity of this reaction has been investigated with α-alkoxy imines,[97] imines with carbohydrate templates,[98] imines with 2-morpholinobutadienes,[99] and imines derived from esters of amino acids.[100] Yamamoto has reported success through the use of nonracemic Lewis acids to promote asymmetric Diels–Alder reactions.[101] In some cases, an addition–cyclization pathway was suggested, while in others a cycloaddition pathway was suggested. Because 1,3-oxazolidines behave as imines toward nucleophilic dienes, that class of heterocycles has recently been investigated for use as substrates in a cycloaddition study.[102]

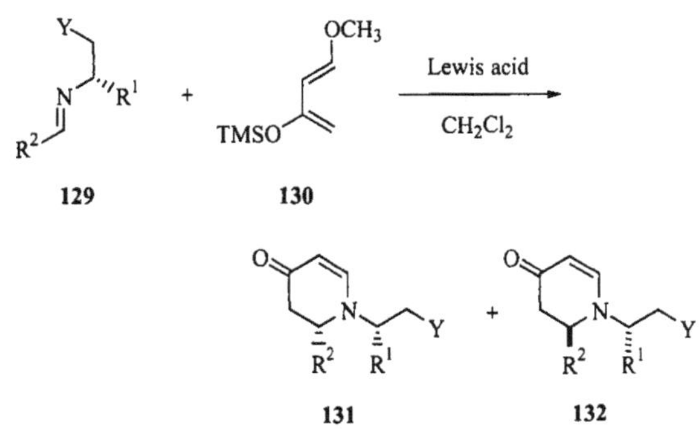

(14)

Reaction of **129** with Danishefsky's diene (**130**)[104] gave interesting results,[102] as reported in Table 11. The reaction (Eq. 14) was normally carried out employing 1–1.5 equiv of Lewis acid ($Zn(OTf)_2$, $BF_3{\cdot}OEt_2$, and $SnCl_2$) in dichloromethane at room temperature.

Several trends in Table 11 were pointed out by the authors.[102] The steric requirements of the stereodirecting R^1 group have a direct influence on the stereoselectivity. As the steric size increases (e.g. methyl, isopropyl, phenyl), the stereoselectivities are enhanced, e.g. entries 3, 4, 7, 8, and 11. Unfortunately, a *t*-butyl or larger group was not included to more fully explore the steric requirements of R^1. For the aldimine, very little difference was observed between aliphatic or aromatic groups (entries 3, 4, 15, and 16) or the electronic nature of the substituents on the aromatic groups (entries 3, 4, 12, and 13).

Binding on oxygen to form a stable cyclic chelated structure (**133**, Figure 8) appears to have a definitive effect on stereoselectivity. For example, imine **129i**, which cannot form a cyclic chelate, gave a lower stereoselectivity than imine

Table 11. Cycloaddition of Imines **129a–n**

Entry	*Imine* (**129**)	R^1	R^2	*Y*	*Lewis Acid*	**131**:**132**	*Yield (%)*
1	**a**	*i*-Pr	Ph	OCH_3	Et_2AlCl	72:28	24
2	**a**	*i*-Pr	Ph	OCH_3	Me_3Al	—	—
3	**a**	*i*-Pr	Ph	OCH_3	$BF_3{\cdot}Et_2O$	79:21	55
4	**a**	*i*-Pr	Ph	OCH_3	$Zn(OTf)_2$	87:13	67
5	**a**	*i*-Pr	Ph	OCH_3	$SnCl_2$	83:17	66
6	**a**	*i*-Pr	Ph	OCH_3	$ZnCl_2$	76:24	33
7	**b**	CH_3	Ph	OCH_3	$BF_3{\cdot}Et_2O$	53:47	54
8	**b**	CH_3	Ph	OCH_3	$Zn(OTf)_2$	59:41	63
9	**c**	CH_2Ph	Ph	OCH_3	$BF_3{\cdot}Et_2O$	73:27	53
10	**c**	CH_2Ph	Ph	OCH_3	$Zn(OTf)_2$	73:27	47
11	**d**	Ph	Ph	OCH_3	$Zn(OTf)_2$	87:13	51
12	**e**	*i*-Pr	*p*-Ph-NO_2	OCH_3	$BF_3{\cdot}Et_2O$	77:23	57
13	**f**	*i*-Pr	*p*-Ph-OCH_3	OCH_3	$Zn(OTf)_2$	89:18	79
14	**g**	*i*-Pr	*t*-Bu	OCH_3	$Zn(OTf)_2$	—	—
15	**h**	*i*-Pr	*i*-Pr	OCH_3	$BF_3{\cdot}Et_2O$	77:23	48
16	**h**	*i*-Pr	*i*-Pr	OCH_3	$Zn(OTf)_2$	72:28	50
17	**i**	*i*-Pr	Ph	H	$Zn(OTf)_2$	72:28	60
18	**j**	Ph	Ph	H	$Zn(OTf)_2$	80:20	69
19	**k**	Ph	*n*-Pr	OTMS	$Zn(OTf)_2$	58:42	45
20	**l**	H	Ph	CO_2CH_3	$Zn(OTf)_2$	—	—
21	**m**	CH_3	Ph	CO_2CH_3	$Zn(OTf)_2$	>95: <5	60
22	**n**	Ph	Ph	OH	$BF_3{\cdot}Et_2O$	>95: <5	60

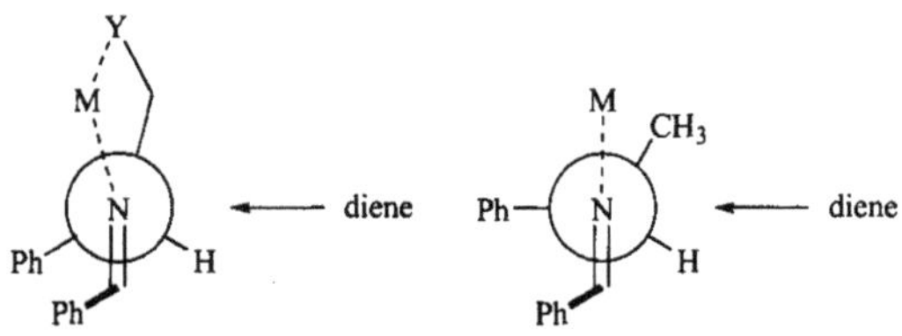

133A 133B

Figure 8. Transition state model for Diels–Alder type additions.

129a (entry 17 vs. 4). Imine **129j**, which also cannot form a cyclic chelate, gives lower stereoselectivity than its counterparts, imines **129d** and **129n** (entry 18 vs. entries 11 and 22). Imines, wherein the oxygen available for chelation is highly Lewis basic (e.g. Y = CO_2CH_3, OH), gave higher diastereoselectivity than imines with a less Lewis basic site (Y = OTMS, entries 21 and 22 vs. 19). Imines with intermediate Lewis basicity (Y = OCH_3) gave corresponding intermediate stereoselectivity. Although the authors report a > 95 : 5 ratio of isomers, Pridgen et al. were only able to obtain, after repeated attempts, a ~ 70/30 ratio of the same isomers when using $ZnCl_2$ or $BF_3{\cdot}OEt_2$ to catalyze the reaction. In this case, **131** and **132** were formed but with phenylglycinol as chiral auxiliary (entry 22).[103]

The chelated structure (**133A**, Figure 8) as proposed by Oh[102] is similar to the previously discussed Figure 2 above. The same arguments as presented above were also used by Oh and coworkers to rationalize their results—namely, that

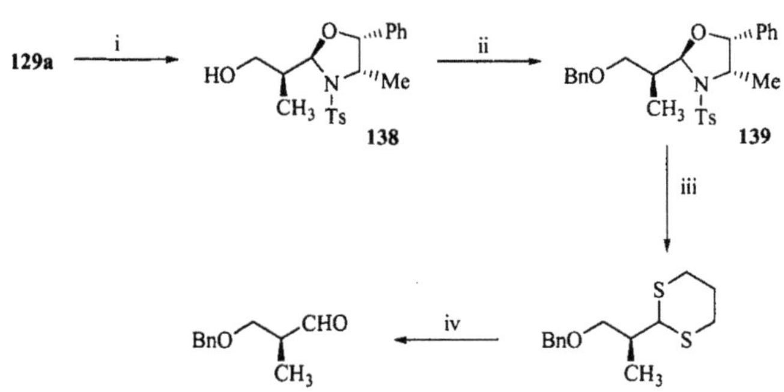

i. $LiAlH_4$ (95%)
ii. $PhCH_2Br$ / NaOH / Bu_4NHSO_4 (86%)
iii. $HS(CH_2)_3SH$ / BF_3Et_2O / CH_2Cl_2 (73%)
iv. $CaCO_3$ / MeI / H_2O / acetone (65%)

Scheme 25.

chelation between the imine and alkoxy group improves the selectivity.[102] Accordingly, there is a direct correlation between the Lewis basicity of the alkoxy functionality and selectivity.

A novel application of nonracemic 1,3-oxazolidines as electrophiles was reported by Scolastico, wherein 2-methoxy oxazolidine **134** (Eq. 15) was shown to be capable of reacting stereoselectively with various nucleophiles such as enamines, silyl enol ethers, and silylketene acetals in the presence of Lewis acids.[105,106] The addition usually occurs with >95% diastereoselectivity. Hydrolysis or removal of the auxiliary leads, in effect, to enantioselective formylations in good yields (Scheme 25).

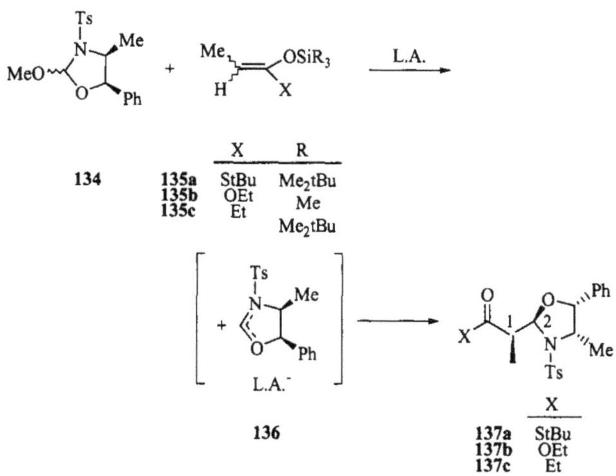

(15)

The cyclic orthoamide **134** was synthesized as a 1:1 mixture of C-2 epimers by condensing trimethylorthoformate and (1*R*,2*S*)-*N*-tosyl-norephedrine in the presence of pyridinium toslyate as catalyst. On treating the 1:1 mixture of **134** with 2 molar equiv of silylketene thiolacetal (**135a**) in the presence of $SnCl_4$, only two isomers were formed in a 94:6 diastereomeric ratio (Table 12 and Eq. 16). The configuration of the major isomer was verified by X-ray diffraction and found to be (2*R*, 2′*R*) as indicated in **137**.[105]

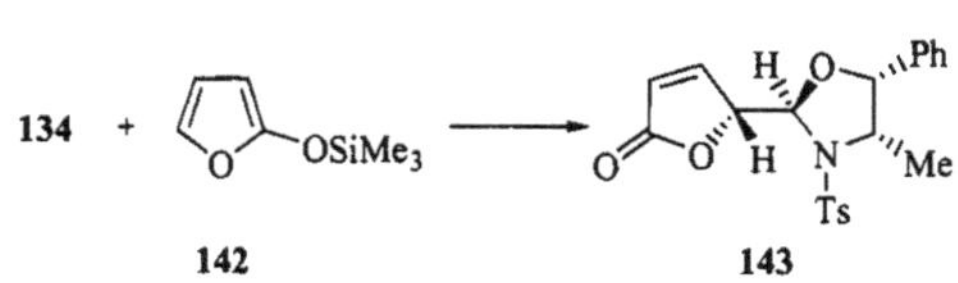

(16)

Table 12. Addition of Silylethers to **134**[a]

Entry	*Nuc.*	*Lewis Acid*[b]	*Diast. Exc. (%)*[c]	*Yield (%)*[d]
1	(*Z*)-**135a**	$SnCl_4$	92	85(83)
2	(*Z*)-**135a**	BF_3OEt_2	92	90
3	(*E*)-**135a**	BF_3OEt_2	92	90
4	**135a**(2:1 *E/Z*)	BF_3OEt_2	92	92(91)
5	(*E*)-**135a**	Me_3SiOTf[e]	84	87(85)
6	**135b**[f]	$SnCl_4$	—	—
7	**135b**[f]	BF_3OEt_2	50	97(95)
8	(*Z*)-**135c**	Me_3SiOTf	78	67
9	(*Z*)-**135c**	$TiCl_4$	88	92
10	(*Z*)-**142c**	BF_3OEt_2	92	56
11	**142**	*t*-$BuMe_2SiOTf$	74	78(75)
12	**142**	*t*-$BuMe_2SiOTf$[g]	88	95(90)

Notes: [a]Unless otherwise stated, reactions were performed by adding the Lewis acid to a CH_2Cl_2 1:2 solution of **134** and **135** at −78 °C. Reaction is usually over within 2 h at this temperature.

[b]The following amounts of Lewis acid were used: $SnCl_4$, 1.4 equiv; BF_3OEt_2, 2.0 equiv, $TiCl_4$, 1.4 equiv; Me_3SiOTf, and *t*-$BuMe_2SiOTf$, 0.25 equiv.

[c]Determined by ^1H and ^{13}C NMR spectroscopy.

[d]Yield as determined by NMR on the crude reaction mixtures; isolated yields in brackets.

[e]Reaction performed at −20 °C.

[f]**135b** was used as a 85:15 *E/Z* mixture.

[g]Reaction performed at −100 °C.

The data as shown in Table 12 indicates that selectivity is largely dependent on silylketene acetal configuration (entries 2–4) and to a lesser extent, the Lewis acid used to promote the addition. Noteworthy are the results of entries 11 and 12 (Table 12) where 2-trimethylsilyloxyfuran (**142**) was reacted with **134** to yield furanone **143** in 74–88 de%. Similar results were obtained by Pridgen using an N-unsubstituted 2-phenyl-phenylglycinol analog of **82a**.[103a] Pelter also reported similar results on tosylated 1,3-oxazolidines.[103b]

The fact that final product selectivity is not dependent on the stereochemistry of the starting material most likely indicates that the reaction is proceeding through a common intermediate since the reaction is conducted under nonequilibrating conditions in most cases. In order to explain the stereoselection, the reaction was proposed to proceed through an oxazolinium ion and the acyclic extended transitional structure **144A** (Figure 9). In this hypothesis, the silyl enol ether adds to the less hindered side of the oxazolinium ion. X-ray structures of both **137a** and **139** have the tosyl group oriented in the same plane as the norephedrine phenyl and methyl groups. Thus, in the transition state, the *N*-tosyl group can contribute to hindering the approach of the nucleophile from this side of the ring. An NMR study was performed to study the interaction of **137** and

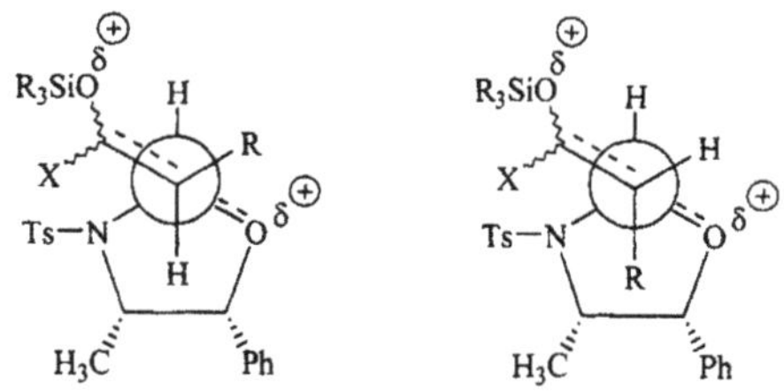

144A **144B**

Figure 9. Silyl enol ether conformational models.

similar structures with Lewis acids to develop a better understanding of the nature of the resulting cationic species.[107]

Structure A in Figure 9, which gives rise to the major 2′(R) epimer, should be lower in energy, because it minimizes steric interactions between the silyl enol ether substituent and the ring.

Concurrently with the reports of Scolastico,[105–107] Hoppe[108] reported his efforts to utilize 2-methoxy-3-(*p*-toluenesulphonyl)-1,3-oxazolidines (**146**) (Scheme 26) as reagents for the asymmetric formylation of silyl enol ethers and allylsilanes. To explain the observed selectivity, it was proposed that the intermediate cation exists as a mixture of rapidly equilibrating conformational isomers **147A** and **147B** that differ in the configuration at the pyramidal nitrogen atom and where **147B** is favored by the influence of the large substituent R^1 (Scheme 26). As a result, the steric differentiation of the nucleophile between the α- and the β-face of the cation is poor. Consequently, it was concluded that the introduction of a small substituent R^1 and a large substituent R^3 should suffice to shift the equilibrium cations in favor of **147B**. The shielding of the β-face is diminished and, thus, the formation of the *cis*-diastereomer **149c** in the nucleophilic addition step is supported.

This assumption was confirmed by the reaction of 2-methoxy-3-(*p*-toluenesulphonyl)-1,3-oxazolidine (**146c**) with 2-trimethylsilyloxy-propene (**153**) to yield 2,4-*cis*-oxazolidine (**154**) (Eq. 17).[108a]. The condensation of **146c** with

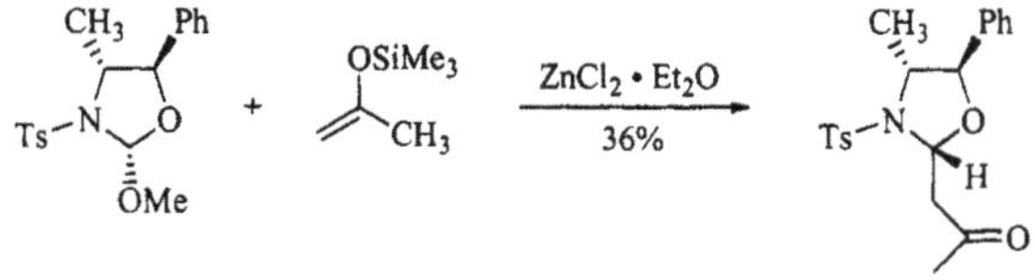

146 **153** **154** (17)

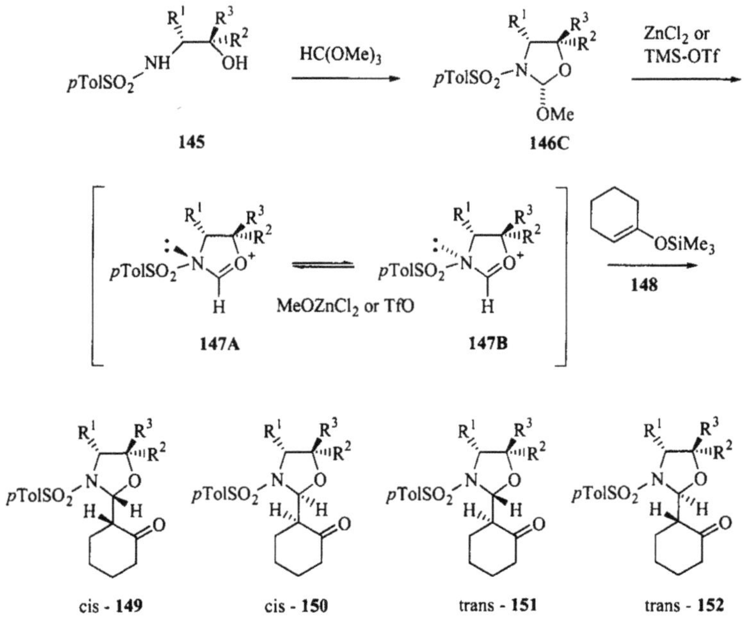

a: R^1=C_6H_5, R^2=R^3=H
b: R^1=CH_3, R^2=C_6H_5, R^3=H
c: R^1=CH_3, R^2=H, R^3=C_6H_5

Scheme 26.

cyclic prostereogenic silyl enol ethers (**148**) also gave rise to the formation of the diastereomer **149c** in 91% yield and > 95% de. Table 13 reports those results

From Table 13 it is apparent that in several cases *cis*- and *trans*-isomers are formed (with respect to the C-2 and C-1′ bond). These arise from the two competing diastereomeric transition states A and B (Scheme 27). Results as presented in Table 13 suggest that the unsubstituted cycloalkenyl trimethylsilyl ethers **148c** or **148d** favor transition state **A** over **B** (Scheme 27). Here, besides minimized steric interaction in the antiperiplanar arrangement of the positively charged double bonds, a favorable dipolar attraction exists between the negatively charged oxygen atoms at the sulfonyl group and the positive carbon atom C-2′. This argument is similar to the one presented above by Scolastico.[77,79] The additional methyl group in enol ether **148e** increases the relative energy of transition state **A** for steric reasons.

The preference of the silyl enol ethers **148f–i** for formation of the diastereomers **149f–i** through transition state **B** is surprising. Presumably, the origin is an

Table 13. Ketones **149** and **151** Prepared from the 2-Methoxy-1,3-oxazolidine **146c**

Educt	*Products*	*Yield (%)*	*Ratio* **149:151**	$[\alpha]_D^{25}$, *Major Product*
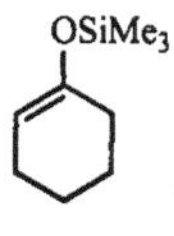 **148c**	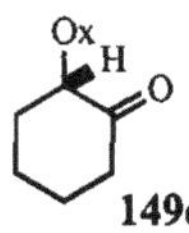 **149c**	91	> 95:5	+2.0
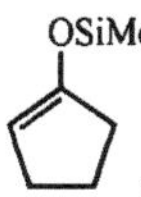 **148d**	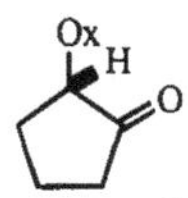 **149d**	75	90:10	+51.5
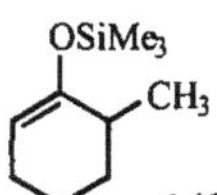 **148e**	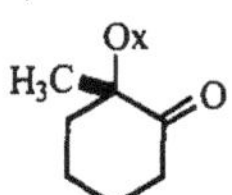 **149e** 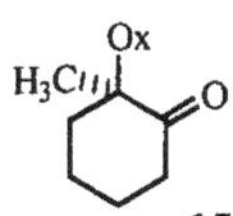 **151e**	78	50:50	**5e:** −19.0 **7e:** −52.5
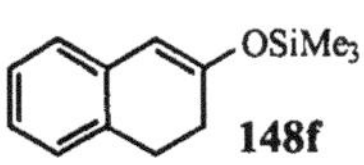 **148f**	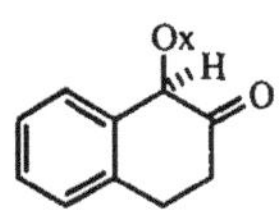 **149f**		20:80	−139.2
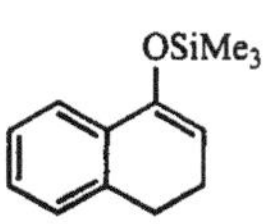 **148g**	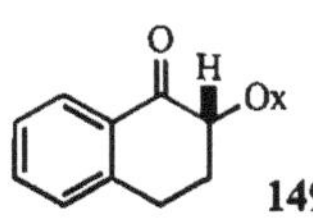 **149g**	52	20:80	+53.0
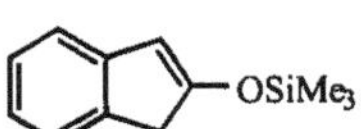 **148h**	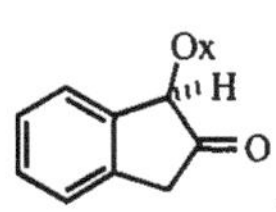 **149g**	46	< 5:95	−68.3
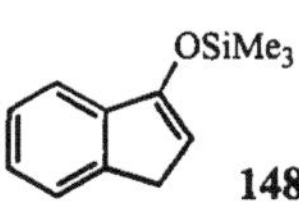 **148i**	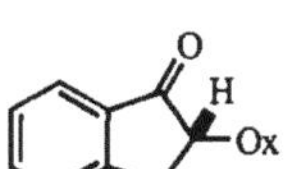 **149j**	51	< 5:95	+63.0

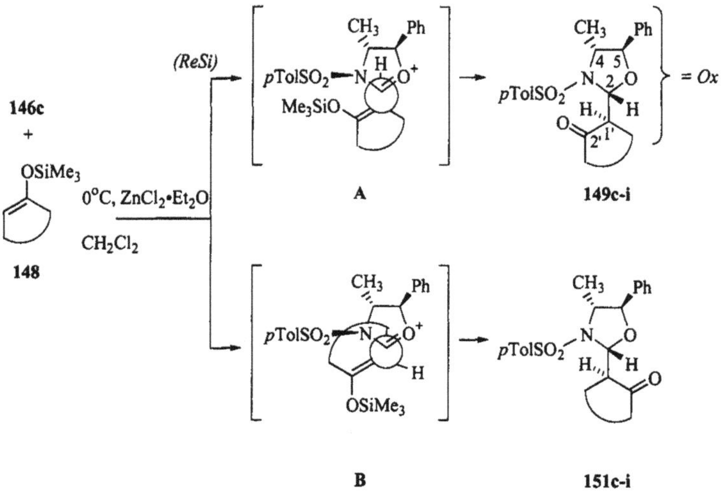

Scheme 27.

attraction interaction of the benzene ring with the tosyl group. Overall, these compounds represent valuable intermediates for the synthesis of homochiral cycloalkane derivatives with few steps and high efficiency.[108c,d]

Hoppe[109a] also reported a versatile approach to 2-acyl-1,3-oxazolidines using the electrophilic synthon **146b** (Scheme 28). The (*R*)-phenylglycinol derived oxazolidine **146a**, under trimethylsilyl cyanide catalysis at low temperature, affords predominately the *trans*-nitrile **153A** (75%, 10:1). Grignard reagents add smoothly to the cyano group, affording the ketones **154–158** after aqueous workup. No epimerization was noted at the labile stereocenter at C-2 under these conditions. It has been further demonstrated that compounds **154–158** may be stereoselectively reacted with other nucleophilic agents.[9g,107b]

Interestingly, it was observed that selectivity of addition was markedly improved when employing Grignard reagents containing the harder counter ions, e.g. bromide or chloride, and using THF as opposed to ether. A chelation-controlled mechanism (structure B of Scheme 29) was invoked to account for the selectivity where *re*-face addition of the carbonyl of **157** or **158** was dominant, yielding **161** and **162**. Conversely, when powerful external complexation of the magnesium ion was possible, the Felkin–Ahn products **159** and **160** from the *si*-face were predominantly obtained.

O-Alkyl-*O*-(trimethylsilyl)ketone acetals are known to react with either imines or iminium ions generated in situ.[110–112] The resulting β-amino esters can be

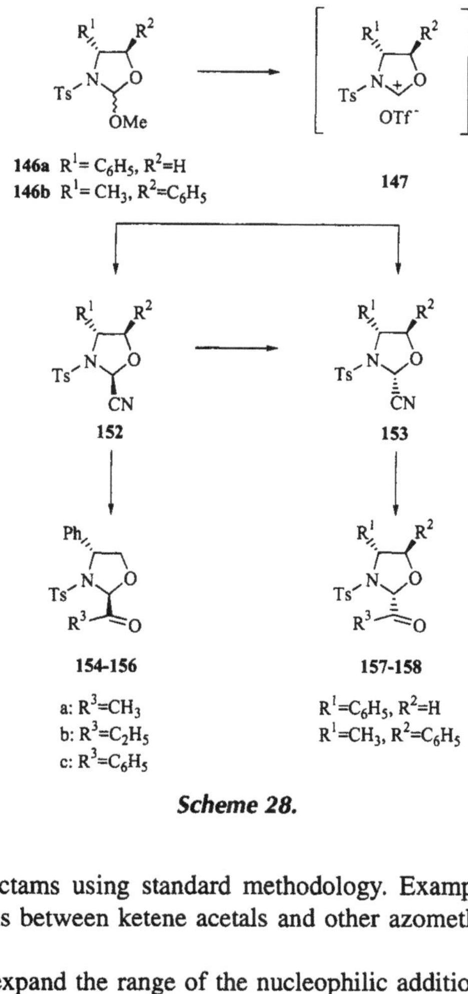

Scheme 28.

converted to β-lactams using standard methodology. Examples of Lewis acid mediated reactions between ketene acetals and other azomethines are also well documented.[110]

In seeking to expand the range of the nucleophilic additions to 2-substituted 1,3-oxazolidines, Pridgen et al.[103a] explored the addition of *O*-methyl-*O*-(trimethylsilyl)ketene acetal (**163**) to **82**. Poor results were obtained employing a variety of catalysts (Eq. 18). The best results were obtained with $TiCl_4$ and $Et_2O{\cdot}BF_3$ as catalysts. For the $TiCl_4$-catalyzed examples, a 41% yield with 53% de was obtained, while with $Et_2O{\cdot}BF_3$ a 88% yield with 22% de was obtained. A similar reaction with *O*-ethyl-*O*-trimethylsilyl ethylketene acetal (**164**) gave results analogous to those of $Et_2O{\cdot}BF_3$.[103a]

Exploitation of the addition of silyl enol ethers to 1,3-oxazolidines offers the possibility of synthesizing a variety of nonracemic β-amines (Scheme 30), e.g. α-

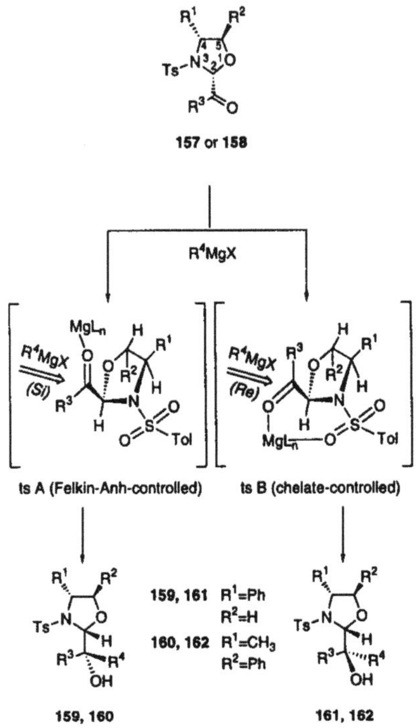

Scheme 29.

hydroxy-β-aminoethers **170** from α-siloxylketene acetal **169** and α-substituted-β-aminoketones **168** from silyl enol ether **167**. In most cases of simple enol ether addition, the reaction was regiospecific. Thus, in addition to studying the parameters that will allow one to control the reaction pathways of addition, it must also be possible to stereoselectively form the desired silyl (or siloxy) enol ether.[a]

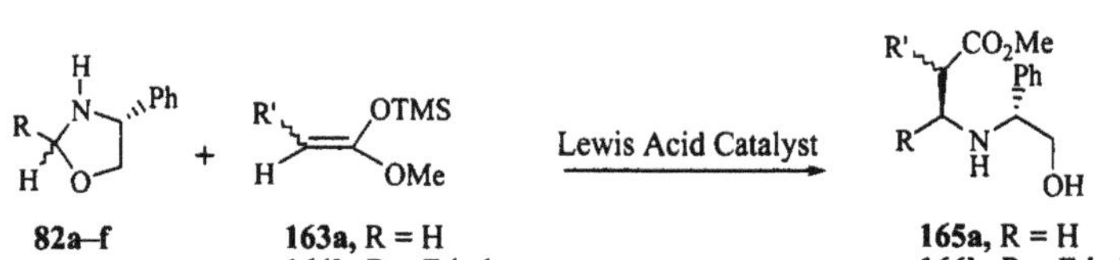

82a–f **163a**, R = H **164b**, R = Ethyl **165a**, R = H **166b**, R = Ethyl (18)

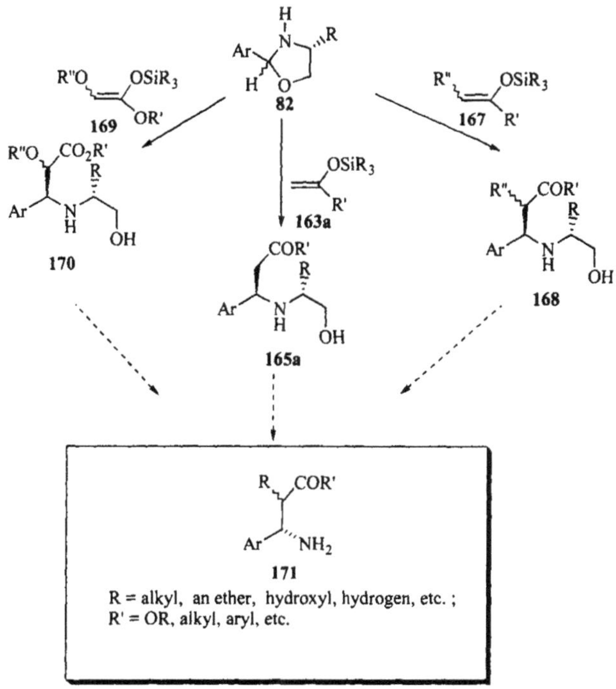

Scheme 30.

Finally, if one is able define the parameters that control addition *and* silyl enol ether geometry, one will be able to control the reaction pathways leading selectively to **170**, **165**, **168**, and finally to **171**. However, because magnesium and/or zinc ions are not involved, the chelation mechanism is not as prominent as before. Therefore, the origin of selectivity will have to be sterically based or involve, as of yet, an unidentified more effective metal mediator or reaction controller.

VII. OXIDATIVE CHIRAL AUXILIARY CLEAVAGE

Unlike the Takahashi examples, which contained only one benzylic amine bond susceptible to reductive cleavage, it was anticipated that a selective hydrogenolysis procedure had to be developed in order to cleave the ethanolamine carbon–nitrogen bond (b) of **14** (see Scheme 3). All initial attempts to hydrogenolyze **14** to **15** were without much success, at least up to hydrogen pressures of 80 psi.

Also attempted were the hydrogenolysis conditions of Bringmann,[114] but mainly the undesired bond cleavage at (a) was obtained. Only the carefully controlled oxidative conditions of Gawley[115] proved to be effective in obtaining the desired nonracemic phenylmethylamines, albeit in moderate yields. This text has summarized the current literature for asymmetric syntheses of secondary *N*-(2′-hydroxy-1′-phenylethyl)aryl and alkyl α-substituted methylamino alcohols **172** by 1,2-addition of nucleophilic agents to nonracemic 2-substituted 1,3-oxazolidines. An initial report described the oxidative cleavage of amino alcohols **172** via the intermediacy of imines **175** to primary amines **176**.[8] Although excellent diastereoselectivity was reported in the organometallic addition step to the oxazolidines, only moderate yields of the primary amines **176** were obtained via the reported lead tetraacetate (LTA) oxidative cleavage procedure.

Several excellent reductive conditions are available for carrying out transformations similar to those for converting **172** to **176**.[116,117] However, they mostly employ palladium-based reducing conditions which nonselectively cleave benzylic amino or ethers bonds. The lithio di-*tert*-butylbiphenyl procedure utilized by Meyers[118] may be considered as an alternative when two benzylic amino groups are present, but its overall efficiency is less than the $Pb(OAc)_4$ procedure. As a consequence, such conditions are inappropriate for substrates containing two differing benzylic amino carbons. The alternative use of lead tetraacetate or periodic acid[6,119] for cleavages of amino alcohols is well documented. However, initial results using the reported conditions for both reagents were capricious, providing only moderate yields. Because the hydroxymethyl functionality offers a manipulative handle, a considerable amount of time was invested in seeking to

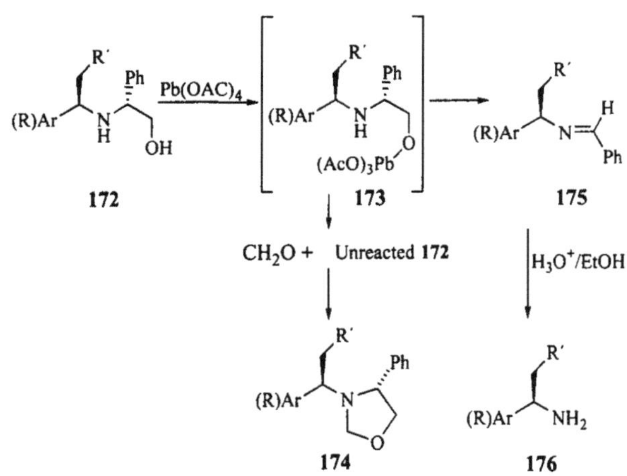

Scheme 31.

Table 14. Lead Tetraacetate Conversion of Amino Alcohols **172** to Amines **176**

Ar (R)	*R'*	*Substrate*	*% de*	*Product*	*Yield, %*	*% ee*
C_6H_5	CO_2Et	**172a**	92	**176a**	86	91
p-BrC_6H_4	CO_2Et	**172b**	94	**176b**	87	95
C_6H_{11}	CO_2Et	**172c**	94	**176c**	73	94
C_6H_5	H	**172d**	>99	**176d**	83	>99
$C_{10}H_7$	C_3H_7	**172e**	23	**176e**	87	24
$C_{10}H_7$	C_6H_6	**172f**	>99	**176f**	88	>99

improve the oxidative cleavage, thereby exploiting the possible advantage of being able to selectively cleave the desired benzylic amino bond of **172**.

Coates[6b] was the first to observe that formaldehyde, which was formed as an expected periodic acid oxidative cleavage product, will survive the reaction conditions long enough to react with an unreacted substrate similar to **172** and furnish an oxazolidine impurity similar to **174** (Scheme 31). By ^{1}H NMR a similar result was observed employing lead tetraacetate as oxidant.[113] An oxazolidine of the type **174** was isolated and identified as a major contaminant initiating a major effort to minimize its formation. After trying several reagents and reaction conditions, it was discovered that by inversely adding the amino alcohol in methylene chloride or methanol to lead tetraacetate in methanol at 0 °C the formation of **174** was minimized with a resulting increase in the formation of **175**. Using this protocol, apparently the formaldehyde was trapped as methylal as it was formed, thereby preventing its reaction with **172**. The resulting phenylimine intermediates **175** were isolated quantitatively and subsequently hydrolyzed under aqueous–ethanolic acidic conditions to aryl(alkyl)methylamines **176** in high yields. In examples where R was aliphatic and R′ contained an ester functionality, some decomposition on chromatography was noted. In no instance was racemization of the chiral center detected. Table 14 outlines the reported results.

This procedure is applicable to both aryl and alkyl amines and appears to be appropriate for other amino acid derived auxiliaries. The ability to isolate **175** quantitatively and cleanly by this procedure also offers the possibility of employing similar molecules as substrates for nucleophilic additions to form other nonracemic amino analogs

VIII. SUMMARY

It is apparent from this brief review that organometallic addition to 1,3-oxazolidines offers an excellent opportunity for synthesis of secondary amines in potentially optically pure form *and* with predictable absolute configuration. Further exploitation should enhance their utility and lead to acceptance of these simple amino acid based substrates as nonracemic amino electrophilic sources.

ACKNOWLEDGMENTS

The author wishes to acknowledge the helpful suggestions of Prof. Larry Overman and the late Dr. Paul Gassman and to my former postdoctoral colleagues Drs. Mohamed K. Mokhallalati, Ming-Jung Wu, and K. Raman Muralidharan for their intellectual contributions to portions of this review; to Dr. Drake Eggleston of the Physical and Analytical Department for X-ray crystallographic data; and to Ms. Kim Parker for administrative support.

REFERENCES AND NOTES

1. (a) Jurczak, J.; Golebiowski, A. *Chem Rev.* **1989**, 89, 149. (b) Whitesell, J.K. *Acc. Chem. Res.* **1985**, *18*, 280. (c) ApSimon, J.W.; Collier, T.L. *Tetrahedron*, **1986**, *42*, 5157. (d) Morrison, J.D., Ed. *Asymmetric Synthesis*; Academic Press: New York, 1983, 1984, Vol. 5.
2. For reviews on organometallic additions to imines and their derivatives see: (a) Volkmann, R.A. In *Comprehensive Organic Synthesis*; Trost, B.M.; Fleming, I; Schreiber, S.L., Eds.; Pergamon Press: Oxford, 1991, Vol. 1, pp. 355–396. (b) Kleinman, E.; Volkmann, R.A. In *Comprehensive Organic Syntheses*; Trost, B.M.; Fleming I., Heathcock, C.H., Eds.; Pergamon Press: Oxford, Vol. 2, pp. 975–1006.
3. (a) Vassella, A. In *Modern Synthetic Methods*; Scheffold, R., Ed.; Salle & Sauerlander: Frankfurt, 1980, p. 173. (b) Hanessian, S. *Total Synthesis of Natural Products: The "Chiron" Approach*; Pergamon Press: Oxford, 1983. (c) Scott, J.W. In *Asymmetric Synthesis*; Morrison, J.D., Ed.; Academic Press: New York, 1984, Vol. 4, p. 1. (d) Inch, T. D. *Tetrahedron* **1984**, *40*, 3161. (e) Jurczak, J.; Pikul, S.; Bauer, T. *Tetrahedron*, **1986**, *42*, 447.
4. (a) Martens, J. *Top. Curr. Chem.* **1984,** *125*, 165. (b) Coppola, G.M.; Schuster, H.F. *Asymmetric Synthesis: Construction of Chiral Molecules Using Amino Acids*; Wiley: New York, 1987.
5. (a) Takahashi, H.; Suzuki, Y.; Hori, T. *Chem. Pharm. Bull.* **1983**, *31*, 2183. (b) Takahashi, H.; Suzuki, Y.; Inagaki, H. *Chem. Pharm. Bull.* **1982**, *30*, 3160. (c) Suzuki, Y.; Takahashi, H. *Chem. Pharm. Bull.* **1983**, *31*, 31. (d) Takahashi, H.; Chida, Y.; Suzuki, S.; Yanaura, Y.; Suzuki, Y.; Masuda, C. *Chem. Pharm. Bull.* **1983**, *31*, 1659. (e) Suzuki, Y.; Takahashi, H. *Chem. Pharm. Bull.* **1983**, *31*, 2895.
6. (a) Chang, Z.-Y.; Coates, R.M. *J. Org. Chem.* **1990**, *55*, 3464. (b) Chang, Z.-Y.; Coates, R.M. *J. Org. Chem.* **1990**, *55*, 3475. (c) Schwartz, M.A.; Hu, X. *Tetrahedron Lett.* **1992**, *33*, 1689.
7. (a) Takahashi, H.; Tomita, K.; Otomasu, H. *J. Chem Soc., Chem. Commun.* **1979**, 668. (b) Takahashi, H.; Tomita, K.; Noguchi, H. *Chem. Pharm. Bull.* **1981**, *29*, 3387. (c) Takahashi, H.; Inagaki, H. *Chem. Pharm. Bull.* **1982**, *30*, 922. (d) Takahashi, H.; Suzuki, Y. *Chem. Pharm. Bull.* **1983**, *31*, 4295. (e) Enders, D.; Schubert, H.; Nubling, C. *Angew. Chem. Int. Ed. Engl.*, **1986**, *25*, 1109. (f) Denmark, S.E.; Weber, T.; Piotrowski, D.W. *J. Am. Chem. Soc.* **1987**, *109*, 2224.
8. Wu, M.-J.; Pridgen, L.N. *J. Org. Chem.* **1991**, *56*, 1340 and references therein.

9. (a) Coote, S.J.; Davies, S.G.; Middlemiss, D.; Naylor, *Tetrahedron Lett.* **1989**, *30*, 3581. (b) Fairhurst, R.A.; Heaney, H.; Papageorgiou, G.; Wilkins, R.F.; Eyley, S.C. *Tetrahedron Lett.* **1989**, *30*, 1433. (c) Marco, J.L. *J. Heterocycl. Chem.* **1987**, *24*, 1235. (d) Singh, H.; Singh, K. *Tetrahedron*, **1988**, *44*, 5897 and references therein. (e) For a general 3-benzazepine review see: Kametani, T.; Fukumoto, K. *Heterocycles* **1975**, *3*, 975. (f) Agami, C.; Couty, F.; Mathieu, H. *Tetrahedron Lett.* **1996**, *37*, 4001. (g) Agami, C.; Couty, F.; Lequesne, C. *Tetrahedron Lett.* **1994**, *35*, 3309.

10. (a) Bergmann, E.D. *Chem. Rev.* **1953**, *53*, 309. (b) Senkus, M. *J. Am. Chem. Soc.* **1945**, *67*, 1515. (c) Goodson, L.H.; Christopher, H. *J. Am. Chem. Soc.* **1950**, *72*, 358. (d) For the earliest example of a Grignard addition to an oxazolidine see: Robinson, G.M.; Robinson, R. *J. Chem. Soc.* **1923**, *123*, 532. (e) Courtois, G.; Miginiac, P. *Bull. Chem. Soc. Fr.* **1982**, 395. (f) Ficini, J.; Normant, H. *Bull. Soc. Chem. Fr.* **1957**, 1454. (g) Germon, C.; Alexakis, A.; Norman, J.F. *Tetrahedron Lett.* **1980**, *21*, 3763. (h) Glacet, C.; Couturier, J.C. *Bull. Soc. Chem. Fr.* **1962**, 2097. (i) Shono, T.; Matsumura, Y.; Inoue, K.; Ohmizu, H.; Kashimura, S. *J. Am. Chem. Soc.* **1982**, *104*, 5753.

11. (a) Fulöp, F.; Bernáth, G.; Mattinen, J.; Pihlaja, K. *Tetrahedron* **1989**, *45*, 4317. (b) Fülöp, F.; Mattinen, J.; Pihlaja, K. *Tetrahedron* **1990**, *46*, 6545. (c) Fülöp, F.; Pihlaja, K.; Neuvonen, K.; Bernáth, G.; Argay, G.; Kálmán, A. *J. Org. Chem.* **1993**, *58*, 1967. (d) Vainiotalo, P.; Ronkanon, S.; Fülöp, F.; Pihlaja, K. *Tetrahedron* **1990**, *46*, 3683. (e) Fülöp, F.; Pihlaja, K. *Tetrahedron* **1993**, *49*, 6701. (f) Fülöp, F.; Dahlquist, M.; Pihlaja, K. *Acta Chem. Scand.* **1991**, *45*, 273.

12. (a) Fülöp, F.; Bernáth, G. *Tetrahedron* **1993**, *49*, 6701. (b) Riddell, F. G.; Arumugam, S.; Fülöp, F.; Bernáth, G. *Tetrahedron* **1992**, *48*, 4979.

13. Astudillo, M.; Chokotho, N.; Jarvis, T.; Johnson, C.; Lewis, C.; McDonnell, P. *Tetrahedron* **1985**, *41*, 5919.

14. Baldwin, J. E. *J. Chem. Soc., Chem. Commun.* **1976**, 734.

15. (a) Pelleteir, S.W.; Mody, N.V. *J. Am. Chem. Soc.* **1979**, *101*, 492. (b) Filer, C.N.; Granchelli, F.E.; Soloway, A.H.; Neumeyer, J. L. *J. Org. Chem.* **1978**, *43*, 672.

16. (a) Lambert, J.B.; Wharry, S.M.; *J. Org. Chem.* **1982**, *47*, 3890 and references therein. (b) Lambert, J.B.; Majchrzak, M.W. *J. Am. Chem. Soc.* **1980**, *102*, 3588.

17. See also: Nishiyama, T.; Nishikawa, T.; Yamada, F. *J. Heterocycl. Chem.* **1989**, *89*, 1687.

18. Takahashi, H.; Suzuki, K.; Kametani, T. *Heterocycles* **1983**, *20*, 607.

19. (a) Takahashi, H.; Chida, K.; Higashiyama, K.; Onishi, H. *Chem Pharm. Bull.* **1985**, *33*, 4662. (b) Eggleton, G.L.; Cooper, B.A.; Sturch, C.L.; Trent, J.C.; Mulevar, S.V. *J. Het. Chem.* **1990**, *27*, 1853.

20. Neelakantan, L. *J. Org. Chem.* **1971**, *36*, 2256.

21. (a) Agami, C.; Rizk, T. *Tetrahedron* **1985**, *41*, 537. (b) Agami, C.; Rizk, T. *J. Chem. Soc., Chem. Commun.* **1983**, 1485. (c) Agami, C.; Meynier, F.; Puchot, C.; Guilhem, J.; Pascard, C. *Tetrahedron* **1984**, *40*, 1031. (d) Alcaide, B.; Lopez-Mardomingo, C.; Perez-Oscorio, R. *J. Heterocycl. Chem.* **1982**, *19*, 45.

22. Eleveld, M.B.; Hogeveen, H.; Schudde, E.P. *J. Org. Chem.* **1986**, *51*, 3685.

23. Miao, C.K.; Sorcek, R.; Jones, P.-J. *Tetrahedron Lett.* **1993**, *34*, 2259.

24. Agami, C.; Rizk, T. *J. Chem. Soc., Chem. Commun.* **1983**, 1485.

25. Neelakantan, L.; Molin-Case, J.A. *J. Org. Chem.* **1971**, *36*, 2261.

26. Just, G.; Potvin, P.; Uggowitzer, P.; Bird, P. *J. Org. Chem.* **1983**, *48*, 2923.

27. (a) Takahashi, H.; Chida, Y.; Suzuki, T.; Onishi, H.; Yanaura, S. *Chem. Pharm. Bull.* **1984**, *32*, 2714. (b) Takahashi, H.; Chida, Y.; Suzuki, T.; Onishi, H.; Yanaura, S. *Chem. Pharm. Bull.* **1986**, *34*, 2071. (c) Johansen, M.; Bungaard, H. *J. Pharm. Sci.* **1983**, *72*, 1294. (d) Young-Har-

vey, J.; Rea, I.; Pitman, I. *Int. J. Pharm.* **1986**, *30*, 151. (e) Andres, C.; Nieto, J.; Pedrosa, R.; Villamán, N. *J. Org. Chem.* **1996**, *61*, 4130. (f) Higashiyama, K.; Nakagawa, K.; Takahashi, H. *Heterocycles* **1995**, *41*, 2007.

28. Higashiyama, K.; Inoue, H.; Takahashi, H. *Tetrahedron Lett.* **1992**, *33*, 235.
29. Takahashi, H.; Hsieh, B.C.; Higashiyama, K. *Chem. Pharm. Bull.* **1990**, *38*, 2429.
30. (a) Rufer, C.; Loser, W. *J. Med. Chem.* **1979**, *22*, 750. (b) Fuller, R.W.; Hemrick, S.K. In *Monoamine Oxidase*; Singer, T.P.; van Korff, R.W.; Murphy, D.L., Eds.; Academic Press: New York, 1979, p. 245.
31. (a) Ingersoll, A.W. *Organic Synthesis*; J. Wiley and Sons: New York, 1943, Collect, Vol. II, p. 506. (b) Nicholson, J.S.; Tantum, J.G. Boots Co. Ltd., U. S. Patent 4, 209, 638, 1980.
32. Nogradi, M. In *Stereoselective Synthesis*; Verlag Chemie: Weinheim, 1986, p. 54.
33. Menachery, M.P.; Lavanier, M.L.; Wetherly, H.; Guinaudeau, H.; Shamma, M. *J. Nat. Prod.* **1986**, *49*, 745.
34. Bringmann, G. *The Alkaloids*; Brossi, A., Ed.; Academic Press: New York, 1986, Vol. 29, p. 141.
35. Lu, S.-T.; Su, T.-L.; Kametani, T.; Ujiie, A.; Ihara, M.; Fukumoto, K. *Heterocycles* **1975**, *3*, 459. (b) Wang, E. C. *Phytochemistry* **1982**, *21*, 809.
36. Danishefsky, S.J.; Harrison, P.J.; Webb, R.R.; O'Neill, B.T. *J. Am. Chem. Soc.* **1985**, *107*, 1421.
37. Kunz, H. *Angew. Chem. Int. Ed. Engl.* **1987**, *26*, 294.
38. Pridgen, L.N.; Prol, J.; Alexander, B.; Gillard, L. *J. Org. Chem.* **1989**, *54*, 3231.
39. Stewart Jr., A.T.; Hauser, C.R. *J. Am. Chem. Soc.* **1955**, *77*, 1098.
40. Imamoto, T.; Sugiura, Y. *J. Organomet. Chem.* **1985**, *285*, C21.
41. Takahashi, H.; Niwa, H.; Higashiyama, K. *Heterocycles* **1988**, *27*, 2099.
42. Davidsen, S.K.; Chu-Moyer, M.Y. *J. Org. Chem.* **1989**, *54*, 5558.
43. For a similar lithium chelated complex see: Barr, D.; Berrisford, D.J.; Jones, R.V.H.; Slawin, A.M.Z.; Snaith, R.; Stoddard, J.F.; Williams, D. *J. Angew Chem. Int. Engl.* **1989**, *28*, 1044.
44. Corey, E.J.; Yuen, P.-W.; Hannon, F.J.; Wierda, D.A. *J. Org. Chem.* **1990**, *55*, 784.
45. Bernardi, A.; Cardani, S.; Pilati, T.; Poli, G.; Scolastico, C.; Villa, R. *J. Org. Chem.* **1988**, *53*, 1600.
46. Miao, C.K.; Sorcek, R.; Jones. P.-J. *Tetrahedron Lett.* **1993**, *34*, 2259.
47. (a) Higashiyama, K.; Nakahata, K.; Hiroshi, T. *Hetrocycles* **1992**, *33*, 17 (b) Amat, M.; Llor, N.; Bosch, J. *Tetrahedron Lett.* **1996**, *37*, 2223.
48. (a) Zhu, J.; Guirion, J.-C.; Husson, H.-P. *J. Org. Chem.* **1993**, *58*, 6451. (b) Zhu, J.; Guirion, J.-C.; Husson, H.-P. *Tetrahedron Lett.* **1989**, *30*, 5137.
49. Guerrier, L.; Royer, J.; Grierson, D.S.; Husson, H.-P. *J. Am. Soc. Soc.* **1983**, *105*, 7754.
50. (a) Marco, J.L.: Royer. J.: Husson, H.-P. *Tetrahedron Lett.* **1985**, *26*, 6345. (b) Delgado, A.; Mauleon, *Synth. Commun.* **1988**, *18*, 823.
51. (a) Royer, J.; Husson, H.-P. *Tetrahedron Lett.* **1985**, *26*, 1515. (b) Grierson, D.S.; Royer, J.; Guerrier, L.; Husson, H.-P. *J. Org. Chem.* **1986**, *51*, 4475. (c) Gosmann, G.; Guillame, D.; Husson, H.-P. *Tetrahedron Lett.* **1996**, *37*, 4369.
52. (a) Yamato, M.; Hashigaki, K.; Ishikawa, S.; Qais, N. *Tetrahedron Lett.* **1988**, *29*, 6949. (b) Takeuchi, Y.; Kamada, Y.; Nishimura, K.; Nishioka, H.; Nishikawa, M.; Hashigaki, K.; Yamato, M.; Harayama, T. *Chem. Pharm. Bull.* **1994**, *42*, 796.
53. Arseniyadis, S.; Huang, P.Q.; Piveteau, D.; Husson, H.P. *Tetrahedron* **1988**, *44*, 2457.
54. (a) Huryn, D.M. In *Comprehensive Organic Synthesis*; Trost, B.M.; Fleming, I.; Schreiber, S.L., Eds.; Pergamon Press: Oxford, 1991, Vol. 1, pp. 49–75. (b) Whitesell, J.K. *Chem. Rev.* **1989**, *89*, 1581.

55. Higashiyama, K.; Inoue, H.; Takahashi, H. *Tetrahedron* **1994**, *50*, 1083.
56. (a) Arseniyadis, S.; Huang, P.Q.; Husson, H.P. *Tetrahedron Lett.* **1988**, *29*, 1391. (b) Arseniyadis, S.; Huang, P.Q.; Husson, H.P. *Tetrahedron Lett.* **1988**, *29*, 631.
57. (a) Meyers, A.I.; Snyder, L. *J. Org. Chem.* **1992**, *57*, 3814. (b) Meyers, A.I.; Snyder, L. *J. Org. Chem.* **1993**, *58*, 36. (c) For an excellent review of the use of this versatile nonracemic synthon see: Romo, D.; Meyers, A.I. *Tetrahedron Lett.* **1991**, *47*, 9503. (d) Meyers, A.I.; Lefker, B.A.; Sowin, T.J.; Westrum, L.J. *J. Org. Chem.* **1989**, *54*, 4243. (e) For the analogous syntheses of nonracemic piperidines and isoquinolines see: Munchhof, M.; Meyers, A.I. *J. Org. Chem.* **1996**, *61*, 4607 and references therein.
58. Meyers, A.I.; Westrum, L.J. *Tetrahedron Lett.* **1994**, *35*, 973.
59. Salzmann, T.N.; Ratcliffe, R.W.; Christensen, B.G.: Bouffard. F.A. *J. Am. Chem. Soc.* **1980**, *102*, 6161.
60. d'Angleo, J.; Maddaluno, J. *J. Am. Chem. Soc.* **1986**, *108*, 8112.
61. Gmeiner, P. *Tetrahedron Lett.* **1990**, *31*, 5717.
62. Estermann, V.H.; Seebach, D. *Helv. Chim. Acta* **1988**, *71*, 1824.
63. Laschat, S.; Kunz, H. *J. Org. Chem.* **1991**, *56*, 5883.
64. Wu, M.-J.; Pridgen, L.N. *Synlett* **1990**, 636.
65. Weber, T.; Edwards, J.P.; Denmark, S.E. *Synlett* **1989**, 20.
66. Mokhallalati, M.K.; Wu, M.-J.; Pridgen, L.N. *Tetrahedron Lett.* **1993**, *34*, 47.
67. Andrés, C.; González, A.; Pedrosa, R.; Pérez-Encabo, A. *Tetrahedron Lett.* **1992**, *33*, 2895.
68. Bocoum. A.: Savoia. D.; Umani-Ronchi. A. *J. Chem. Soc., Chem. Commun.* **1993**, 1542.
69. Chakraborty, T.K.; Reddy, G.V.; Hussain, K.A. *Tetrahedron Lett.* **1991**, *32*, 7597.
70. (a) Anh, N.T.; Eisenstein, O. *Nouv. J. Chem.* **1977**, *1*, 6. (b) Broeker, J.L.; Hoffmann, R.W.; Houk, K.N. *J. Am. Chem. Soc.* **1991**, *113*, 5006.
71. (a) Chakraborty, T.K.; Reddy, G.V. *J. Org. Chem.* **1992**, *57*, 5462. (b) Zhu, J.; Bouillon, J.-P.; Singh, G.P.; Chastanet, J.; Beugelmans, R. *Tetrahedron Lett.* **1995**, *37*, 7081. (c) Zhu, J.; Beugelmans, R.; Bourdet, S.; Chastanet, J.; Roussi, G. *J. Org. Chem.* **1995**, *60*, 6389.
72. Andrés, C.; Maestro, A.; Pedrosa, R.; Pérez-Encabo, A. *Synlett* **1992**, 45.
73. Mori, A.; Fujiwara, J.; Maruoka, K.; Yamamoto, H. *J. Organomet. Chem.* **1985**, *285*, 83.
74. Ishihara, K.; Hanaki, N.; Yamamoto, H. *J. Am. Chem. Soc.* **1991**, *113*, 7074.
75. (a) Andrés, C., Gonzales, A.; Pedrosa, R.; Pérez-Encabo, A.; Garcia-Granda, S.; Salvado, M.A.; Gomez-Beltran, F. *Tetrahedron Lett.* **1992**, *33*, 4743. (b) Mori, I.; Ishihara, K.; Flippin, L.A.; Nozaki, K.; Yamamoto, H.; Bartlett, P.A.; Heathcock, C.H. *J. Org. Chem.* **1990**, *55*, 6107. (c) Denmark. S.E.; Almstead, N.G. *J. Org. Chem.* **1991**, *113*, 7074.
76. (a) Hucke, M.; Auboult, J.; Pourcelot, G.; Berlan, J. *Tetrahedron Lett.* **1983**, *24*, 585. (b) Mongeney, P.; Alexakis, A.; Normant, J.F. *Tetrahedron*, **1984**, *40*, 1803. (c) Berlan, J.; Besace, J.; Stephen, E.; Cressan, P. *Tetrahedron Lett.* **1985**, *26*, 5459.
77. Bernadi, A.: Cardani. S.: Poli. G.; Scolastico, C. *J. Org. Chem.* **1986**, *51*, 5041.
78. Hutchinson, D.K.; Fuchs, P.L.; *Tetrahedron Lett.* **1986**, *27*, 1429.
79. Cardani, S.; Poli, G.; Scolastico, C. *Tetrahedron* **1988**, *44*, 5929.
80. Ahn, N.T. *Top. Curr. Chem.* **1980**, *88*, 145.
81. (a) Isobe, M.; Hirose, Y.; Shimokawa, K.-I.; Nishikawa, T.; Goto, T. *Tetrahedron Lett.* **1990**, *31*, 5499. (b) Langois, N.; Dahuran, N. *Tetrahedron Lett.* **1996**, *37*, 3993.
82. Pridgen, L.N.; Mokhallalati, M.K.; Wu, M.-J. *J. Org. Chem.* **1992**, *57*, 1237.
83. Andrews, R.C.; Teague, S.J.; Meyers, A.I. *J. Am. Chem. Soc.* **1988**, *110*, 7854.
84. Meyers, A.I.; Roth, G.P.; Hoyer, D.; Garner, B.A.; Laucher, D. *J. Am. Chem. Soc.* **1988**, *110*, 4611.
85. Meyers, A.I.; Higashiyama, K. *J. Org. Chem.* **1987**, *52*, 4592.

86. Tomioka, K., Shindo, M.; Koga, K. *J. Am. Chem. Soc.* **1989**, *111*, 8266.
87. Tomioka, K.; Koga, K. In *Asymmetric Synthesis*; Morrison, J.D., Ed.; Academic Press: Orlando, 1983, Vol. 2, Chapter 7.
88. Meyers, A.I., Brown, J.D.; Laucher, D. *Tetrahedron Lett.* **1987**, *28*, 5279.
89. Meyers, A.I., Brown, J.D.; Laucher, D. *Tetrahedron Lett.* **1987**, *28*, 5283.
90. Lutomski, K.A.; Meyers, A.I. In *Asymmetric Synthesis*; Morrison, J.D., Ed.; Academic Press: Orlando, 1984, Vol. 3, Chapter 3.
91. Tomioka, K.; Inoue, I.; Shindo, M.; Koga, K. *Tetrahedron Lett.* **1989**, *31*, 6681.
92. (a) Barner, B.A.; Meyers, A.I. *J. Am. Chem. Soc.* **1984**, *106*, 1865. (b) Meyers, A.I.; Hoyer, D. *Tetrahedron Lett.* **1984**, *25*, 3667.
93. Mokhallalati, M.K.; Muralidharan, K.R.; Pridgen, L.N. *Synthetic Commun.* **1993**, *23*, 2055.
94. Rawson, D.J.; Meyers, A.I. *J. Org. Chem.* **1991**, *56*, 2292.
95. Meyers, A.I.; Schmidt, W.; McKennon, M.J. *Synthesis* **1993**, 250.
96. For reviews see: (a) Weinreb, S.M. *Chem. Rev.* **1989**, *89*, 1234. (b) Weinreb, S.M.; Boger, D.L. *Hetero Diels-Alder Methodolgy in Organic Synthesis*; Academic Press: New York, 1987. (c) Weinreb, S.M.; Staib, R.R. *Tetrahedron* **1982**, *38*, 3087. (d) Hopman, J.; Van der Berg, E.; Oilero, L.; Hiemstra, H.; Speckamp, W.N. *Tetrahedron Lett.* **1995**, *36*, 4315.
97. Midland, M.M.; Koops, R.W. *J. Org. Chem.* **1992**, *57*, 1158.
98. (a) Pfrengle, W.; Kunz, H. *J. Org. Chem.* **1989**, *54*, 4261. (b) Kunz, H.; Pfrengle, W. *Angew. Chem. Int. Ed. Engl.* **1989**, *29*, 1067.
99. Barluenga, J.; Aznar, F.; Valdes, C.; Cabal, M.-P. *J. Org. Chem.* **1993**, *58*, 3391.
100. (a) Waldemann, H.; Braun, M. *J. Org. Chem.* **1992**, *57*, 4444. (b) Waldemann, H.; Braun, M. *Gazz. Chim. Ital.* **1991**, *121*, 277.
101. (a) Hattori, K.; Yamamoto, H. *J. Org. Chem.* **1992**, *57*, 3264. (b) Hattori, K.; Yamamoto, H. *Tetrahedron* **1993**, *49*, 1749.
102. Devine, P.N.; Reilly, M.; Oh, T. *Tetrahedron Lett.* **1993**, *34*, 5827.
103. (a) Pridgen, L.N., an unpublished observation. (b) Pelter, A.; Ward, R.S.; Sirit, A. *Tetrahedron:Asymmetry* **1994**, *5*, 1745.
104. Danishefsky, S. *Acc. Chem. Res.* **1981**, *14*, 400.
105. Bernardi, A; Cardani, S.; Carugo, O.; Colombo, L; Scolastico, C.; Villa, R. *Tetrahedron Lett.* **1990**, *31*, 2779.
106. Palazzi, C.; Poli, G.; Scolastico, C; Villa, R. *Tetrahedron Lett.* **1990**, *31*, 4223.
107. (a) Bernardi, A.; Cardani, S.; Poli, G.O.; Potenza, D.; Scholastico, C. *Tetrahedron* **1990**, *48*, 1343. (b) Poli, G.; Manzoni, L.; Pilati, T.; Scolastico, C. *SynLett* **1995**, 71.
108. (a) Conde-Friebos, K.; Hoppe, D. *Tetrahedron* **1992**, *48*, 6011. (b) Conde-Friebos, K.; Hoppe, D. *Synlett* **1990**, 99. (c) Prien, O.; Hoffmann, H.; Conde-Friebos, K.; Krettek, T.; Berger, B.; Wagner, K.; Bolte, M.; Hoppe, D. *Synthesis* **1994**, 1313. (d) Prien, O.; Hoppe, D. *Synthesis* **1996**, 27.
109. (a) Conde-Friebos, K.; Harder, T.; Aulbert, D.; Strahringer, C.L.; Bolte, M.; Hoppe, D. *Synlett* **1993**, 921. (b) Hoppe, I.; Hoffmann, H.; Gartner, I.; Krettek, T.; Hoppe, D. *Synthesis* **1991**, 1157. (c) Hoppe, I.; Hoppe, D.; Herbst-Irmer, R.; Egert, E. *Tetrahedron Lett.* **1990**, *31*, 6859.
110. Hart, D.J.; Ha, D.-C. *Chem. Rev.* **1989**, *89*, 1447.
111. Martin, S.F.; Corbett, J. W. *Synthesis* **1992**, 55.
112. Harding, K.E.; Coleman, M.T. Liu, L.T. *Tetrahedron Lett.* **1991**, *32*, 3795.
113. Mokhallalati, M.K.; Pridgen, L.N. *Synth. Commun.* **1993**, *23*, 2055.
114. Bringmann, G.; Geisler, J.-P. *Tetrahedron Lett.* **1989**, *30*, 317.
115. Gawley, R.E.; Rein, K.; Chemburkar, S. *J. Org. Chem.* **1989**, *54*, 3002.

116. Quirion, J.-C.; Grierson, D.S.; Royer, J.; Husson, H.-P. *Tetrahedron Lett.* **1988**, *29*, 3311 and references therein.
117. Ram, S.; Ehrenkaufer, R.E. *Synthesis* **1988**, 91.
118. Meyers, A.I.; Burgess. L.E. *J. Org. Chem.* **1991**, *56*, 2294.
119. Narukawa, Y.; Juneau, K.N.; Snustad, D.; Miller, D.B.; Hegedus, L.S. *J. Org. Chem.* **1992**, *57*, 5453.

ASYMMETRIC SYNTHESES OF α,α-DISUBSTITUTED β-DIKETONES AND β-KETO ESTERS

André Guingant[1]

Advances in Asymmetric Synthesis
Volume 2, pages 119–188

ISBN 1-55938-797-1

I. INTRODUCTION

α,α-Disubstituted β-dicarbonyl compounds are versatile precursors in the synthesis of a wide range of natural products. The sustained interest devoted to these compounds stems from the two following characteristics:

1. They possess a stereogenic quaternary carbon center, a common structural feature present in numerous classes of biologically significant natural products and pharmaceutical compounds. Moreover, the development of effective methods for the construction of such structural units still remains a major challenge in organic synthesis in spite of the good results obtained in recent years.[2]
2. They possess several nucleophilic and electrophilic reactive sites which give them sufficient structural flexibility to permit diversion into several families of compounds.

From the above considerations it is not surprising that the development of effective methods for the preparation of α,α-disubstituted β-dicarbonyl compounds in optically pure form has been the subject of intense research over the last 20 years or so.

The present article is primarily concerned with the methods of formation of α,α-disubstituted β-diketones and β-keto esters which feature, as starting material, the corresponding α-monosubstituted β-dicarbonyl compounds or their

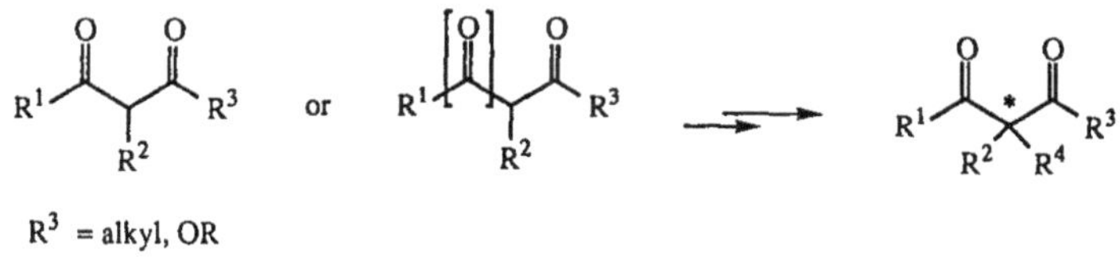

Scheme 1.

ketone carbonyl protected forms as it may be represented by the generalized outline in Scheme 1. (The reader will find in the Addendum a brief description of alternative methods that do not follow the above scheme.)

This review is divided into three main sections. The first is devoted to enantioselective syntheses (mainly of α,α-disubstituted β-diketones) and the second and third ones, which not unexpectedly form the main body, cover diastereoselective syntheses (mainly of α,α-disubstituted β-keto esters). Because the overall transformation, as reported in Scheme 1, was achieved in most examples by resorting to alkylation and conjugate addition reactions, a clear distinction between these two processes has been made within each section.

Finally, I should add that, when writing this chapter, my objective has been to bring out the salient features of each method and to emphasize its scope and limitations, particularly with regard to selectivity.

II. ENANTIOSELECTIVE SYNTHESES OF α,α-DISUBSTITUTED β-DIKETONES

A. Michael Addition Reaction

Catalysis by Cinchona Alkaloids

The first example of the use of a chiral catalyst in a Michael reaction was due to Långström and Bergson[3] who reported the addition of methyl indane-1-one-2-carboxylate **1** to acrolein and α-isopropyl acrolein in the presence of a catalytic amount of partially resolved (57% ee) 2-(hydroxymethyl)-quinuclidine **2** in benzene solution at 22 °C (Scheme 2). Unfortunately, although the reaction products did not show insignificant specific rotations, the enantiomeric excesses were not determined.

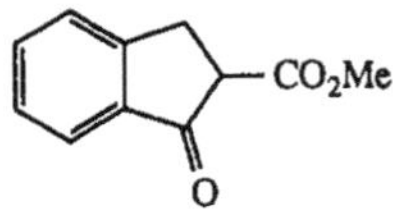

1

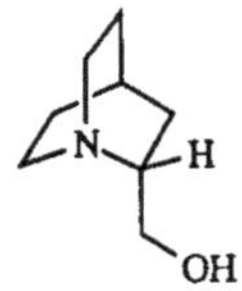

2

Scheme 2.

Two years later, the possibility of preparing optically active Michael adducts by use of cinchona alkaloids as catalysts was disclosed by Wynberg and co-workers.[4,5] These authors discovered that the indanone derivative **1** reacts with methylvinylketone (MVK) under the catalytic influence of (–)-quinine **3** or (+)-quinidine **4** (Scheme 3) to give the optically active Michael adduct **5** (Scheme 4). Enantiomeric excesses as high as 76% could be achieved in CCl_4 at relatively low temperature (Table 1).

As previously suggested by Långström and Bergson[3] it is probable that a tight ion pair is formed between the enolate of **1** and the protonated quinuclidine ring of the catalyst. It was also assumed that the hydroxyl at C-9 forms a hydrogen bond with the carbonyl group of the MVK, enhancing its electrophilic properties and bringing it into position for stereoselective attack by the β-keto ester enolate. These proposals are in agreement with the two following experimental observations:

- Acetylation of the hydroxyl at C-9 caused a strong decrease in both the reaction velocity and the optical purity of Michael adduct **5**.
- Addition of 2% ethanol to CCl_4, which disfavors both the ion pair and the

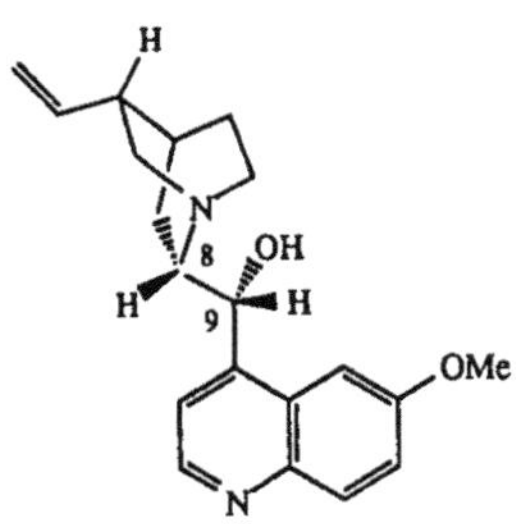

3

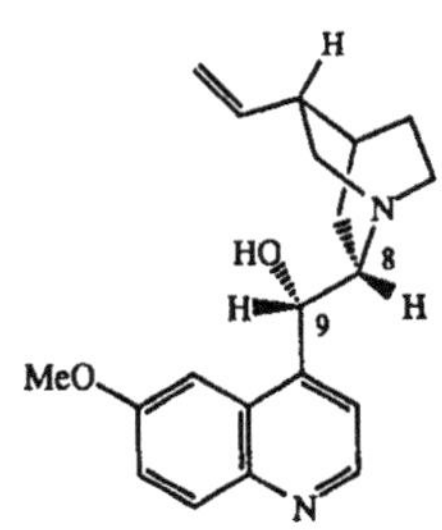

4

Scheme 3.

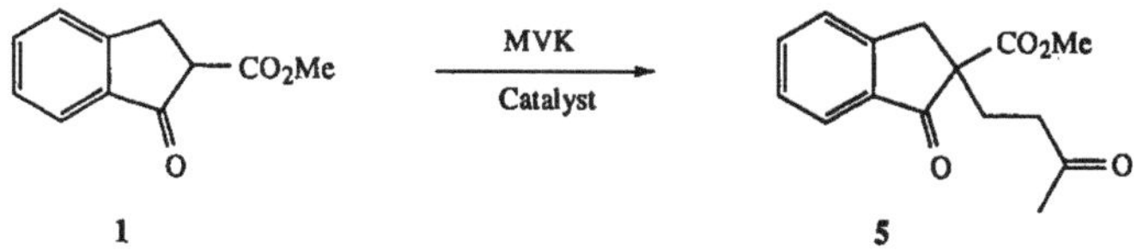

Scheme 4.

hydrogen bond formations, resulted in a half diminution of the enantiomeric purity of **5** (entries 1 and 3, Table 1).

It is worthwhile mentioning that the absolute configuration of the created quaternary carbon center in **5** is closely related to the spatial arrangement of the catalyst substituents at C-8 and C-9. Thus, (–)-quinine **3**—C-8*S*, C-9*R*—leads to the preferential formation of (*S*)-**5**, whereas (+)-quinidine **4**—C-8*R*, C-9*S*—leads to (*R*)-**5** in slightly less enantiomeric purity (entries 1 and 4, Table 1). This reversal of enantioselectivity is clearly the result of the enantiomeric relationship that exists at the crucial amino alcohol part of quinine and quinidine.

Later on, deeper insight into the mechanistic details of the reaction was provided by extensive NMR studies of quinine and quinidine. Thus, it could be demonstrated that both of these alkaloids exist in solution as the only "open" conformation (schematically represented in Scheme 5 for quinidine **6**) and that the latter, contrary to previous assumptions,[6] is not switched to the "closed" conformation **7** by protonation at the quinuclidine nitrogen.[7,8] These findings, added to the results of a molecular docking study, led Dijkstra et al.[7] to propose a transition-state model for the quinine-catalyzed Michael addition of thiophenol to cyclohexenone. However, no reasonable transition-state model has yet been put forward to account for the good enantioselectivity of the addition of **1** to MVK.[9]

The Michael addition of β-keto ester **8** to MVK was also examined.[5] Use of quinine methohydroxide, a stronger base than quinine, was necessary for deprotonation. Compared to the preceding example, Michael adduct **9** was formed with a disappointingly low selectivity (Scheme 6), possibly because the reaction had to be performed in the presence of a small amount of ethanol.

Table 1.

Entry	*Catalyst*	*Conditions*	*% Yield*	*% ee*
1	(–)-Quinine	CCl_4, –21 °C	99	76 (*S*)
2	(–)-Quinine	CCl_4, –25 °C	98	60 (*S*)
3	(–)-Quinine	2% EtOH/CCl_4 25 °C	97	33 (*S*)
4	(+)-Quinine	CCl_4, –21 °C	100	69 (*R*)

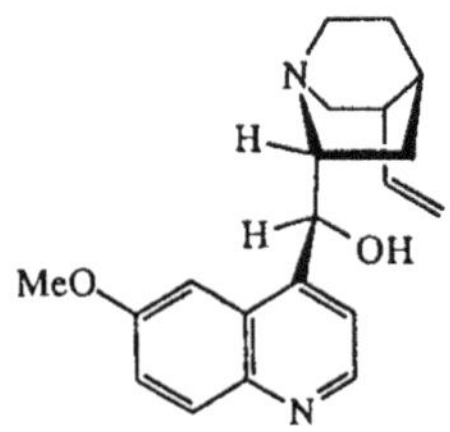

"open" quinidine **6**

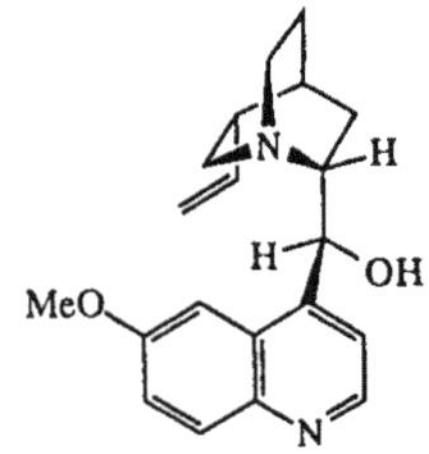

"closed" quinidine **7**

Scheme 5.

Using acrolein or MVK as Michael acceptors and quinine as base catalyst, some other β-diketo compounds gave optically active adducts.[4] Unfortunately, in none of the cases examined were the enantiomeric excesses determined.

Use of polymer-supported cinchona alkaloids to catalyze the Michael reaction is particularly attractive because the catalysts are more easily removed at the end of the reaction and can be eventually reused. Such a mode of catalysis has been investigated by a few authors and three different strategies have been devised:

1. Formation of quaternary ammonium salts by reaction of alkaloids with chloromethylated cross-linked polystyrenes (Hodge's method).[10]
2. Acylation of the alkaloid hydroxyl at C-9 by reaction with functionalized copoly(styrene-2% divinylbenzene) (Wynberg's method).[11]
3. AIBN radical-induced copolymerization of alkaloids with acrylonitrile (Kobayashi's method).[12]

The first two methods gave low enantioselectivities and are of limited synthetic value. However, the method of Kobayashi, which incorporates the alkaloid in the polymer matrix through its vinyl group and without chemical modification of the crucial amino alcohol moiety, gave somewhat better results. Thus, a 42%

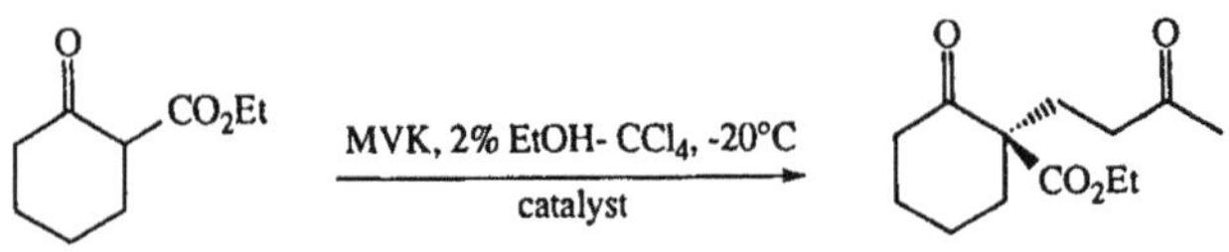

8 (*R*)-**9** (ee = 22%)

Scheme 6.

ee was achieved for (*R*)-**5** by reacting a mixture of indanone derivative **1** with two equivalents of MVK in toluene at room temperature for 48 h in the presence of a (+)-quinidine–acrylonitrile copolymer. The efficiency of the catalyst depends on the nature of the alkaloid, its content in the copolymer, the reaction temperature, and the applied pressure.[13] Finally, it is noteworthy that modification of this strategy, by inserting a spacer group between the carbon C-3 of the alkaloid and the polymer matrix, led Oda's group[14] to substantially improve the selectivity of the reaction. Michael adduct (*S*)-**5** was thus obtained in 65% ee after addition of **1** to MVK (2 equiv in toluene, –48 °C for 28 h, 88% chemical yield) in the presence of a (–)-quinine–acrylonitrile copolymer armed with a 15-atom spacer group.

Transition-Metal Catalysis

Transition-metal catalysts are known to promote the Michael addition of 1,3-dicarbonyl compounds to α,β-unsaturated systems. The reaction generally proceeds under mild conditions and avoids the formation of by-products sometimes associated with the classical base-catalyzed version, thus providing high yields of Michael adducts. Another obvious advantage of this method is that chiral catalysts can be easily generated in situ by complexation with a large array of homochiral polydentate ligands. The development of such chiral catalysts and their use in the enantioselective Michael addition have been the subject of recent reports by the groups of Brunner and Desimoni.

The first example[15] to be reported involved the octahedral cobalt (II) complex obtained by addition of 1 equiv of Co $(acac)_2$ to C_2-symmetric (+)-(*S*,*S*)-1,2-diphenylethylene diamine. In the presence of a catalytic amount of this complex in toluene at –50 °C, methyl indane-1-one-2-carboxylate (**1**) added to MVK gave adduct (*R*)-**5** in 50% chemical yield and 66% ee. A model has been offered by the authors to account for the preferred formation of the (*R*)-enantiomer of **5**. They first assumed that the Δ-trischelate octahedral complex **10**, in which the ketone carbonyl group occupies an axial position and the ester carbonyl group an equatorial one, is formed in situ and represents the most stable configuration. It was therefore postulated that the MVK is directed to the enantiotopic *si*-face of the complexed substrate by hydrogen bonding assistance of one of the equatorial NH_2 ligands (Scheme 7; for clarity the phenyl substituents of the diamine ligand are not shown and one of the two 1,3-dicarbonyl ligands is schematically represented).

Later on, the same reaction was also investigated by the groups of Botteghi and Boga using a variety of Ni (II), Co (II), and Cu (II) chiral complexes. The cobalt complexes proved to be better catalysts than the nickel ones and a 38% ee could be obtained for adduct **5** with the tetradentate (4*R*,5*R*)-3,4-dihydroxy-1,6-

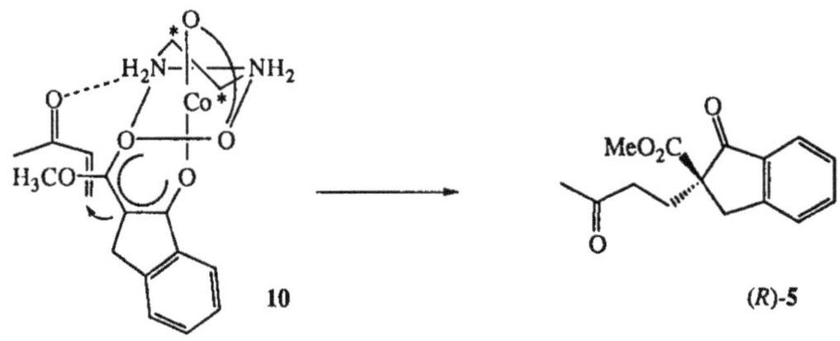

Scheme 7.

diaminohexane.[16] However, the most interesting results were obtained with tetradentate ligand-derived copper complexes.[17] Thus, the two dimeric Cu(II) complexes **11** and **12**, prepared from salicylaldehyde and (*S*)-2-amino-1,5-pentane diol or (*S,S*)-2-amino-3-phenyl-1,3-propane diol, respectively, catalyze the addition of **1** to MVK in CCl_4 at −20 °C to give adduct (*S*)-**5** in up to 70% ee. By analogy with the work of Brunner,[15] the efficiency of catalyst **11** was ascribed to the in situ formation of a Δ-octahedral complex **13** which is well suited to direct the MVK on the *re*-face of the complexed substrate by the hydrogen bonding assistance of the hydroxy equatorial ligand (Scheme 8). A similar complex, in which a water molecule occupies an axial position, was postulated to account for

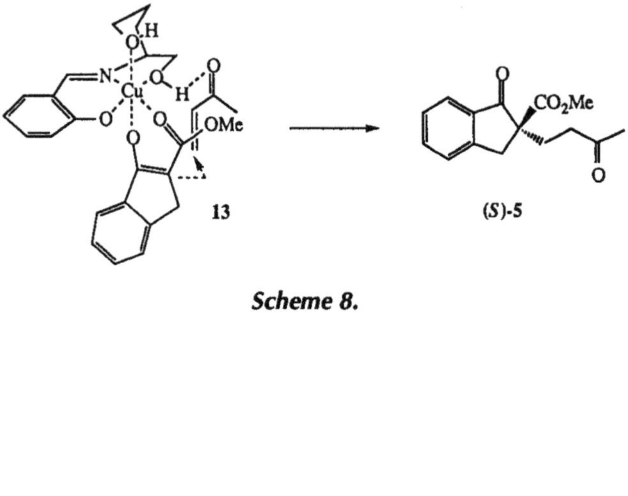

Scheme 8.

the nearly identical efficiency of catalyst **12**. Less rigid complexes, prepared from salicylaldehyde and bi- or tridentate ligands, led to less satisfying results.

Brunner and Kraus[18] have carried on with their earlier investigations in studying the effectiveness of their earlier-mentioned cobalt catalyst in some other examples of donor–acceptor systems. Methyl 3,4-dihydro-1(2*H*)-naphthalenone-2-carboxylate, methyl indane-1-one-2-carboxylate, 2-acetylcyclohexanone, and ethyl cyclohexanone-2-carboxylate were chosen as the donors, and methylvinylketone, di-*tert*-butyl methylenemalonate and acrolein as the Michael acceptors. All of the reactions were performed in toluene solution in the presence of 3 mol% of the catalyst. Some of the results of this study are displayed in Scheme 9 and certainly did not come up to the authors expectations, giving 39% ee at best for the ethyl cyclohexanone-2-carboxylate/acrolein system.

Although a final discussion concerning the enantioselective metal-catalyzed Michael addition reaction must certainly await further experimentation and a more systematic optimization with regard to ligands, some of the following trends can be drawn from the previously mentioned data:

- For a given catalyst, the enantioselection is strongly dependent on both the Michael donor and the Michael acceptor.
- For a given reaction the results are highly sensitive to structural changes in the catalyst.
- The formation of a rigid complex capable of anchoring both the donor and the acceptor is necessary to obtain significant levels of enantioselection. Accordingly, polydentate ligands led to the best results.
- Hydrogen bonding is of paramount importance to selectively direct the acceptor on one enantiotopic face of the complexed substrate.

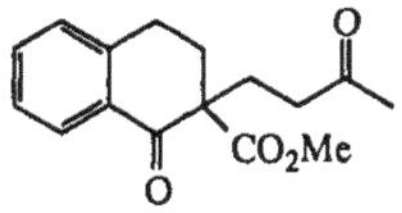

14 (ee 18%)

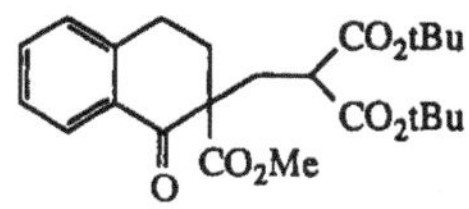

15 (ee 27%)

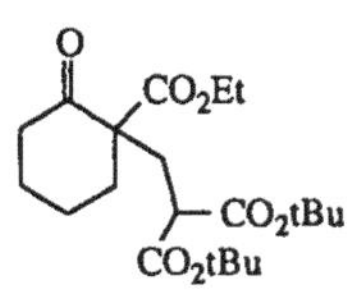

16 (ee 12%)

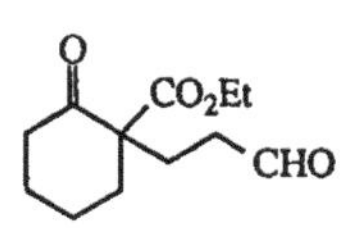

17 (ee 39%)

Scheme 9.

- Oxygenated solvents are to be avoided as they may act as ligands or may be involved in hydrogen bonding with a hydroxyl group. Accordingly, best results were obtained in toluene and carbon tetrachloride solutions.
- Finally, it is noteworthy that most of the above observations are also valid for the Cinchona alkaloid-mediated Michael addition reaction.

Catalysis by Chiral Crown Complexes

Complexation of organic molecules with optically active macrocyclic hosts are of considerable interest for stereoselective syntheses. In this area, two examples of chiral crown complexes catalyzing the Michael addition reaction of methyl indane-1-one-2-carboxylate (**1**) to methylvinylketone have been reported to date.

By using 4 mol% of chiral macrocyclic compound (*S*,*S*)-**18** (Scheme 10) associated with *t*-BuOK as a cocatalyst and working in toluene at –78 °C for 120 h, Cram and Sogah[19] were able to obtain 48% yield of adduct (*R*)-**5** in virtually enantiomerically pure form. Unfortunately, the reaction is sensitive to temperature change and a moderate 67% ee was obtained at 25 °C.

The way in which the potassium enolate of **1** fits into the chiral cavity of **18** is assumed to be governed by steric factors originating from the two dinaphthyl units.

Later, Raguse and Ridley[20] investigated a series of chiral podands incorporating ether and sulfoxide functionalities. The selectivities observed were quite low [6.5% ee at best for podand **19** (Scheme 10)] probably because of the conformational flexibility of the host molecules.

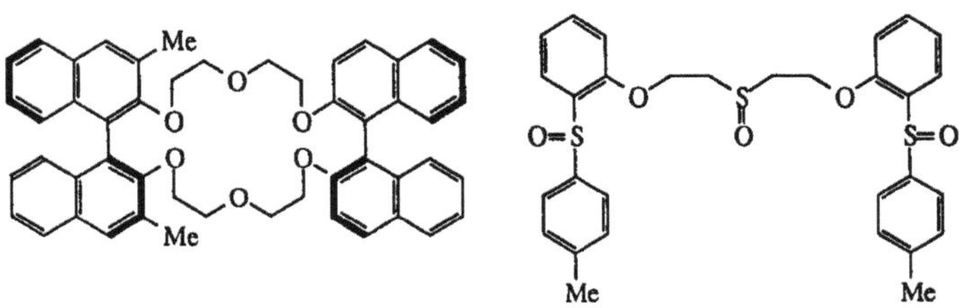

(*S*,*S*)-**18** (*R*,*R*)-**19**

Scheme 10.

B. Alkylation Reactions

Catalysis by Transition-Metal Complexes

Although asymmetric carbon–carbon bond-forming reactions catalyzed by chiral transition-metal complexes have been widely developed over the years, only a few examples are known concerning the alkylation reactions of 1,3-dicarbonyl systems. The first example to be reported was due to Kagan's group[21] who examined the allylation of cyclic β-diketones with chiral palladium complexes. Using 1 mol% of chiral complex $PdCl_2$·(+)DIOP together with sodium phenoxide as a cocatalyst, the catalytic transfer of an allyl group from allyl phenyl ether to 2-acetyl-1-tetralone (**20**) and 2-acetyl-cyclohexanone (**21**) could be effected in excellent chemical yields but with poor enantioselectivities (10 and 7% ee, respectively) (Scheme 11).

Taking into account that the nucleophile attacks the face of the Π-allylpalladium(II) complex opposite to the palladium, the low asymmetric induction was ascribed by the authors to the large distance between the chiral-inducing ligand and the prostereogenic quaternary carbon center of the β-diketonate. Later, a somewhat different approach was devised by Kumada's group[22] to circumvent such a problem. New phosphine ligands such as **23**, bearing a chiral functional group at an appropriate distance from the coordinated phosphino groups, were designed and prepared. The hope was that the terminal chiral functional group would be able to coordinate to the sodium enolate of **21** and, thereby, help to differentiate between its two enantiotopic faces as the reaction is proceeding (Scheme 11).

The results displayed in Table 2 demonstrate the pertinence of the working hypothesis as up to 52% ee could be observed in case ligand **23a** (n = 2) is used. This hypothesis is also supported by the lower efficiency of ligand **23a** (n = 3), which points out the importance of the distance between the chiral ligand and the diphenylphosphino-groups coordinated to palladium, and by the poor enantiose-

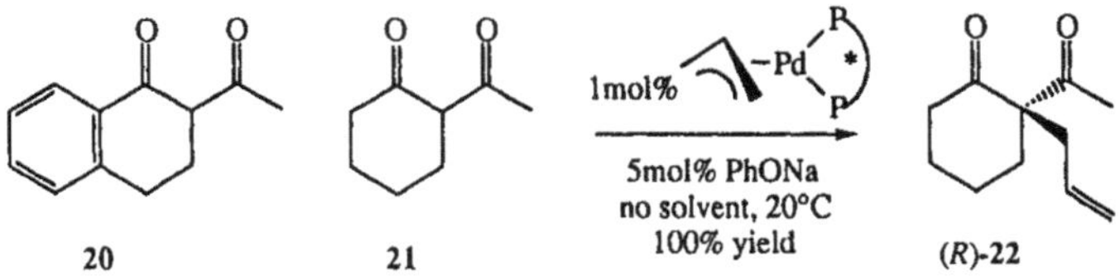

20 21 (*R*)-22

Scheme 11.

Table 2.

Chiral Ligand	*T* °C	*% Isolated Yield*	*% ee*
23a (*n* = 2)	−50	52	52 (*R*)
23b (*n* = 2)	−30	64	31 (*R*)
23c (*n* = 2)	−30	86	15 (*R*)
23a (*n* = 3)	20	82	5 (*R*)

lectivity recorded with ligand **23c** (*n* = 2) lacking a methoxycarboxyl or a hydroxy chelating group (Scheme 12).

Refinement of this methodology could be later achieved by Hayashi's group[23] who discovered that allylation of β-diketones with allyl acetate proceeds with good enantioselectivity in the presence of palladium complexes bearing functionalized chiral ferrocenylphosphine ligands.

Among the chiral ligands prepared, ligand **24** armed with a *N*-methyl-(2-hydroxyethyl)-amino group was found the most effective. For example the Pd(II) complex prepared in situ by mixing di-μ-chlorobis(Π-allyl)dipalladium with ligand **24** catalyzed the allylation of the sodium enolate of **21** in THF at −50 °C to give the β-diketone (*S*)-**22** in 82% ee. The absolute configuration at the quater-

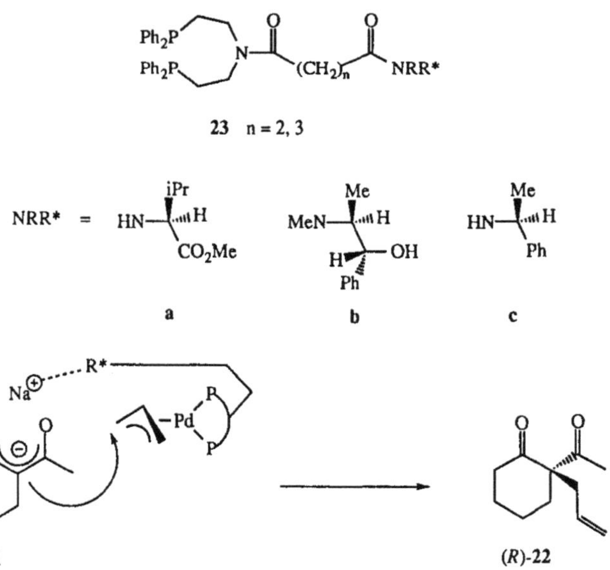

Scheme 12.

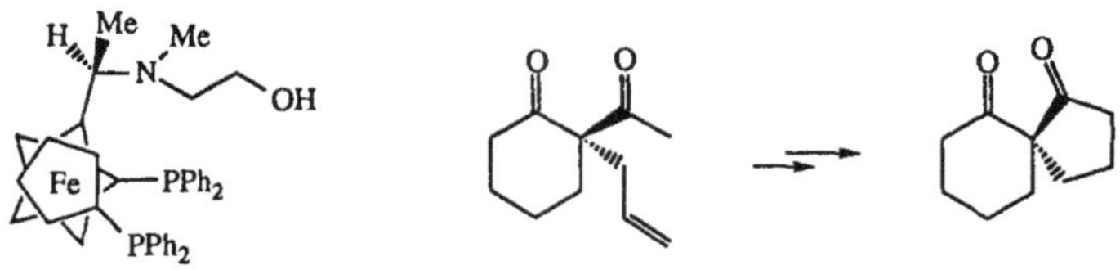

24 (S)-22 (R)-25

Scheme 13.

nary carbon center was assigned to be *S* after conversion of **22** into the known spirodione (*R*)-**25** (Scheme 13).

Other experiments conducted with structurally different ligands led to the conclusion that the presence of a terminal hydroxy group in the ligand is a key feature to obtain good enantioselectivities. Furthermore, this hydroxy group must be located at an appropriate distance from the ferrocenylbiphosphine moiety and not sterically hindered at the hydroxymethyl carbon. A model which accounts for the stereochemical results has been proposed by the authors. They assumed that the enolate of **21** is anchored to the complex by hydrogen bonding to the terminal hydroxy group of the ligand and is thus spatially well-disposed to react with the Π-allyl moiety. Steric interactions between the enolate and the ferrocenyl moiety are assumed to play a major role in differentiating the enantiotopic faces of the prochiral enolate. In contrast to the reaction reported by Kumada's group (*vide supra*), coordination to the sodium cation of the enolate here is unlikely because the replacement of the terminal ligand hydroxy group by amino or alkoxy groups resulted in the formation of **22** in low selectivities.

The same chiral catalyst was also examined in the allylation reaction of other 1,3-dicarbonyl compounds. As can be seen from Scheme 14 variable results were obtained with a series of cyclic and acyclic β-diketones, the highest enantioselectivity (82% ee) being attained in case of 2-acetyl-1-tetralone (**20**). As for the allylation of **21**, the selectivities were strongly temperature-dependent and best results were obtained at low temperature for a long reaction period. The results also demonstrate that good selectivities were achieved with β-diketones displaying different steric bulkiness around the two carbonyls. Unfortunately, no such satisfying results could be obtained with β-keto esters.

More recently, homogeneous palladium-catalyzed allylation of β-diketone **20** was reported by Lemaire's group.[24] Reacting the sodium enolate of **20** to the Π-allyl complex prepared in situ by adding 5 mol% of allyl palladium chloride dimer to allyl acetate in the presence of 10 mol% of the C_2 symmetric (*S,S*)-*N,N'*-dimethyl-1,2-diphenylethylene diamine led to the expected allylation product

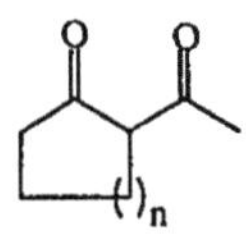

26

n = 1 ee = <5% (-50°C, 17h)
n = 4 ee = 58% (-60°C, 20h)

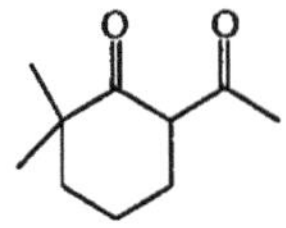

27

ee = 70% (-60°C, 68h)

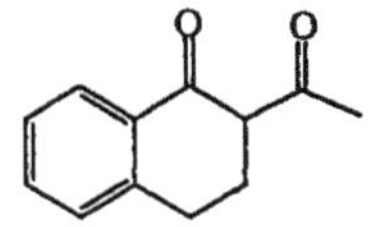

20

ee = 82% (-60°C, 45h)

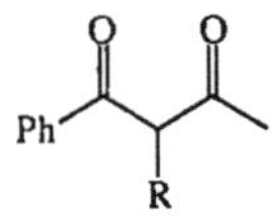

28

R = Me ee = 60% (-60°C, 45h)
R = iPr ee = 32% (-30°C, 21h)

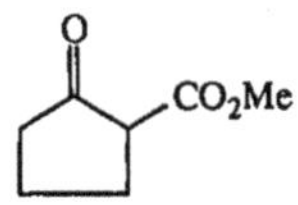

29

ee = 22% (-50°C, 18h)

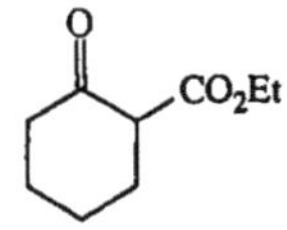

8

ee = 3% (-50°C, 44h)

Scheme 14.

with 51% chemical yield and 30% ee. Although less efficient than the Hayashi method in case of β-diketone **20**, it is worth mentioning that this simple protocol proved to be very efficient (up to 95% ee) in case of sodiomalonate allylation.

One single example of enantioselective telomerization was reported by Keim's group.[25] They studied the telomerization reaction of butadiene with β-keto ester **29** catalyzed by palladium acetate in combination with a chiral bidentate phosphine to give alkylated β-keto ester **30** (absolute configuration unknown) (Scheme 15). The best result (ee = 41%) was recorded at −10 °C with (2*S*,4*S*)-*N*-*tert*-butoxycarbonyl-2,4-bis(diphenylphosphino)-methylpyrrolidine (BPPM) as the ligand (catalyst/substrate ratio = 1/350).

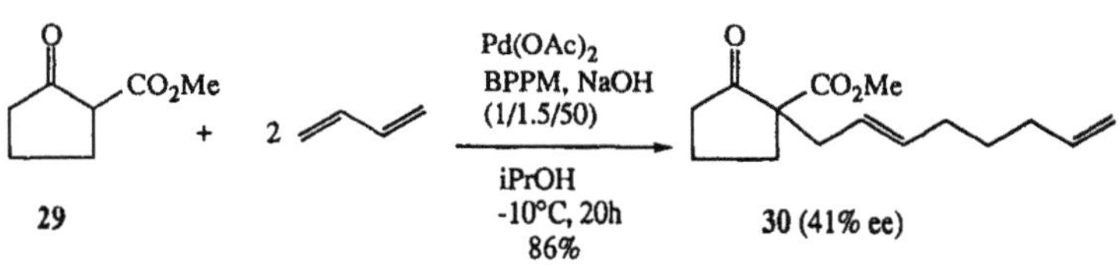

29 **30** (41% ee)

Scheme 15.

Phase Transfer Catalysis

Phase transfer catalysis using chiral catalysts offers a potentially simple way to achieve asymmetric alkylation reactions. The first example to be reported[26] involved β-keto esters as substrates and ephedrinium halide as catalysts. The estimated ee were very low (6–7%) and subsequent work did not achieve better results.[27] It must be pointed out here that recent progress has been made in the area of chiral phase transfer catalysis, allowing the efficient asymmetric synthesis of cyclic compounds bearing a quaternary carbon centers to be achieved by means of substituted *N*-benzylcinchonidium halides as catalysts.[28] However, these findings have not yet been applied to the alkylation reaction of 1,3-dicarbonyl compounds.

III. DIASTEREOSELECTIVE SYNTHESES OF α,α-DISUBSTITUTED β-KETO ESTERS FROM α-MONOSUBSTITUTED β-KETO ESTERS

In relation with the synthesis of natural products of biological importance, a number of elegant and, in some examples, efficient strategies have been devised to achieve the asymmetric transformation of α-substituted β-keto esters to chiral α,α-disubstituted β-keto esters. Four distinct processes have been used to reach this goal: (1) Michael addition reaction of chiral β-keto ester enolates to achiral electrophilic olefin, (2) alkylation of chiral β-keto ester enolates with achiral electrophiles, (3) alkylation of achiral β-keto ester enolates with chiral electrophiles, and (4) intramolecular metal-induced cyclopropanation of chiral α-diazo-β-keto esters. These processes may be represented by the generalized outline in Scheme 16.

A number of diverse secondary alcohols derived from natural sources have been employed as chiral auxiliaries in the above reaction sequences. These (**A** to G) are represented in Scheme 17.

A. Michael Addition Reaction of Chiral β-Keto Ester Enolates to Achiral Electrophilic Olefins

Stork and Saccomano[29] have demonstrated that intramolecular capture of chiral α,β-unsaturated esters by sodium β-keto ester enolates provides an effective method for the diastereoselective construction of vicinally substituted *trans*-cyclopentanones which are useful synthons in steroid total synthesis. In their study, the easily-prepared substrates **31** were cyclized in benzene at room temper-

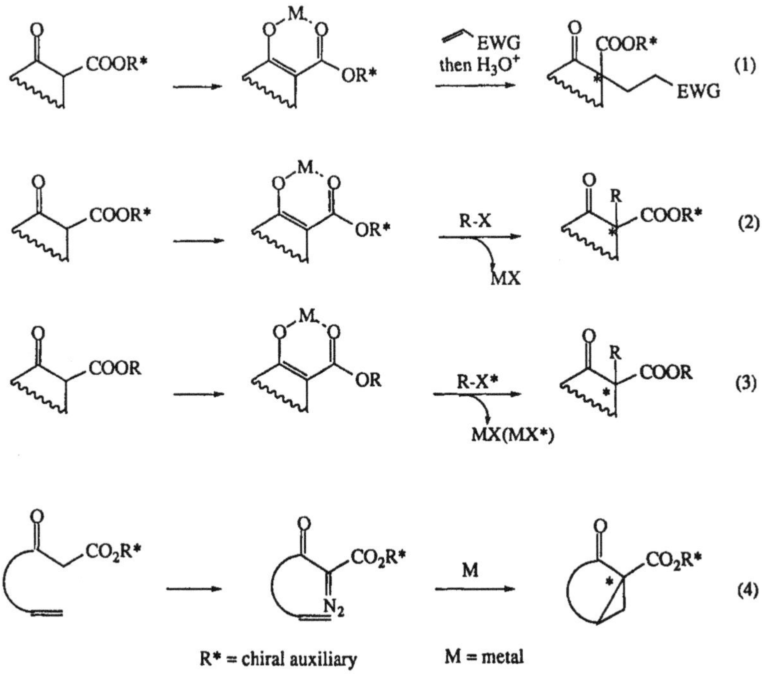

Scheme 16.

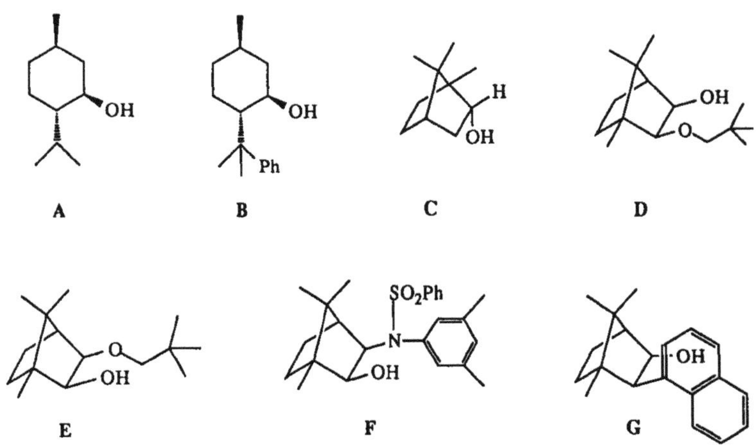

Scheme 17.

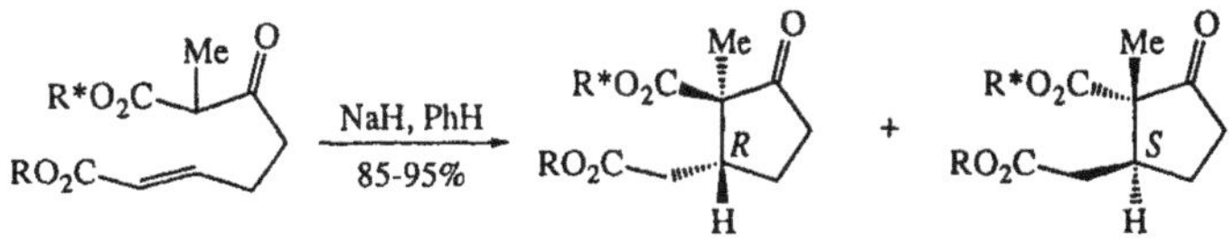

31 32 33

Scheme 18.

ature with a catalytic amount of sodium hydride to give the Michael adducts **32/33** in excellent chemical yields (Scheme 18). The diastereoselectivities, reported in Table 3, were greatly dependent on both the selected chiral auxiliaries (alcohols **A**, **B**, **D** and **E**; cf. Scheme 17) and the size of the ester grouping in the α,β-unsaturated ester moiety. 8-Phenylmenthol **B** and borneols **D** and **E** are good diastereotopic Π-facial directing groups, and *tert*-butyl esters always provide higher diastereomeric ratios than the corresponding methyl esters.

The observed Π-face discrimination of the above addition reactions was attributed to transition-state models that feature (1) a planar chelate ring for the sodium enolate, and (2) a *syn*-planar arrangement $C{=}O/H_a$ (Scheme 19).

For the cyclization of the 8-phenylmenthol derivatives **34**, the dimethylbenzyl group is well disposed to shield the *si*-face of the Π-enolate system and, hence, to direct the internal electrophile to the opposite *re*-face (Scheme 19). With regard to the cyclization of the neopentyloxy borneol derivatives, the neopentyl ether grouping sterically blocks the *si*-face of the borneol ester **35** and the *re*-face of the borneol ester **36**, thus leading to the major diastereomers which are observed. It is to be noted that the preferred transition state conformations **35** and **36** are enantiomeric, with the exception of the position of the bridgehead methyl group. The higher topological control provided by **36** can be accounted for by severe conformational restrictions on the reactive side chain moiety. Indeed, rotation of the chelate ring would result in the establishment of steric interactions with the bridgehead methyl group. Such a buttressing situation is no longer found in geometry **35** to which the shielding effect of the chiral auxiliary is less pronounced (Table 3).

Table 3.

*R**, *R*	**A**, *Me*	**A**, *t-Bu*	**B**, *Me*	**B**, *t-Bu*	**D**, *Me*	**D**, *t-Bu*	**E**, *Me*	**E**, *t-Bu*
abs. conf.	(*R*)	(*R*)	(*R*)	(*R*)	(*R*)	(*R*)	(*S*)	(*S*)
32/33	2.2/1	5.3/1	4/1	14/1	3.3/1	4.2/1	1/15	1/40

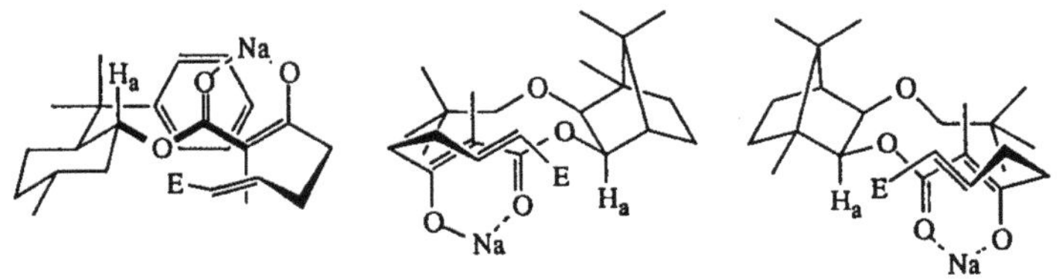

34 35 36

Scheme 19. (E = COOR)

Starting from β-keto ester **37**, which could be isolated in enantiomerically pure form after chromatographic purification of its dioxolane derivative, the synthetic potential of this cyclization process has been elegantly applied to the synthesis of adrenosterone **38** (Scheme 20).[30]

The preceding intramolecular Michael addition reaction allowed the diastereoselectively preparation of five-membered ring β-keto esters. Later on, Taber's group[31] showed that intermolecular Michael addition of an acyclic chiral β-keto aldehyde to methylvinylketone followed by Robinson annulation could nicely serve to prepare six-membered ring β-keto esters of synthetic importance with a high stereocontrol. In close relation with their work on the synthesis of the mesembrane alkaloid *O*-methyl Joubertiamine (**45**), they have studied the base-induced addition of β-aldehyde ester **40**, prepared from **39** by 4-(dimethylamino)pyridine-catalyzed ester exchange with chiral alcohol **G** (Scheme 21), to methylvinylketone as the acceptor.

Of several solvents and bases screened, potassium carbonate in dry DME, to which was added 0.1% H_2O, afforded the optimum balance between reaction stereoselection and product yield. The preferred formation of diastereomer **41** has

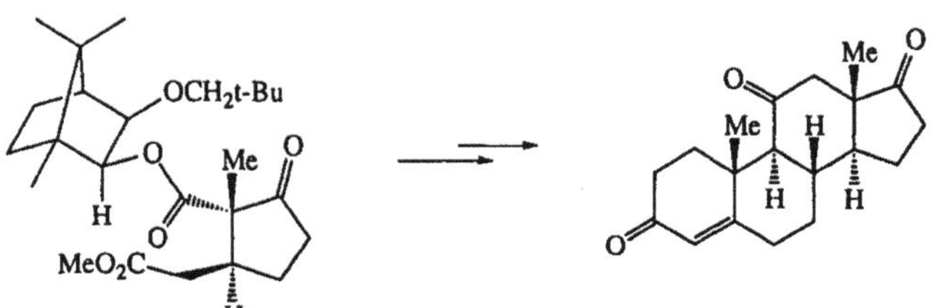

37 38

Scheme 20.

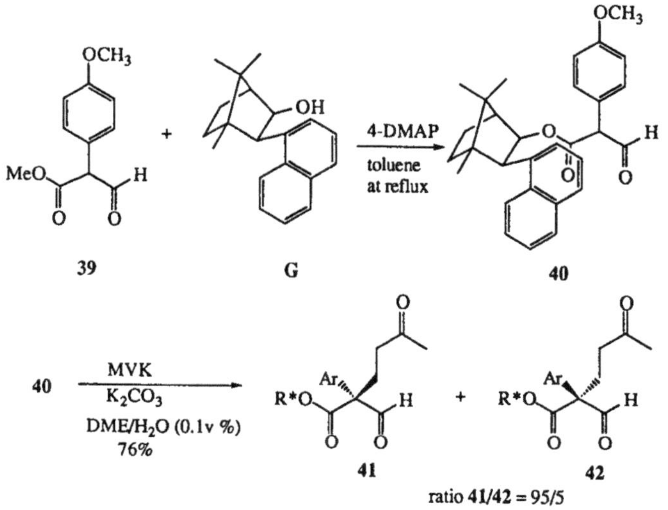

Scheme 21.

been interpreted on the assumption that, in agreement with earlier work, (1) the potassium enolate forms a planar chelate ring, (2) there exist a *syn*-planar arrangement between the ester carbonyl and the H_a proton, and (3) one of the diastereotopic faces of the chelate ring is shielded by the naphthyl ring, lying in a parallel plane. If one accepts all of these assumptions it appears, rather unexpectedly, that diastereomer **41** originates from enolate **44** rather than its rotamer **43** (Scheme 22).

The synthetic potential of this reaction has been illustrated by the transformation of diastereomer **41** into the unnatural enantiomer of *O*-methyl Joubertiamine

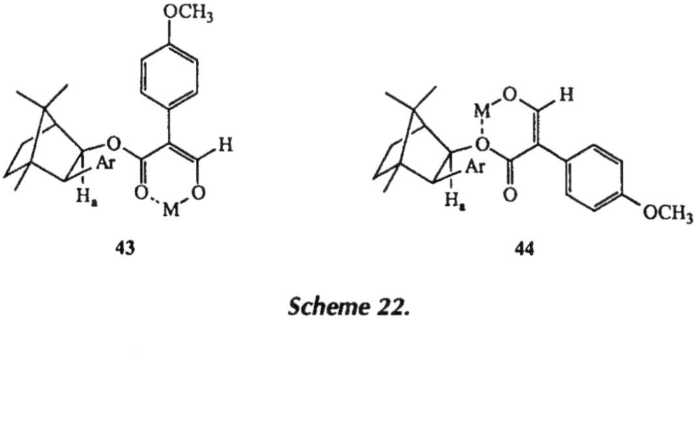

Scheme 22.

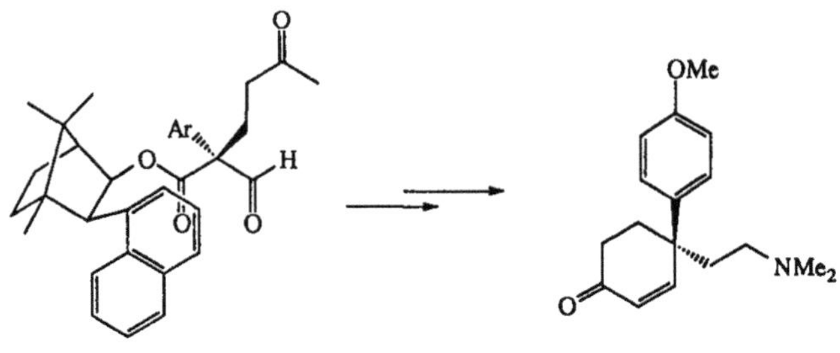

44 45

Scheme 23.

(**45**) (Scheme 23). However, this synthesis was made difficult by the high propensity of **41** to be deformylated and an 11-step sequence was necessary to finally reach the target.

Another example of diastereoselective Michael addition reaction of a chiral β-keto ester to an achiral acceptor is found in the work of Kozikowski's group[32] relative to the total synthesis of the alkaloid (−)-Huperzine A **49**, a potent acetylcholinesterase inhibitor (Scheme 24). These authors have shown that the 8-phenylmenthol ester derivative **46** could be added to methacrolein in the presence of tetramethylguanidine at room temperature over 2 days to give a mixture of **47**

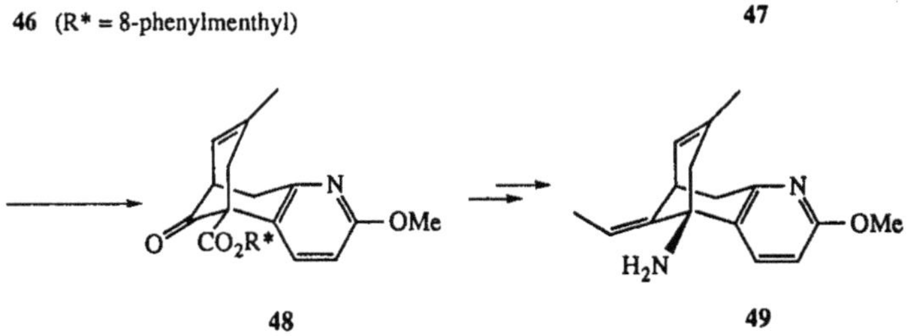

46 (R* = 8-phenylmenthyl) 47

48 49

Scheme 24.

resulting from a tandem Michael–aldolization reaction. Compound **47** was subsequently dehydrated to a mixture of chromatographically separable olefins in a ratio of 4/1 (9/1 if the reaction was carried out at –20 °C) in favor of **48**.

B. Alkylation Reactions of Chiral β-Keto Ester Enolates with Achiral Electrophiles

In order to search for new analgesic and antipyretic agents possessing the indazole backbone, Ito's group[33] has studied the alkylation reaction of (–)-menthyl-derived β-keto ester anion of **50**. The reaction was carried out in ether and generated optically active β-keto esters **51** which were subsequently transformed to the desired heterocycles **52** by the action of hydrazine hydrate in ethanol at reflux (Scheme 25). Although no effort was made to determine the diastereoselection, this example is interesting since it constitutes the first report of an asymmetric alkylation reaction of a β-keto ester bearing a chiral inductor at the ester grouping.

Following this work, Welzel's group[34] reported the diastereoselective methylation of enolates of chiral α-substituted β-keto esters **54** (Scheme 26) prepared by photolysis of diazodimedone **53** in the presence of the chiral alcohols **A**, **C**, **D**, and **F** (Scheme 26; cf. Scheme 17). The results in Table 4 show that low to moderate diastereoselection was obtained depending on the chiral auxiliary and conditions used.

In agreement with a large body of earlier literature data, it was assumed that the sodium enolates are held in planar six-membered chelate rings with the carbonyl ester group *syn*-planar with the C–H bond at C-2 (**F**) or C-3 (**D**) of the chiral auxiliaries. The buttressing effect of the neopentyl chain or of the sulfonylamino group then directs the electrophile to the *re*-face of **56** and to the *si*-face of **57** leading to the major diastereomers **55D** and **55F**, respectively, as shown in Scheme 27.

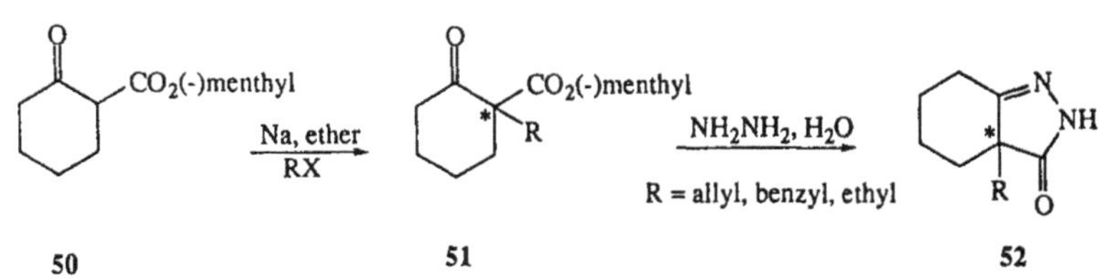

Scheme 25.

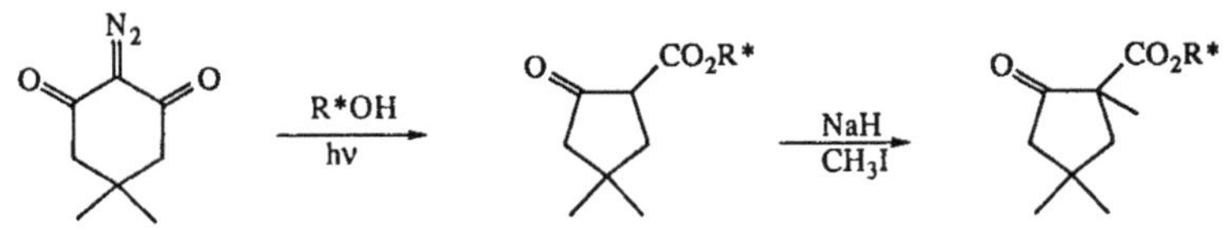

53 54 55

Scheme 26.

C. Alkylation Reactions of Achiral β-Keto Ester Enolates with Chiral Electrophiles

In a series of papers, Kobayashi, Matsuyama, and co-workers[35a–d] have reported some examples of asymmetric alkylations of achiral α-substituted β-keto esters with chiral sulfonium and selenonium salts. The method suffers from synthetic limitations due to very low asymmetric inductions and formation of several alkylation products. As an example, alkylation of β-keto ester **1** with homochiral (*R*)-**58** in the presence of anhydrous potassium carbonate in dichloromethane gave a mixture of C-methylated (*S*)-**59** (30%, 6% ee) and C-ethylated (*R*)-**60** (44%, 10% ee) products along with some O-alkylated products (Scheme 28). It was proposed that the reaction proceeds via a S–O sulfurane intermediate.

An example of the use of a chiral leaving group in a substitution reaction was reported by Differding and Lang.[36] They explored the fluorination of the sodium enolate of β-keto ester **63** with the *N*-fluoro sultams **61** and **62** acting as enantioselective "F^+" transfer reagents. As can be seen in Scheme 29, enantioselectivity of up to 70% was obtained using a slight excess of **61** in ether as the solvent. In contrast, **62** failed to give satisfactory results probably because of steric reasons. It proved also difficult to extend the moderate efficiency of this method to other β-keto esters, as exemplified by the fluorination of **65**.

Following this work, Davies's group[37] introduced a new fluorinating reagent, namely (–)-*N*-fluoro 2,10-(3,3-dichloro)-camphorsultam (**68**), prepared in 68% yield by passing 10% F_2/N_2 through a solution of (+)-**67** in chloroform at –40 °C

Table 4.

Chiral Auxilliary	*Solvent*	*Reaction Product*	*% Yield*	*Diastereoselection*
A	DMF	55A	42	18% (*R*)
C	DMF-toluene 3/2	55C	81	16%
D	DMF-toluene 3/2	55D	70	58% (*S*)
F	THF-HMPA 3/2	55F	60	62% (*R*)

56 ⟶ (S)-55D 57 ⟶ (R)-55F

Scheme 27.

for 1 h. Although this new sultam is more reactive and often gives better chemical yields than the parent sultam **61**, its superiority, as far as enantioselection is concerned, is not clearly manifest as judged from the reported two examples of β-keto ester enolate fluorination (Scheme 30).

D. Intramolecular Metal-Induced Cyclopropanation of Chiral α-Diazo-β-Keto Esters

Cyclopropanation methodologies have already been found to be quite useful in the synthesis of natural products and one of the most challenging issues to be pursued is to develop an efficient chiral version to control the absolute stereochemistry of cyclopropane formation. Taber's group[38] has recently developed a new method allowing preparation of α,α-disubstituted β-keto esters **72/73** with a reasonable level of diastereoselection. Practically, they have prepared a series of chiral α-diazo-β-keto esters **71** (Scheme 31) derived from the naphthyl borneol **G** (cf. Scheme 17) and studied their metal-induced decomposition to **72/73** via an intramolecular cyclopropanation process. The level and the sense of diastereoselection proved to be a function of both the metal–ligand association and the substitution on the olefins, as summarized in Table 5 for rhodium and palladium catalysts. The stereochemical outcomes of this reaction have been interpreted, in the palladium series, on the following assumptions: (1) the starting β-keto esters

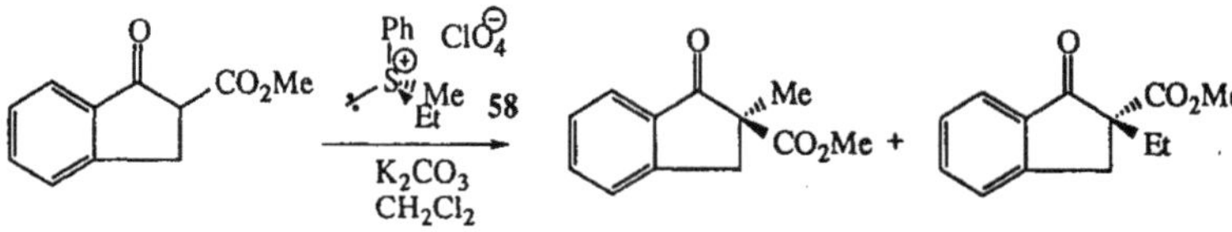

1 (S)-59 (R)-60

Scheme 28.

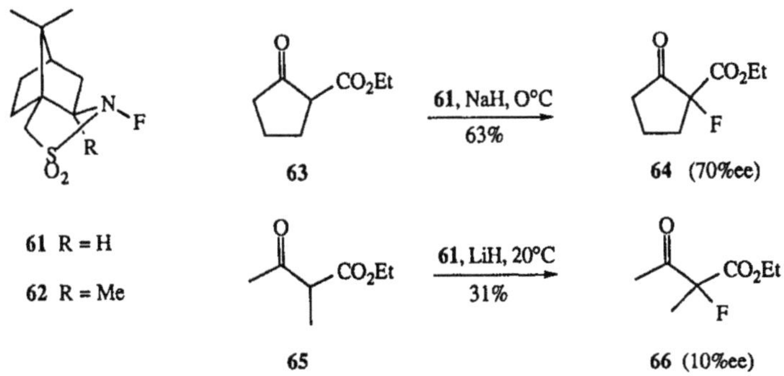

Scheme 29.

exist in the preferred conformation **74** (Scheme 32) in which, and in agreement with results of ab initio calculations, the carbonyls are *syn*-disposed, and (2) the olefin approaches the more hindered side of the transient metallocarbene, thus pushing out the bulky ligands on the opposite face. Inspection of molecular models reveals that the *cis*-olefin **71b** will fit into the cleft between the naphthalene ring and the β-keto ester moieties more easily than the *trans*-olefin **71c**. Accordingly, better chemical yields of addition products were obtained for the *cis*-olefin (compare entries 3 and 5, Table 5).

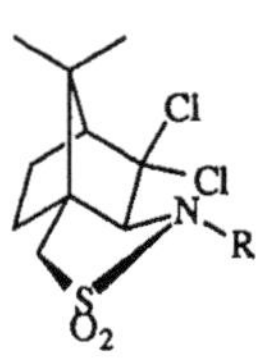

67: R = H
68 R = F

Sultam	Product	% yield	% ee
61	**64**	63	70
68	**64**	71	34
61	**70**	25	28
68	**70**	46	95

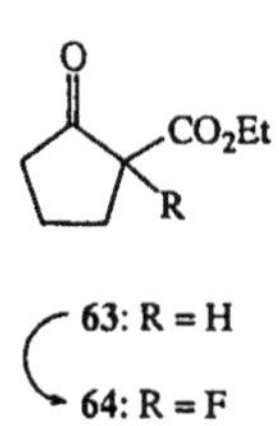

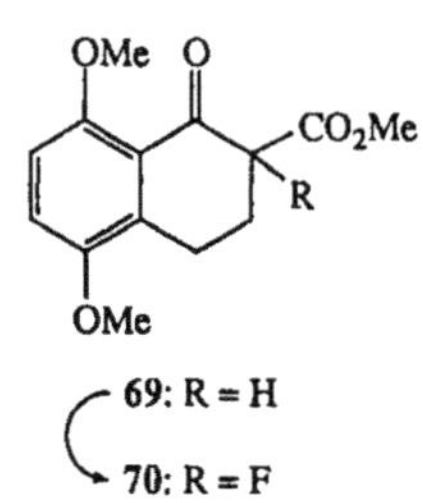

Scheme 30.

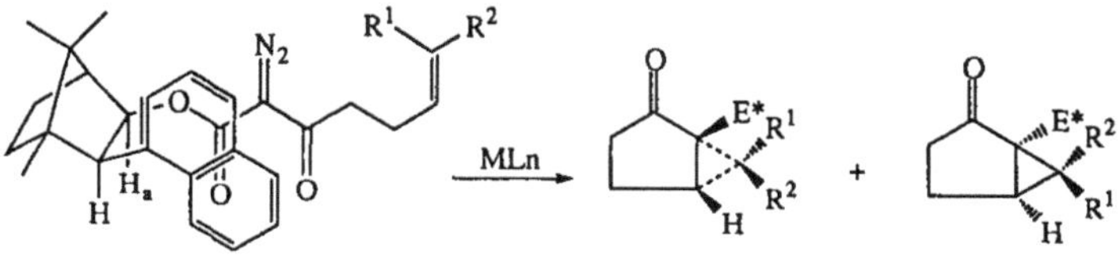

71

a: $R^1 = R^2 = H$

b: $R^1 = Me, R^2 = H$

c: $R^1 = H, R^2 = Me$

72

73

Scheme 31.

Table 5.

Entry	**71**	*Catalyst*	**72/73**	*% Yield*
1	$R^1 = R^2 = H$	(dppp))PdCl$_2$	28/72	57
2		(TPP)RhCl	89/11	64
3	$R^1 = Me$	(dppp)PdCl$_2$	80/20	64
4	$R^2 = H$	(TPP)RhCl	68/32	43
5	$R^1 = H$	(dppp)PdCl$_2$	80/20	38
6	$R^2 = Me$	(TPP)RhCl	70/30	55

As a synthetic demonstration of this method, diastereomer **72b** was converted to (+)-isonepetalactone **75** (Scheme 32). The initially controlled quaternary carbon center was lost during the course of this transformation which features, as the key step, the opening of the cyclopropane ring under the action of an organocuprate reagent.

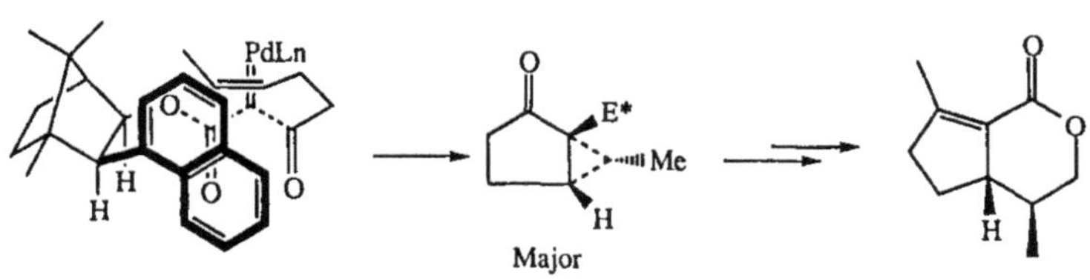

74

72b

75

Scheme 32.

IV. DIASTEREOSELECTIVE SYNTHESES OF α-α-DISUBSTITUTED β-KETO ESTERS FROM ACETAL AND β-ENAMINOESTER DERIVATIVES

We have previously reported and commented on the different modes of preparation of α,α-disubstituted β-diketones and β-keto esters starting from the α-mono-substituted β-dicarbonyl analogs as precursors. As far as the diastereoselective approaches to α,α-disubstituted β-keto esters were concerned, the chiral inductor was borne by the nucleophile at the ester grouping. A potentially useful alternative to this general strategy may be to exploit the chirality of a temporary ketone carbonyl protective group to induce asymmetry at the latent prostereogenic quaternary carbon, as depicted in Scheme 33.

Because protection and deprotection of the ketone carbonyl of a β-keto ester may be readily and efficiently achieved, such an "indirect" route to chiral α,α-disubstituted β-keto esters does not suffer from real disadvantages. A few research groups have recently been involved exploring this route and good to excellent levels of diastereoselection have been attained. The different strategies that have been considered until now make use of the chiral precursors—acetals or β-enaminoesters—represented in Scheme 34.

A. Alkylation and Oxidation Reactions using Chiral Acetals

Sakai and co-workers[39a,b] have recently developed an enzymatic preparation of both enantiomers of the C_2-symmetric 1,2-cyclohexane and cycloheptane diols and have used them as chiral controllers in a number of asymmetric reactions. In particular, they found that the lithium enolates of chiral β-keto ester acetals derived from these alcohols could be alkylated with diastereoselectivity of up to 99% in both cyclic and acyclic series as shown in Schemes 35 and 36 (only reaction products are shown). The anion **77** resulting from a deprotonation–deconju-

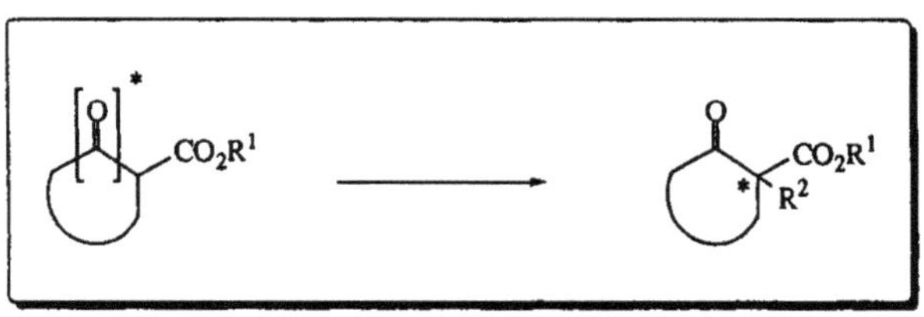

Scheme 33.

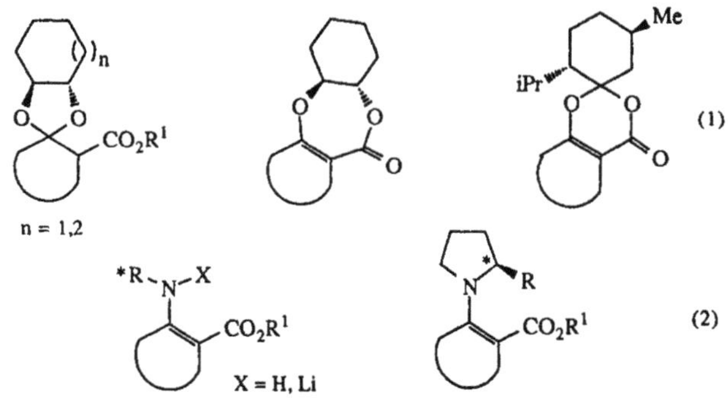

Scheme 34.

gation sequence initiated by an excess of base (5 equiv) is believed, in its sterically least congested form (**77** rather than **78**, Scheme 35), to be the reactive key intermediate. Note that greatly reduced diastereoselectivities were obtained using acyclic chiral diols such as (*R,R*)-2,3-butanediol and (*R,R*)-2,4-pentanediol as chiral auxiliaries.

The α,α-disubstituted β-keto esters could be obtained by simply treating the enol ethers with dilute acetic acid as exemplified by the preparation of **83** (Scheme 37), a known precursor of the marine antibiotic (–)-Malyngolide **84**.

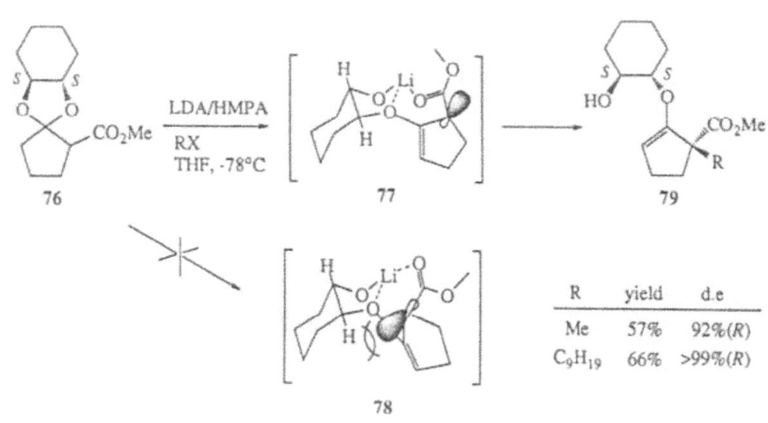

R	yield	d.e
Me	57%	92%(*R*)
C_9H_{19}	66%	>99%(*R*)

Scheme 35.

80 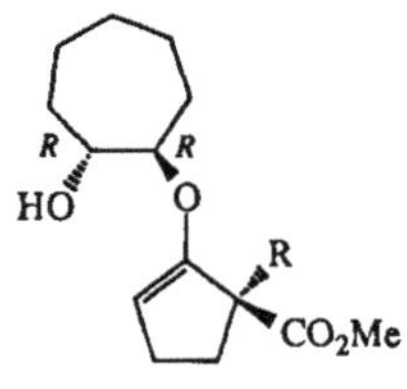

R	yield	d.e
Me	54%	>99%(*S*)
C_9H_{19}	74%	>99%(*S*)

81 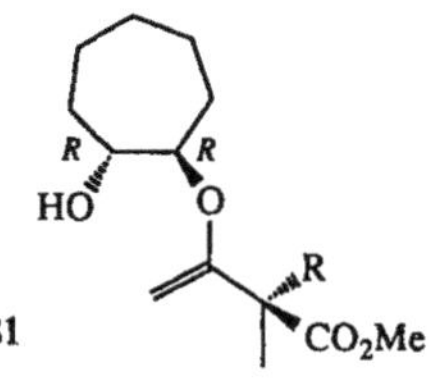

R	yield	d.e
Benzyl	78%	> 99%(*R*)
Allyl	48%	> 99%(*R*)

Scheme 36.

The same group[40] has also demonstrated that the above chiral ketals could be diastereoselectively transformed by asymmetric oxidation of "open" derivatives into protected β-keto esters bearing a hydroxyl group at the newly created stereogenic quaternary carbon.

For example, ketal **76** when submitted to the action of LDA (2.2 equiv in THF at –40 °C) in the presence of TMSCl (3.5 equiv) afforded trimethylsilyloxy enol ether **85** which could be subsequently deprotected into enol ether **86** by treatment with $ZnBr_2$ or HF/pyridine (Scheme 38). Oxidation of trimethylsilyloxy enol ether **85** with MCPBA proceeded in CH_2Cl_2 at –60 °C and was controlled sterically to form the hydroxy ester **87** with exceedingly high diastereoselectivity (>99%). In striking contrast to the above result, MCPBA oxidation of enol ether **86**, in the presence of a basic additive, led to hydroxy ester **88** having the opposite absolute configuration at the quaternary carbon center. The same trend was also observed with acyclic analogs of **85** and **86** which were oxidized under the same experimental conditions into the corresponding hydroxy esters with good to excellent diastereoselectivities.

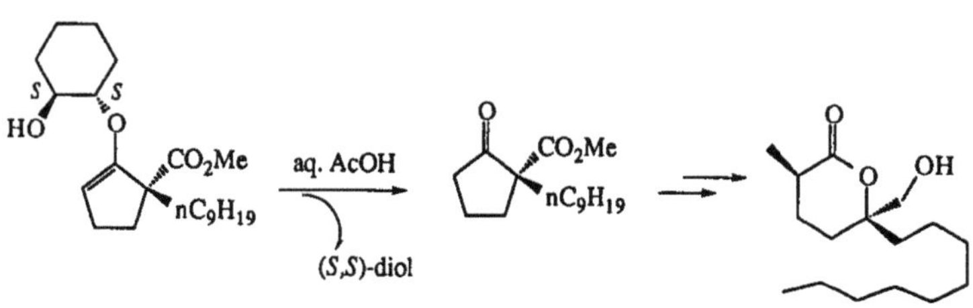

82 83 84

Scheme 37.

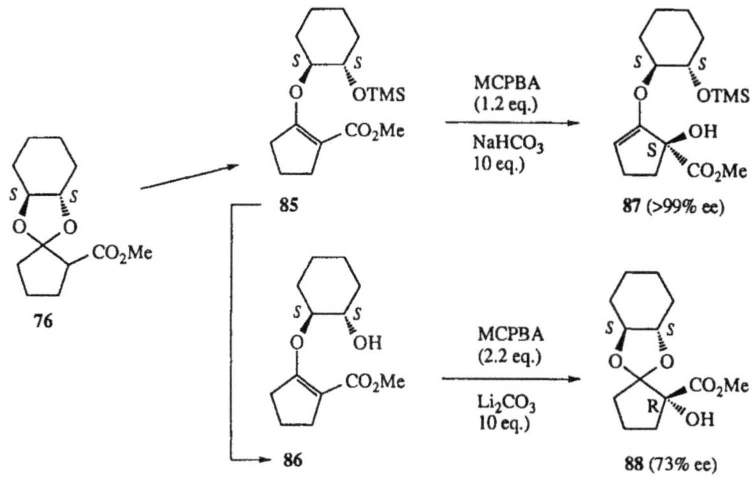

Scheme 38.

The authors proposed that the observed diastereoselectivities could be accounted for by assuming that the reactions proceed through the intermediates **89** and **90** shown for each case (Scheme 39).

As a means of achieving the synthesis of the spiro [4,5]decane-1,6-dione skeleton **97**, a valuable building block for the synthesis of perhydrohistrionicotoxin, Thebtaranonth's group[41] has examined the diasteroselective allylation of chiral acetals **91** derived from tartaric acid. Treatment of these acetals with an excess of base in THF at low temperature followed by addition of allylbromide afforded allylated acetals **95** or allylated β-keto esters **96** depending upon the mode of treatment of the initially formed enol ethers **94**. Although the underlying causes of the good diastereoselectivity are still not clear, the key step in the reaction sequence is believed to be the allylation of dienolates **93**, as shown in Scheme 40.

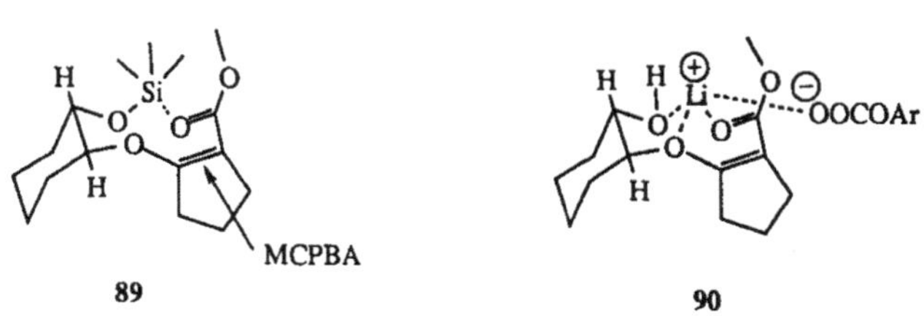

Scheme 39.

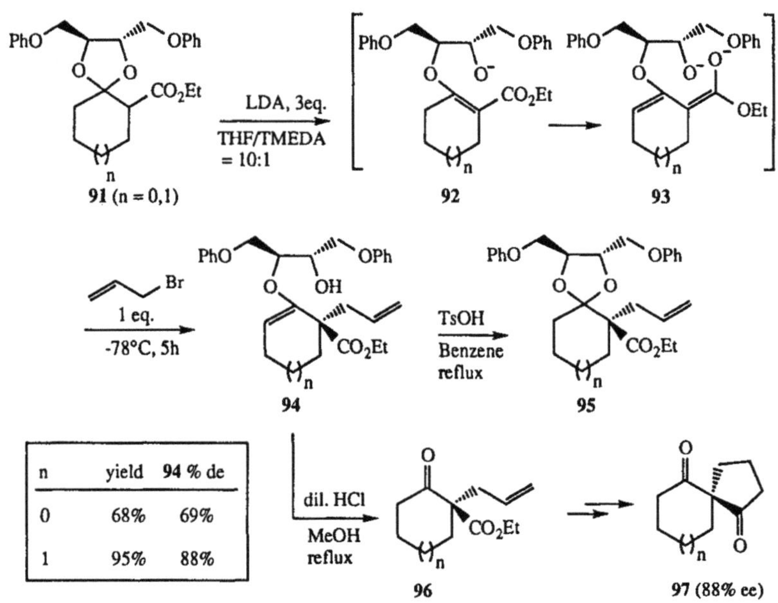

n	yield	94 % de
0	68%	69%
1	95%	88%

Scheme 40.

The best results (up to 88% ee) were obtained in the presence of TMEDA as cosolvent. The use of the OMe ketal analogs gave slightly lower diastereoselectivity.

The same group has also studied the base-induced allylation of ketal ester **98** in which the chirality is conveyed by the ester grouping. Following the mechanism depicted in Scheme 40, the ketal ester **99** was formed in 46% ee at best, provided the reaction was performed in THF as the solvent (Scheme 41). In contrast

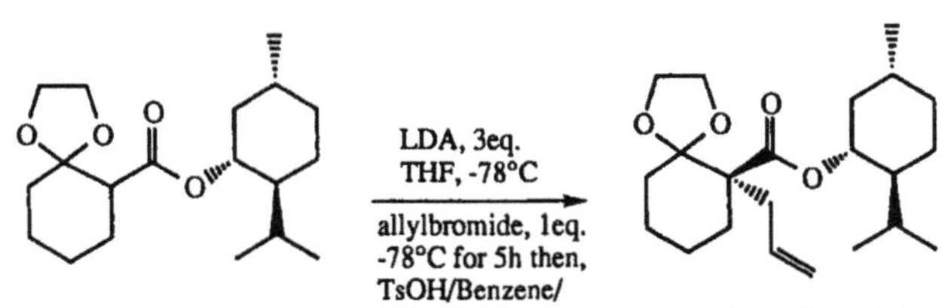

98 **99** (73% yield, 46% ee)

Scheme 41.

to the preceding example, addition of TMEDA or HMPA as cosolvents resulted in marked decrease of the diastereoselectivity.

B. Alkylation Reactions of Chiral α,β-Unsaturated Lactones

Pursuing their research on asymmetric alkylation using cyclic 1,2-diols as chiral auxiliaries, Sakai's group[42] has disclosed an interesting variation to this theme involving the alkylation of the lithium enolate of lactone **100** in place of the ester enolate **77** seen earlier. The reaction was carried out with high level of diastereoselection to afford fair to good yields of lactones **101** which could be further transformed into β-keto esters **103** as shown in Scheme 42. The reaction mechanism is not entirely clear but it is noteworthy, by comparison with the alkylation of **76** (Scheme 35), that products of opposite configuration at the quaternary carbon center are formed in the present reaction.

C. Fluorination of Chiral Spirocyclic Dioxinones

Chiral spirodioxinones derived from (–)-menthone have been shown to display appreciable diastereofacial selection in several different reaction types. Relevant to this chapter is the fluorination reaction of such spirodioxinones which proved to be a more efficient method than those already reported to prepare chiral α,α-disubstituted β-keto esters bearing a fluorine atom at the quaternary carbon center.

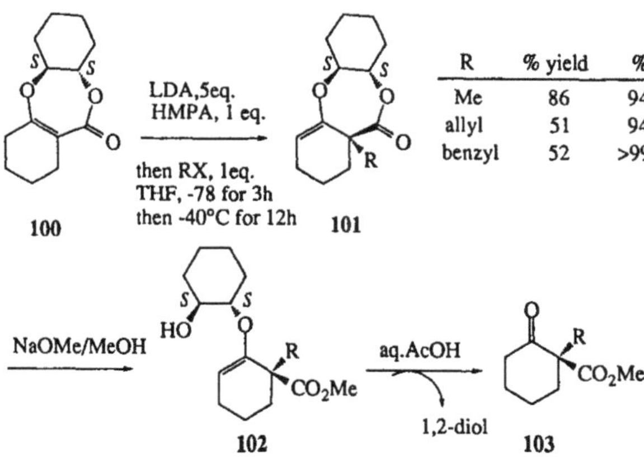

R	% yield	% d.e
Me	86	94(*S*)
allyl	51	94(*R*)
benzyl	52	>99(*R*)

Scheme 42.

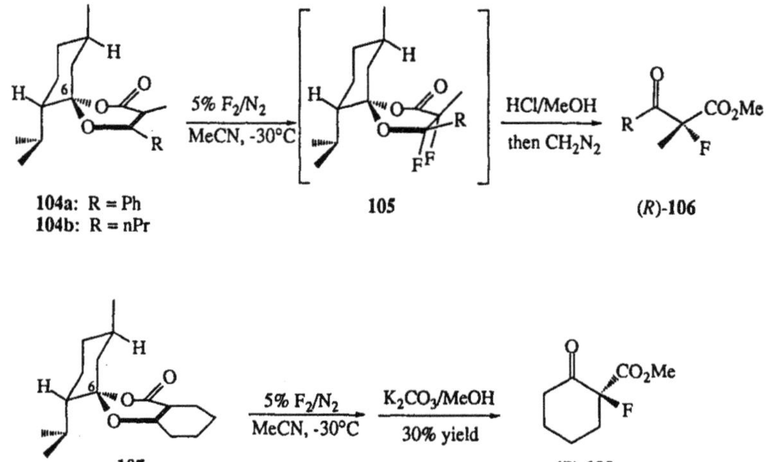

104a: R = Ph
104b: R = nPr
105
(*R*)-106
107
(*R*)-108

Scheme 43.

Thus, Iwaoka's group[43] reported that (6*S*)-dioxinone **104a** reacted with molecular fluorine in acetonitrile at –30 °C to afford the addition product **105a** which, without isolation, was further transformed, by sequential treatment with methanolic HCl and diazomethane, into acyclic β-keto ester (*R*)-**106a** in 43% overall yield and almost 100% ee. Similarly (6*S*)-dioxin-one **104b** was transformed along the same reaction sequence into β-keto ester (*R*)-**106b** in 52% overall yield and 73% ee.

This method was also efficiently applied to prepare cyclic β-keto ester **108** (>98% ee) from (6*S*)-dioxinone **107** (Scheme 43).

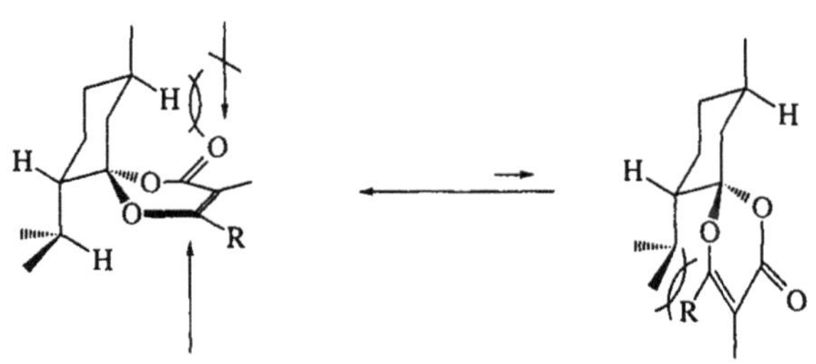

109 110

Scheme 44.

The diasterofacial selectivity has been rationalized assuming that the ground state conformer population is a governing factor in the determination of the stereochemical course of the reaction. Thus, fluorine addition is believed to preferentially occur to the less hindered bottom face of the most stable dioxinone conformer **109** (Scheme 44). However, this interpretation does not hold in other reaction types where the electrophilic partner is bulky and it cannot be excluded, in those cases, that factors other than steric have an influence on diastereoselection.

D. Alkylation Reactions of Chiral Pyrrolidine-Derived β-Enaminoesters

Gallagher's group[44] has investigated the possibility of preparing chiral α,α-disubstituted β-keto esters by regioselective α-alkylation of the lithium enolate derived from cyclic β-enaminoester **111**. This latter was prepared by condensation of the corresponding β-keto ester with (*S*)-2-methoxymethylpyrrolidine and was isolated as a 9:1 mixture with the "conjugated" form (as shown) predominating. Deprotonation of this isomeric mixture afforded a single extended enolate **112** which reacted with alkyl and allylic halides at the α-site exclusively to give, after acidic hydrolysis, the target β-keto esters **113**. As can be seen from Scheme 45, the level of asymmetric induction is quite low and this was attributed to the preponderance of conformer **115** over conformer **114** (which would certainly provide a higher diastereofacial selectivity) and to the slow equilibration rate **114** ⇌ **115** compared to the alkylation rate. As mentioned by the authors themselves, an appropriate means to circumvent that problem would have been to start from a

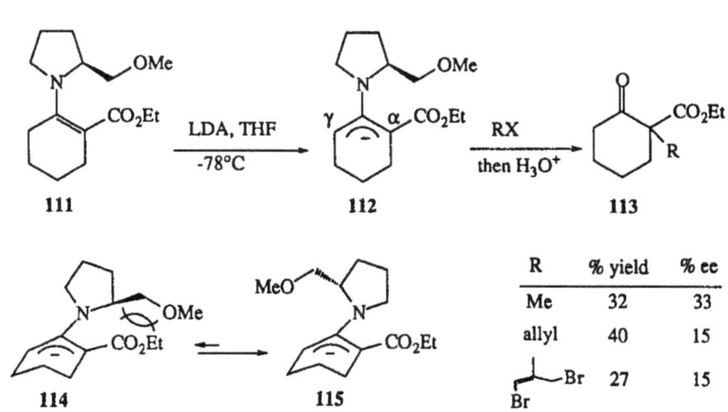

R	% yield	% ee
Me	32	33
allyl	40	15
Br, Br	27	15

Scheme 45.

C_2-symmetric pyrrolidine-derived β-enaminoester. Unfortunately, this could not be attempted in view of the difficulties associated with its generation.

As previously pointed out, asymmetric allylation of certain α-substituted β-diketones can be achieved with reasonable level of enantioselection under the catalytic influence of a palladium complex bearing a ferrocenyl-derived ligand. Unfortunately, the efficiency of the catalytic system did not extend to α-substituted β-keto esters. In a different approach to the problem, Hiroi and Abe later reported[45] that acyclic β-enaminoester **116**, derived from (*S*)-2-(diphenylphosphinomethyl)-pyrrolidine, could be reacted with Π-allyl complexes to generate, after acidic hydrolysis, β-keto ester **117** with enantioselectivity of up to 84%.

Selectivity was highly dependent on the ester substituent in **116** (Me vs. *t*Bu) and, to a lesser extent, on the nature of the leaving group of the allylating partner (Scheme 46). The results were interpreted by assuming that β-enaminoester **116**, in its ground state most preferred conformation, forms a π-allylpalladium complex **118** chelated with both the phosphine group and the allylating reagent. The allyl moiety is then intramolecularly delivered onto the bottom face of the β-enaminoester double bond to give, after acidic hydrolysis, the (*S*)-β-keto ester **117** as the major enantiomer. It is somewhat unfortunate that no work was done to expand the synthetic usefulness of this new method to β-enaminoesters other than **116**. It is worthy of note, however, that allylation reaction of cyclic and acyclic enamines could be performed with similar efficiency by using the same procedure.

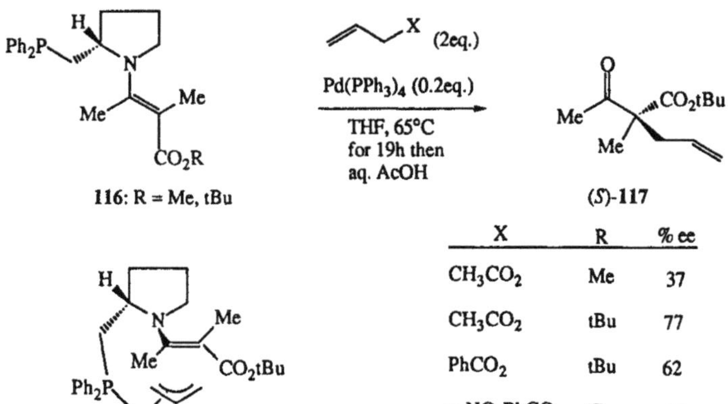

116: R = Me, tBu

(*S*)-117

118

X	R	% ee
CH_3CO_2	Me	37
CH_3CO_2	tBu	77
$PhCO_2$	tBu	62
p-NO_2PhCO_2	tBu	67
p-CH_3PhSO_3	tBu	84
Br	tBu	78

Scheme 46.

E. Alkylation, Conjugate Addition, and Oxidation Reactions of (*S*)-Valine-Derived β-Enaminoesters

One of the major synthetic strategies for the synthesis of chiral α,α-disubstituted β-keto esters was pioneered by Koga's group and involves alkylation and Lewis acid mediated conjugate addition reactions of lithio aza enolates of β-enaminoesters derived from the corresponding β-keto esters and (*S*)-valine *tert*-butylester.

Alkylation Reactions

In a series of papers, Koga's group[46a–c] reported that cyclic and acyclic lithio aza enolates **120**, prepared by low temperature LDA-mediated deprotonation of the β-enaminoester precursors **119**, could be reacted with certain alkyl halides to give, after acidic hydrolysis, α,α-disubstituted β-keto esters **121** in moderate to excellent diastereoselectivity (Scheme 47).

One of the most notable features of this alkylation process is the relation that exists between the solvent composition and the observed diastereofacial selectivity. As generally indicated in Scheme 47 when the reaction is performed in toluene in the presence of 1 equiv of HMPA, β-keto esters **121** are formed predominantly whereas inclusion of 1 to 3 equiv of an additive such as THF, dioxane, or NEt_3 leads to the formation of the β-keto ester enantiomers. In other words, HMPA strongly favors alkylation to occur at the front side of **120** whereas the other additives favor the opposite back side attack as exemplified in Scheme 48 for the cyclic aza enolate **122**. It is noteworthy that, in the latter case, an appropriate combination of alkylating agent and additive is of paramount importance to achieve appreciable levels of diastereoselection (Table 6).

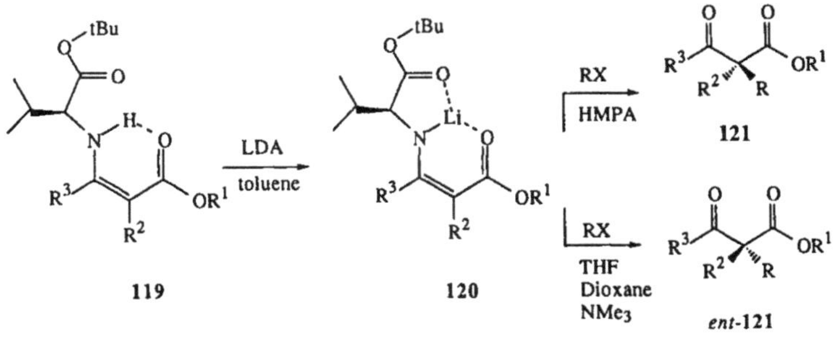

Scheme 47.

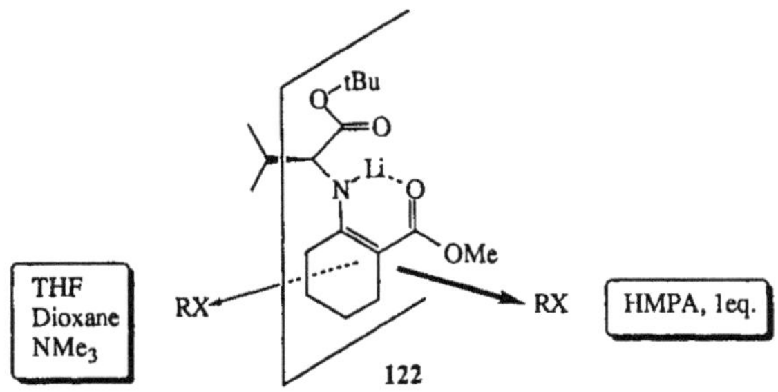

Scheme 48.

From consideration of the aforementioned reaction features, the authors have proposed working models which could help to rationalize all the experimental observations. These models rest on the assumption that, in addition to steric and stereoelectronic effects, the aggregation state of the reagent and chelation of the metal play important roles in controlling the steric course of the reaction. Thus, azaenolate **122** is probably existing as a mixture of higher-order structures in toluene, a structural feature that may explain the observation that, in the absence of an additive, alkylation did not proceed at low temperature. Inclusion in the reaction mixture of HMPA or other electron-donating additives results in "de-aggregation" and formation of the monomeric species **123A** in which a molecule of the additive is coordinated to the lithium cation to satisfy its usually tetracoordination state (Scheme 49). In structure **123A** and for steric reasons, the nitrogen lone pair is expected to be exclusively *cis* to the bulky isopropyl substituent on the stereogenic center. The bulky and powerful ligand HMPA is then believed to both increase the nucleophilicity of the aza enolate and screen its bottom face, hence forcing the alkylation to occur mainly from the less hindered top face. If HMPA is replaced by a weaker coordinating ligand, the latter is no longer assumed to increase the aza enolate nucleophilicity. Instead, this ligand is supposed to be replaced by the alkylating agent which is thus well disposed to react at the bottom face of **123A**.

Table 6.

RX	*Additive (equiv)*	*% Yield*	*% ee*	*RX*	*% Yield*	*% ee*
MeI	THF (2)	63	92 (*S*)	MeI	57	>99 (*R*)
Allyl–Br	Dioxane (1.2)	56	56 (*R*)	Allyl–Br	71	76 (*S*)
Benzyl–Br	Dioxane (1.6)	48	71 (*R*)	Benzyl–Br	77	>99 (*S*)
Acetyl–Br	NMe_3 (3)	78	74 (*R*)	Acetyl–Br	59	70 (*S*)

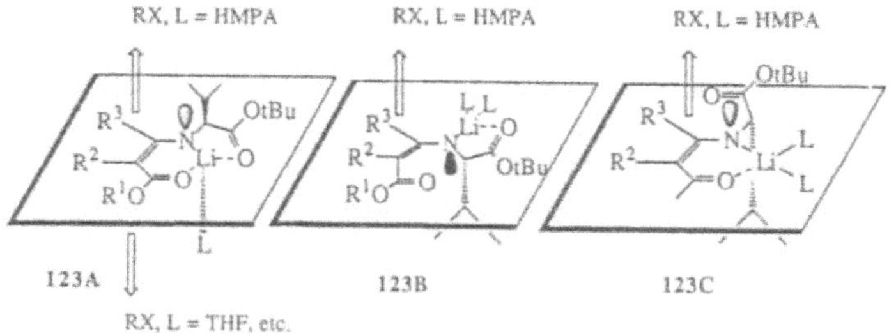

Scheme 49.

In an alternative model, the authors have also considered that HMPA may destroy the chelate structure **123A** to form the species **123B** or **123C** in which the bulky isopropyl group prevents the alkylating agent from approaching to the aza enolate bottom face (Scheme 49).

The utility of this method, as a starting point for the preparation of potentially useful synthetic intermediates, has been already demonstrated. The compounds currently so far synthesized are represented in Scheme 50.

Conjugate Addition Reactions

As a further demonstration of the synthetic possibilities of their method, Koga and co-workers later reported that lithium aza enolates **120** add in a conjugate fashion to electrophilic olefins to diastereoselectively afford α,α-disubstituted β-

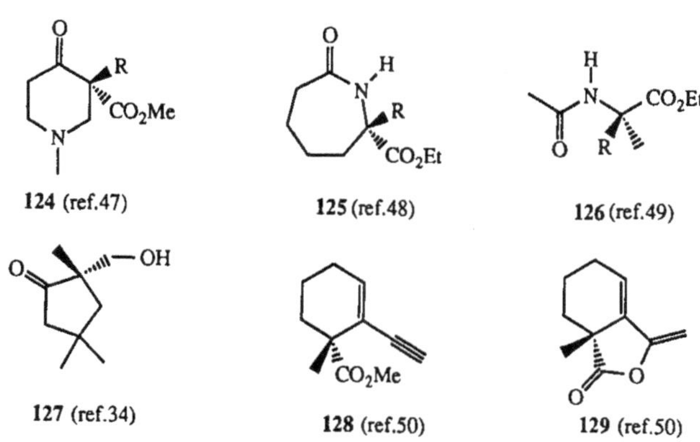

Scheme 50.

keto esters bearing a functionalized side chain at the quaternary carbon center (Scheme 51). Examples of electrophilic olefins used include di-*tert*-butyl methylenemalonate and methyl ethylidenemalonate. Some selected results are reported in Table 7 for the case in which di-*tert*-butyl methylenemalonate is the electrophile.[51] It can be seen that, as for alkylation reactions reported above, a complete reversal of diastereoselectivity may be achieved by simply changing the nature of the additive. Thus, running the reaction in the presence of HMPA (4 equiv) leads to predominant formation of adducts **130**, whereas changing HMPA for THF (8 equiv) promotes the formation of enantiomers. The reaction is temperature dependent and quite useful levels of stereocontrol may be achieved at –95 °C.

The use of methyl ethylidenemalonate in place of di-*tert*-butyl methylenemalonate as electrophilic partner also allowed the highly diastereoselective preparation of β-keto esters having contiguous quaternary and tertiary carbon centers.[52]

Later on, the same group discovered that lithium aza enolates **120** could add to less reactive electrophilic olefins, e.g. MVK and ethyl acrylate (EA), provided that a Lewis acid is added to the reaction mixture.[53] Of the various Lewis acid examined for their ability to promote the reaction, chlorotrimethylsilane proved to be the most useful allowing to achieve moderate to good levels of selectivity in favor of adducts (*R*)-**131**. The reactions were run in THF at very low temperature and it was essential, in order to maximize the selectivity, to add the aza enolate to a preformed mixture of TMSCl and MVK (or EA) (Scheme 52).

As before, either of the enantiomers of **131** could be obtained by varying the solvent system. Thus, working in toluene instead of THF and adding HMPA as an additive reversed the steric course of the reaction, leading to the preponderant formation of (*S*)-**131**. However, the chemical yields and selectivities were not very high (compare the results in Table 8 and 9).

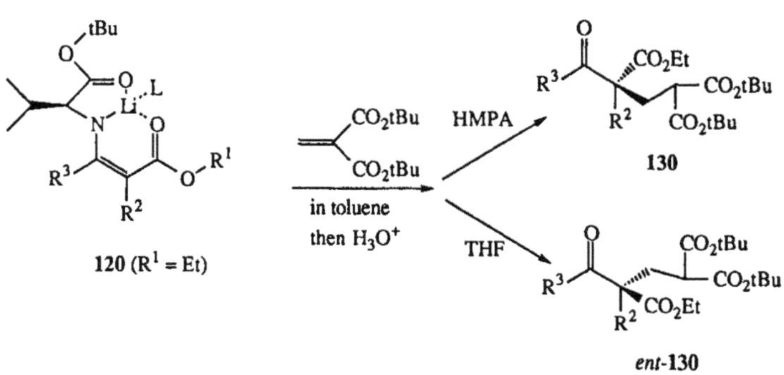

Scheme 51.

Table 7 (see Scheme 52).

R^2, R^3	*L (equiv)*	*T °C*	*% Yield*	*% ee*
Me	HMPA (4)	−95	82	87 (*S*)
Me	THF (8)	−95	58	66 (*R*)[a]
$-(CH_2)_4-$	HMPA (4)	−95	73	92 (*S*)
$-(CH_2)_4-$	THF (8)	−95	87	76 (*R*)[b]

Notes: [a]84% ee in THF at −105 °C.
[b]95% ee in THF at −105 °C.

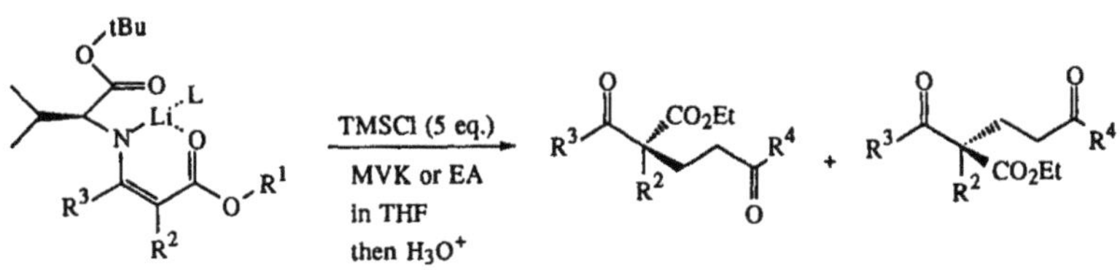

120 (R^1 = Et) (*R*)-**131** (*S*)-**131**
(R^4 = Me, OEt)

Scheme 52.

All these results can be accommodated, with minor variations, by the working model previously invoked for the alkylation reactions. In particular, it was postulated that the role of the TMSCl consists:

- In activating the electrophilic olefin by coordination at its carbonyl site in both THF and toluene–HMPA solvent systems (Lewis acid behavior).
- In displacing the weak THF ligand from the lithium cation.

As a result, this directive effect could cause the C–C bond formation to occur, in THF, from the bottom face of the chelate structure **123A** (Scheme 53).

Asymmetric Oxidation

The asymmetric oxidation of a β-keto ester at its α-position to deliver a tertiary alcohol is an important synthetic transformation in the context of natural product chemistry; we have already mentioned one example of such a reaction (cf. Scheme 38).

Table 8 (see Scheme 52).

R^2, R^3	*Electrophile*[a]	*T °C*	*% Yield*	*% ee*
Me	MVK	−78	65	73 (*R*)
Me	MVK	−100	66	87 (*R*)
$-(CH_2)_4-$	MVK	−78	60	55 (*R*)
$-(CH_2)_4-$	MVK	−100	87	90 (*R*)
Me	EA	−100	43	79 (*R*)
$-(CH_2)_4-$	EA	−100	53	57 (*R*)

Note: [a]The lithium aza enolate **120** was added to a mixture of Lewis acid and MVK (or EA).

Table 9 (see Scheme 52).

R^2, R^3	*Electrophile*[a]	*T °C*	*% Yield*	*% ee*
Me	MVK	−95	38	50 (*S*)
$-(CH_2)_4-$	MVK	−95	48	60 (*S*)
Me	EA	−95	16	41 (*S*)
$-(CH_2)_4-$	EA	−95	23	77 (*S*)

Note: [a]The lithium aza enolate **120** in toluene-HMPA (1 equiv) was added to a mixture of Lewis acid and MVK (or EA).

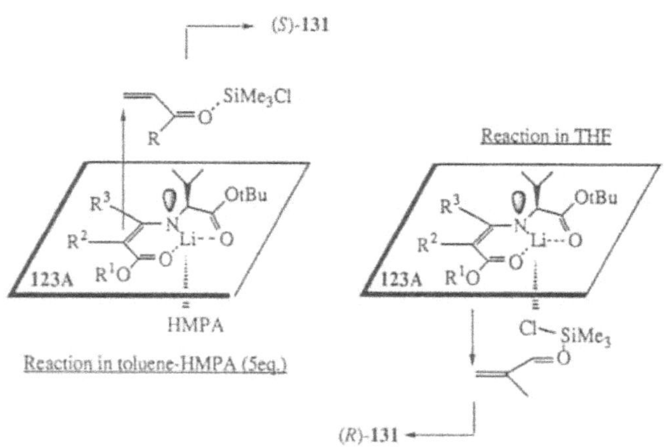

Scheme 53.

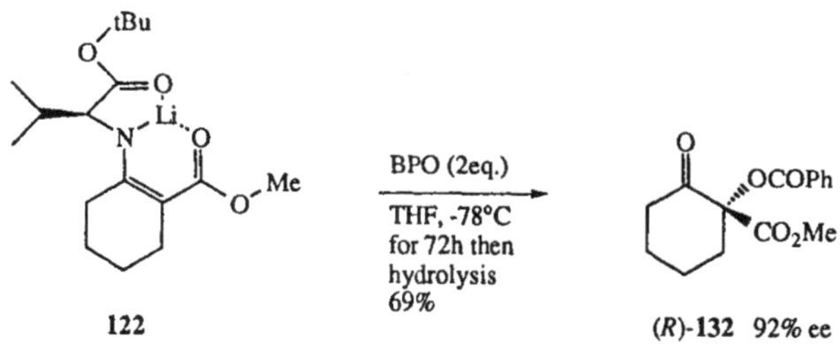

Scheme 54.

A second example was later reported by Snyder's group[54] in the course of their synthesis of the naturally occurring sesquiterpene tanshindiol A. These authors made the interesting discovery that lithium aza enolate **122** could be oxidized to the protected tertiary alcohol **132** by the action of benzoylperoxide. Good chemical yield and excellent stereoselectivity were best achieved when running the reaction in THF as solvent and at low temperature (Scheme 54). Interestingly and contrary to both the alkylation and addition reactions studied by Koga, recourse to toluene–HMPA as the solvent system no longer reversed the steric course of the reaction. This could possibly be realized by working with aluminum instead of lithium aza enolates, though in rather low selectivities.

Application of the procedure to other cyclic and acyclic (*S*)-valine-derived β-enaminoesters was also carried out with similar efficiency.

F. Alkylation and Conjugate Addition Reactions of Chiral 1-Phenylethylamine-Derived β-Enaminoesters

Alkylation Reaction

Thebtaranonth's group[41] has extended, with comparatively less success, the chemistry of Koga's group on the alkylation reaction of aza enolates to include enolates of 1-phenylethylamine-derived β-enaminoesters. These are generated by treatment of the corresponding β-enaminoesters with an excess of LDA in THF at low temperature. Trapping by allylbromide and subsequent hydrolysis afforded the β-keto esters **134** in low to moderate enantioselectivity (Scheme 55).

Conjugate Addition Reactions

In the context of the synthesis of natural products we and others became interested, some years ago, to determine whether or not cyclic β-enaminoesters, noto-

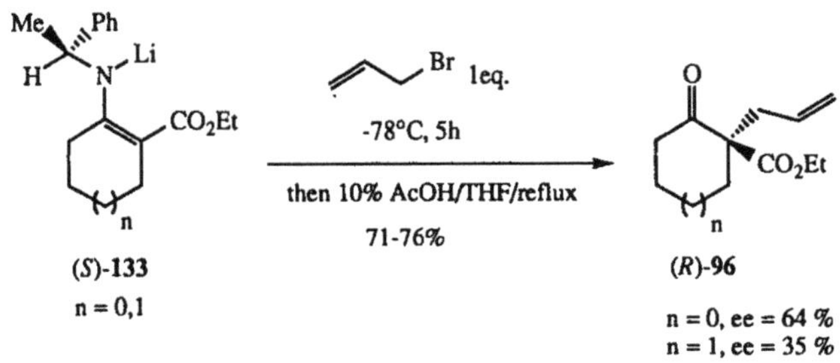

Scheme 55.

riously poorer nucleophiles than enamines, could be conjugatively added, *in their neutral form*, to electrophilic alkenes to give α,α-disubstituted β-keto esters bearing a functionalized side chain at the quaternary carbon center. Using this strategy, incorporation of chirality in the adducts could be approached from two directions: (1) addition of chiral β-enaminoesters to achiral electrophilic alkenes or, conversely, (2) addition of achiral β-enaminoesters to chiral electrophilic alkenes, as illustrated in general terms in Scheme 56.

Both routes have been found to be viable and each has unique advantages. This section will describe the development of the former approach. The latter one will be the subject of the next section.

For the purposes of our study, 1-phenylethylamine-derived β-enaminoesters **135**, easily prepared from the corresponding β-keto esters and (*S*)- or (*R*)-1-phenylethylamine (Scheme 57), were chosen as the nucleophilic components. The choice of 1-phenylethylamine as the chiral inductor is rooted in early work related to the addition of 1-phenylethylamine-derived enamines to electrophilic alkenes, a reaction which is generally achieved with a high degree of diastereose-

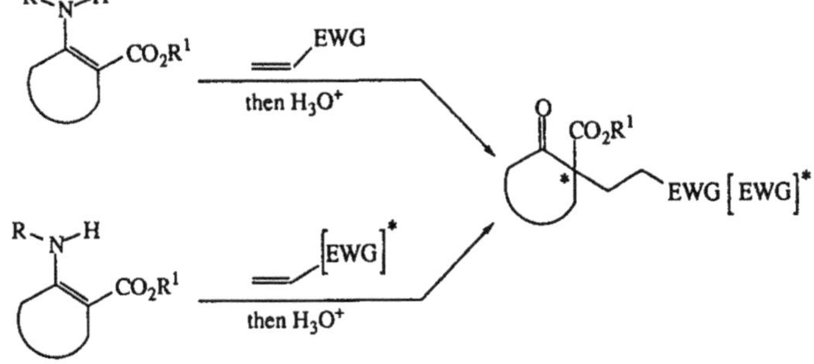

Scheme 56.

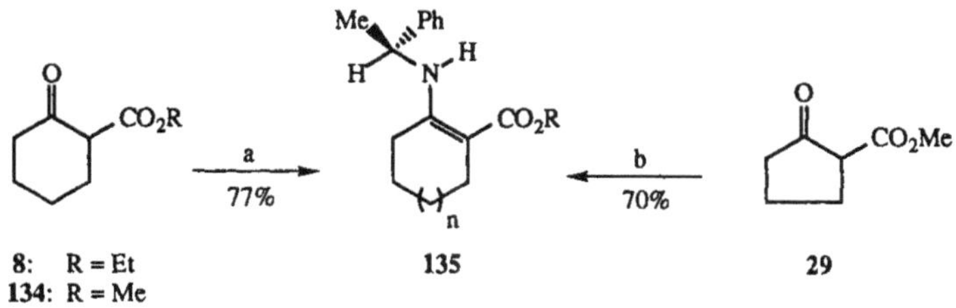

Scheme 57.

lectivity.[55] By way of comparison, we were rather confident that this chiral auxiliary would display a similar conformational bias in case of β-enaminoesters **135** and, thereby, would allow highly diastereoselective conjugate additions to be carried out. Additionally, our choice was also guided by the fact that both 1-phenylethylamine enantiomers are commercially available at relatively low prices.

Conjugate Addition Reaction to MVK

To initially explore the feasibility of our strategy, MVK was chosen as the electrophilic partner.[56] As one would expect from the poor nucleophilicity of β-enaminoesters, mixing 135 and MVK in solvents such as ether or THF at room temperature failed to provide any evidence of the expected conjugate adducts (Scheme 58).

However, we further discovered that the above transformation could be considerably accelerated in the presence of a Lewis acid. The most efficient Lewis acid we have surveyed thus far is zinc chloride and some selected results are reported in Table 10. Several aspects of the reaction merit comments:

1. Reasonable levels of enantioselectivity are attained both in ether and THF provided, in the latter solvent, the reaction time is not too long (*vide infra*).
2. Because of the insolubility of the complex presumably formed between the primary adduct **136** and the Lewis acid, 1 equiv of the latter is necessary in ether to afford β-keto ester **137** in good yield. This is no longer the case in THF where a catalytic amount of $ZnCl_2$ suffices to drive the reaction to completion.

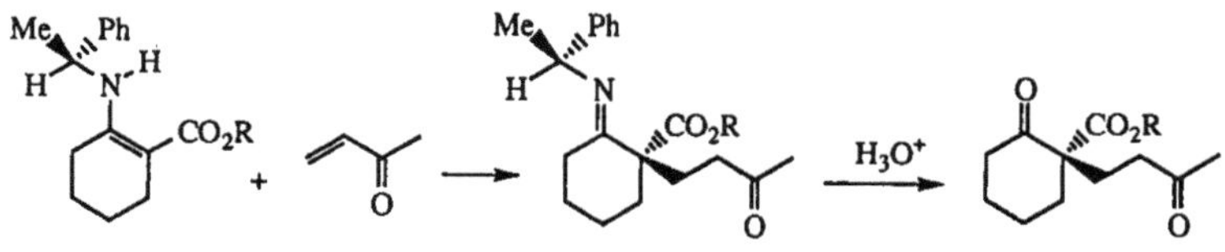

135
a: R = Et
b: R = Me

136
(not isolated)

9: R = Et
137: R = Me

Scheme 58.

Table 10.

			Reaction Conditions					
Entry	*Substrate*	*Solvent*	*temp. °C*	*time (h)*	*$ZnCl_2$ equiv*	*Adduct*	*% Yield*	*% ee*
1	**135a**	ether	0	2	1	**9**	90	72 (*S*)
2	**135a**	ether	–78	1.5	1	**9**	80	79 (*S*)
3	**135b**	THF	20	0.75	1	**137**	80	75 (*R*)
4	**135b**	THF	20	0.75	0.05	**137**	70	72 (*R*)
5	**135b**	THF	20	18	1	**137**	40	30 (*R*)

3. From a mechanistic viewpoint, it is worthy of note that a prolonged reaction time in THF resulted in a dramatic decrease of the ee value. A reasonable interpretation of this behavior is that, under the Lewis acid influence, the primary adducts, once formed, undergo reversal to the starting β-enaminoester **135**, thus allowing thermodynamic factors to be responsible for the stereochemical result.

During the course of our work, Brunner's group[57a] reported that synthesis of β-keto ester (*R*)-**9** could be thermally accomplished in 50% chemical yield and in a claimed ee value[57b] of 79% by reacting β-enaminoester (*R*)-**135a** and MVK in toluene at 40 °C for a period of 12 days! In the same paper, they also reported that β-enaminoketone (*R*)-**138** reacted with MVK in toluene at 60 °C for almost 3 days to give the β-diketone **139** in 65% ee (Scheme 59).

Conjugate Addition Reactions to Acrylates

Pursuant to our study, the conjugate addition reaction of β-enaminoesters (*S*)-**135a** and (*S*)-**135c** to a set of acrylates was next investigated as a means of preparing chiral β-keto esters of the type **140** (Scheme 60).

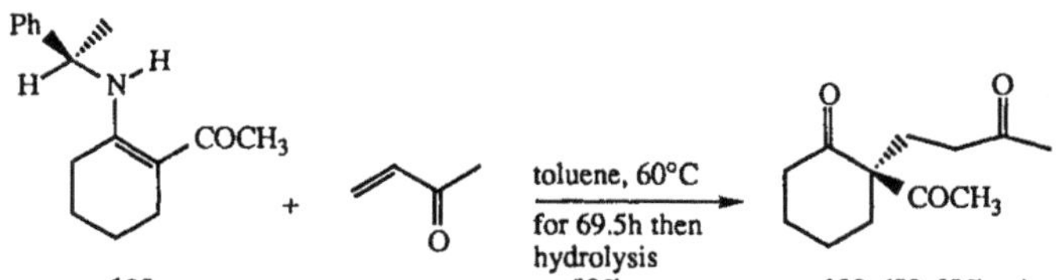

138 139 (59-65% ee)

Scheme 59.

Selected results are presented in Table 11. As can be seen, moderate to good levels of enantioselectivity can be reached by using a stoichiometric amount of magnesium bromide as the Lewis acid promotor. A particularly striking feature of the reaction is the fact that enantioselectivity appears to significantly increase with increasing steric bulk of the acrylates. Thus, the ee values increase from 65 to 90% when going from methyl or ethyl acrylate to *tert*-butyl acrylate. As for the addition reaction to MVK, ether is the solvent of choice although insolubility of complexed material can reduce the chemical yields. Entry 7 reveals that both the velocity and the selectivity of the addition process are reduced in THF.

In an extension of this work, we next investigated the possibility of achieving this reaction under high pressure. Taking into account that the Michael reaction is characterized by a large negative activation volume, it was anticipated that the conjugate addition reaction of β-enaminoesters to acrylates would be accelerated by pressure. Indeed, **135a** and **135c** reacted with ethyl and *tert*-butylacrylates under pressure ranging from 11 to 14 kbar to give, after hydrolysis of the crude primary adducts, β-ketoesters **140a** and **140c** in fair yields but with good enantioselectivity (84–88% ee) (Table 12). It is noteworthy that, in sharp contrast with the $MgBr_2$ catalyzed reaction, the size of the ester grouping R^1 has only a subordinate effect on the enantioselectivity.

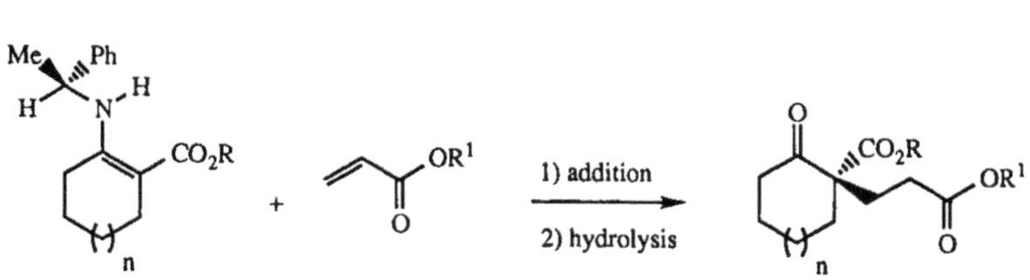

135
a: n = 1, R = Et
c: n= 0, R = Me

140
a: n = 1, R = Et
c: n= 0, R = Me

Scheme 60.

Table 11.[a]

Entry	Substrate	R^1	Reaction Conditions			Adduct (R^1)	% Yield	% ee
			T °C	Solvent	t (h)			
1	**135a**	Me	20	ether	2	**140a** (Me)	76	80 (*S*)
2	**135a**	Et	20	ether	3	**140a** (Et)	80	79 (*S*)
3	**135a**	*t*-Bu	20	ether	3	**140a** (*t*-Bu)	60	90 (*S*)
4	**135c**	Me	20	ether	1	**140c** (Me)	50	70 (*S*)
5	**135c**	Et	20	ether	1	**140c** (Et)	50	65 (*S*)
6	**135c**	*t*-Bu	20	ether	1	**140c** (*t*-Bu)	45	85 (*S*)
7	**135c**	*t*-Bu	65	THF	72	**140c** (*t*-Bu)	65	73 (*S*)

Note: [a]All of the experiments were performed in the presence of a stoichiometric amount of freshly prepared $MgBr_2$.

Table 12.

			Reaction Conditions					
Entry	Substrate	R^1	Pressure	T °C	t (h)	Adduct (R^1)	% Yield	% ee
1	**135a**	Et	11 kbar	40	65	**140a** (Et)	63	88 (*S*)
2	**135a**	*t*-Bu	14 kbar	45	72	**140a** (*t*-Bu)	31	84 (*S*)
3	**135c**	Et	14 kbar	20	67	**140c** (Et)	52	84 (*S*)
4	**135c**	*t*-Bu	14 kbar	45	72	**140c** (*t*-Bu)	45	86 (*S*)

The good results achieved in both the Lewis acid and high pressure-mediated addition of β-enaminoester **135c** to *tert*-butylacrylate have enabled us to use the β-keto ester **140c** (R^1 = tBu) in the synthesis[58] of the (−)-malyngolide precursor **83** (Scheme 61).

Conjugate Addition to Other Electrophilic Olefins

With the encouraging results of good enantioselectivity in the conjugate addition reactions of β-enaminoesters **135** to MVK and acrylates, studies toward the use of other electrophilic olefins such as acrylonitrile and phenylvinylsulfone were undertaken.

By reacting β-enaminoesters **135b** and **135c** with acrylonitrile in ether at −18 °C under the Lewis acid influence of a stoichiometric amount of zinc chloride, β-keto esters (*S*)-**141b** and **141c**, bearing a nitrile functionality at the pendant chain terminus, were obtained in 85% chemical yield and 87–89% ee, following acidic hydrolysis of the nonisolated primary adducts (Scheme 62). Functional group manipulation allowed the facile transformation of β-keto ester **141b** into the

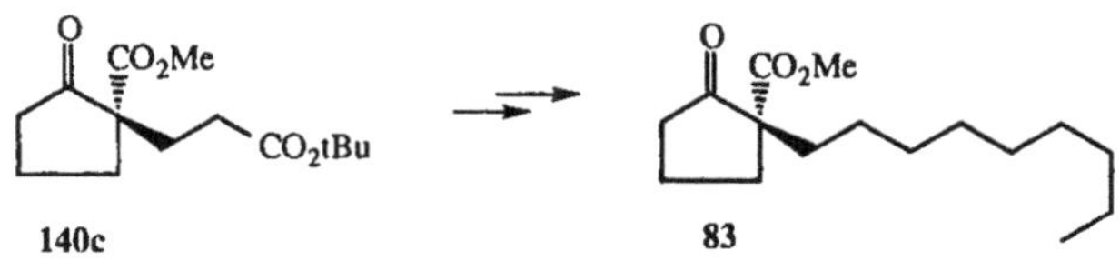

Scheme 61.

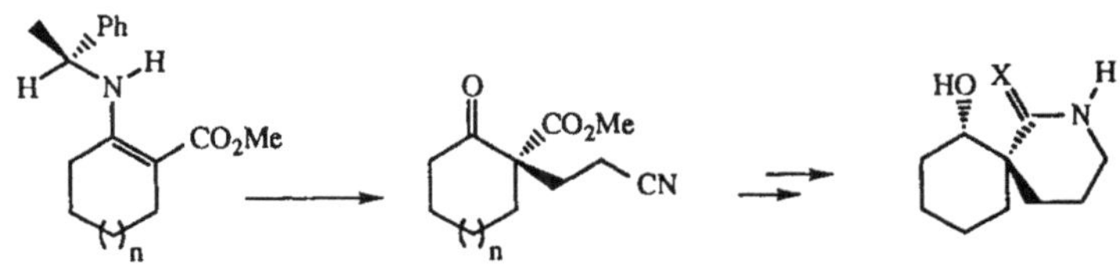

Scheme 62.

spirolactam **142**, a direct precursor of the spiro alkaloid (–)-nitramine **143** (unnatural enantiomer).[59]

The reaction of β-enaminoester (*S*)-**135b** with phenylvinylsulfone was next examined.[60] Lewis acids were unable to promote the reaction which could eventually be achieved by employing high pressure activation (14 kbar in THF at 40 °C for 66 h). Following acidic hydrolysis of the primary adduct, β-keto ester (*S*)-**144** could be isolated in 85% yield and 94% ee. Two synthetic important transformations of **144** were subsequently explored, i.e. desulfonylation into **145** and introduction of an alkyl substituent at the sulfone α-position to give **146**. As outlined in Scheme 63, both of them could be easily accomplished in good overall yield.

Independently, Brunner's group[57] has examined the conjugate addition of β-enaminoester (*R*)-**135a** to di-*tert*-butylmethylenemalonate. The high electrophilic properties of this olefin made it possible to carry out the reaction without Lewis acid assistance. Excellent chemical yield and selectivity were achieved for β-keto ester adduct (*R*)-**147** provided the reaction was conducted in toluene as the solvent (Scheme 64). It is noteworthy that a one pot decarboxylation–transesterification sequence allowed the transformation of **147** into β-keto ester (*R*)-**140a** (R^1 = Et) in 88% yield.

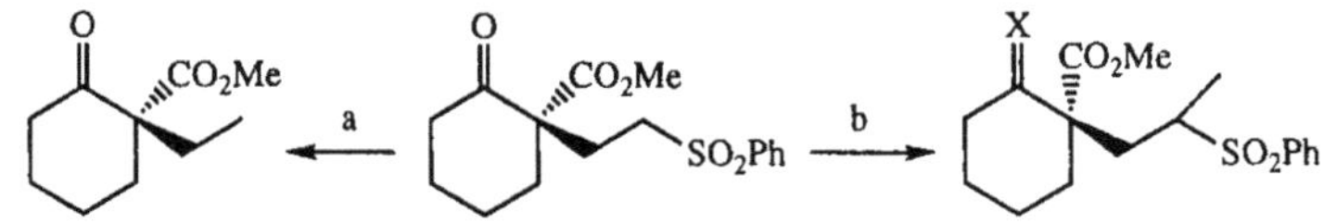

145 **144** **146** X = O-$(CH_2)_2$-O

Conditions: **a** (1) 10 equiv ethylene glycol, APTS, toluene reflux, 85% yield.(2) Mg, MeOH, 50 °C then 3N HCl, 80% yield. **b** (1) 10 equiv ethylene glycol, APTS, toluene reflux, 85% yield. (2) LDA, THF, -78 °C then MeI, 85%.

Scheme 63.

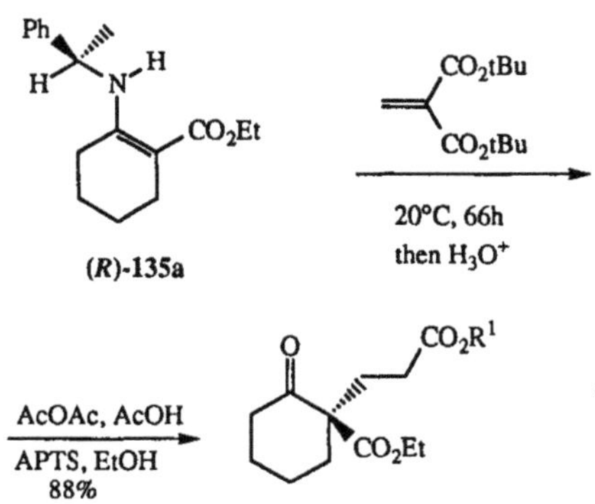

(*R*)-135a **147**

(*R*)-140a (R^1 = Et)

Solvent	% Yield	%ee
toluene	67	>95 (*R*)
THF	77	40 (*R*)

Scheme 64.

Chiral 1-Phenylethylamine-Derived β-Enaminolactone

The chemistry initially developed in case of 1-phenylethylamine-derived β-enaminoesters was later extended by Pfau's group[61] to include the use of (*R*)-1-phenylethylamine-derived β-enamino lactone **148** as the nucleophilic species. Its reaction with acrolein in THF at room temperature afforded the spiro lactone **150a** (de: 96/4, 100/0 after recrystallization) resulting from cyclization of the primary adduct **149** (Scheme 65). Similarly, reaction of **148** with MVK resulted in the formation of spiro lactone **150b** (de: 93/7, 100/0 after recrystallization). Following acidic hydrolysis, both adducts were transformed into spiro lactones **151a** and **151b**, respectively. The conjugate addition of β-enaminolactone (*S*)-**148** to methylacrylate was also reported to occur by heating the mixture at 80 °C for 12 h.

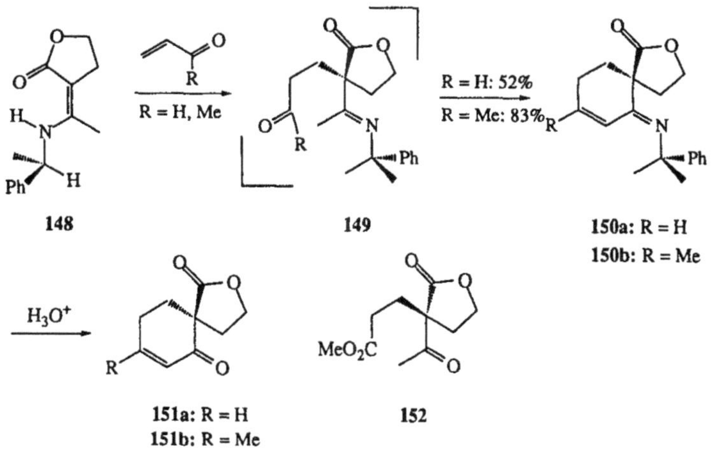

Scheme 65.

By comparison with the preceding examples, the level of chiral induction was significantly lower with the β-keto lactone (*R*)-**152** being finally obtained in 78% ee.

Mechanistic Considerations

As mentioned earlier, conjugate addition of (*S*)- or (*R*)-1-phenylethylamine-derived β-enaminoesters to certain electrophilic olefins can be carried out with good to excellent levels of diastereoselection to afford chiral α,α-disubstituted β-keto esters in good yields. In addition to being a quite simple and effective process, this reaction is also of particular interest since it provides a rare example of 1,4-asymmetric induction. One important stereochemical trend that can be drawn from all of the above results is that the (*S*)-1-phenylethylamine moiety induces the *S* configuration at the stereogenic quaternary carbon center of β-keto ester adducts and vice versa (Scheme 66).

The high quality of asymmetric induction imparted by the chiral 1-phenylethylamine moiety most certainly signifies that the conformational freedom of the substrate is limited and that the chiral auxiliary is proximal to the prostereogenic center; that is, the respective steric environments of the diasterotopic faces of the β-enaminoester are quite distinct. The fact that steric factors contribute to lock the β-enaminoester system in a particular conformation can be easily deduced from conformational analysis based on allylic strain consideration. In order to

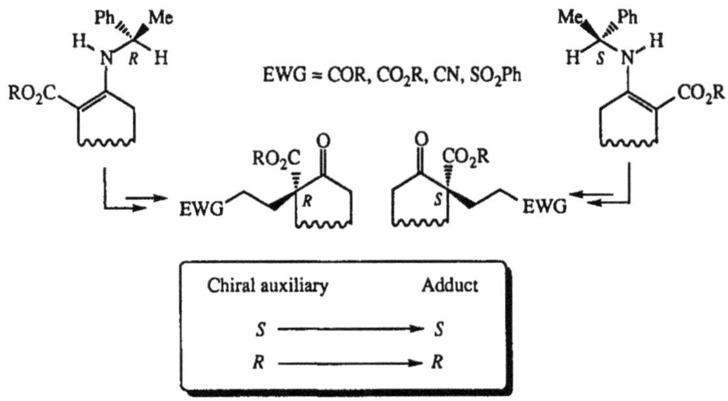

Scheme 66.

avoid both $A^{(1,3)}$ and $A^{(1,2)}$ strains[62] (the azaallyl system is the one considered) it seems reasonable to assume that β-enaminoesters **135** will be forced to adopt the conformation shown in Scheme 67, which also takes advantage of an extra stabilization arising from hydrogen bonding.

Interestingly, such a conformation was also displayed by β-enaminoester **153** in the crystalline state as revealed by an X-ray diffraction analysis.[63] Crystal structure of **153** also revealed the orientation of the phenyl ring relative to the β-enaminoester plane (C_a–N–C_b–C_c = 75.5°) and to the N–C_b bond (N–C_b–C_c–C_d = 44.7°). As a result, it is immediately seen that the upper face of **153** is severely hindered by the protruding phenyl ring. Provided the more stable ground state conformation of β-enaminoesters **135** is not greatly disturbed by the approaching electrophile, it can be safely predicted that the steric barrier imposed

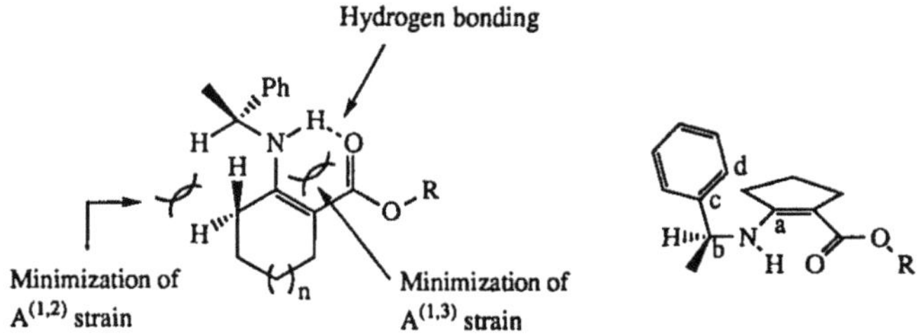

135 (n =0,1)

153 (R = pNO2PhCH2)

Scheme 67.

by the phenyl substituent will force the reaction to occur from the opposite face. All of the experimental results recorded so far are in full agreement with this model (Scheme 68).

Though the above model is able to interpret and also to allow a secure prediction of the stereochemical outcome of β-enaminoester conjugate addition reactions to electrophilic olefins, it provides no information about the transition state topology. In order to gain more insights into the mechanism, ab initio calculations were undertaken using ethyleneamine and propenal as simplified structures.[64] First of all, one must realize that a compact arrangement of both reactant partners is a prerequisite to place the chiral auxiliary in the vicinity of the reacting center. By comparison with a loose approach which minimizes all steric factors, such an arrangement should certainly be disfavored. However, one of the important conclusions arising from the theoretical study is that, in a chair-like structure such as **154** (Scheme 69), important attractive interactions developing mainly between the HOMO of the enamine and the LUMO of the electrophile are able to compensate, to a large extent, the nonbonded repulsive interactions. Another conclusion of fundamental importance which emerged from the theoretical study is that a compact structure also allows the more or less concerted transfer of the N–H hydrogen to the spatially close C–α carbon of the electrophile, thus avoiding the energetically disfavored formation of zwitterionic species.

If we assume that the conclusions of the theoretical study are also valid in case of the addition of β-enaminoesters **135** to electrophilic olefins, the activated complex is best represented in that case by the compact structure **155** (methyl acrylate

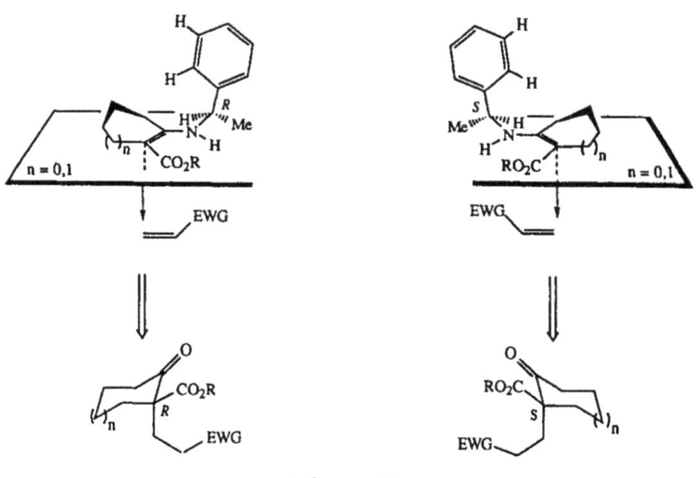

Scheme 68.

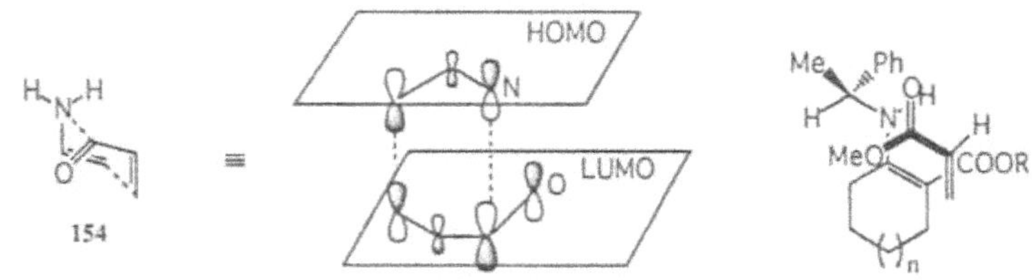

154 155 (n = 0,1)

Scheme 69.

has been chosen as the representative electrophilic olefin) which features a *gauche* relationship of the donor and the acceptor Π-systems.[65]

Note that convincing arguments in favor of such a transition state model had been experimentally found in the conjugate addition reactions of β-enaminoester **135b** and enamine analogs to certain α-substituted acrylates.[66]

G. Conjugate Addition Reactions of Achiral Benzylamine-Derived β-Enaminoesters to Chiral Electrophilic Olefins

As we have already mentioned in the introduction to the preceding section, the conjugate addition of achiral β-enaminoesters to chiral electrophilic olefins represents an alternative and unprecedented approach to the problem of synthesizing chiral α,α-disubstituted β-keto esters. This approach seems particularly attractive as it allows many and structurally different olefins to be used as electrophile partners.

At the outset of our study and in the hope of improving upon the results reported so far, we elected to investigate the reaction of cyclic β-enaminoesters **156** with a set of (*S*)-proline-derived acrylamides.[67]

β-Enaminoesters **156** are not sufficiently reactive to add to acrylamides **157** and the presence of a Lewis acid was required to promote the reaction (Scheme 70). A stoichiometric amount of titanium tetrachloride at 20 °C in methylene chloride was found the best way to promote the reaction and some selected results obtained by using these preferred experimental conditions are reported in Table 13.

From the data (entries 1 to 3) it is evident that conjugate addition of β-enaminoesters **156** to chiral acrylamides **157a** and **157b** is regulated with much higher diastereoselectivity than that previously obtained with chiral β-enaminoesters **135** and acrylates. Concerning the reaction mechanism (*vide infra*) it is noteworthy that acrylamide **157c**, though having two oxygen atoms capable of anchoring

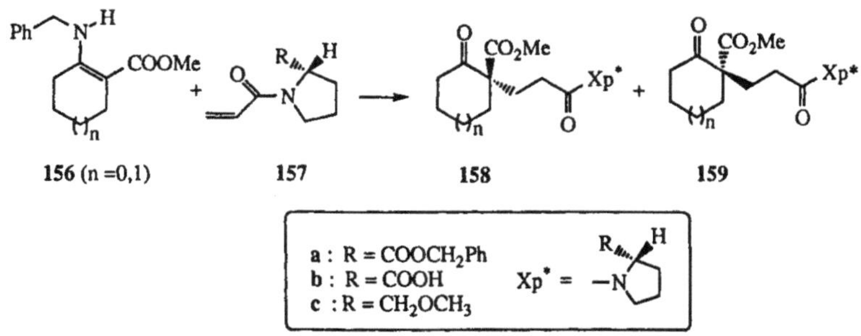

Scheme 70.

$TiCl_4$, led comparatively to disappointing results (entry 4), suggesting that the presence of two carbonyl groups, as in **157a** and **157b**, is a prerequisite for obtaining good selectivity. Recovery of the chiral auxiliary could be effected by treating the adducts with an excess of Meerwein reagent followed by hydrolysis of the resulting O-methylated intermediates. In this way, β-keto esters (*R*)-**140b** and (*R*)-**140c** could be isolated in about 50–60% overall yield and in 96 and 90% ee, respectively (Scheme 71).

Not unexpectedly, cleaving the chiral auxiliary amide bond proved to be a rather difficult and poor-yielding process. In order to circumvent the aforementioned limitation and broaden the synthetic utility of this approach, we have next considered[68] the possibility of adding β-enaminoester **156** to a set of three chiral acrylates, namely **161a,b,c** derived from (*S*)-ethyl lactate, (*R*)-pantolactone and (*S*)-*N*-methyl-2-hydroxysuccinimide, respectively (Scheme 72). By comparison with **158**, it is expected that recovery of the chiral auxiliaries would be greatly facilitated as it requires the cleavage of an ester bond.

Table 13.

Entry	*β-Enamino-ester*	*Proline Derivative*	*Experimental Conditions*[a] temp. (°C)	time (h)	*% Yield*	*Ratio* **158/159**
1	**156** (n = 1)	**157a**	20	5	85	98/2
2	**156** (n = 0)	**157a**	20	5	70	95/5
3	**156** (n = 1)	**157b**	20	5	57	98/2
4	**156** (n = 1)	**157c**	20	3	37	69/31[b]

Notes: [a]Reactions were conducted in CH_2Cl_2 with molar ratio of β-enamino ester:chiral acrylamide = 1:1.
[b]The absolute configuration of the major diastereomer is unknown.

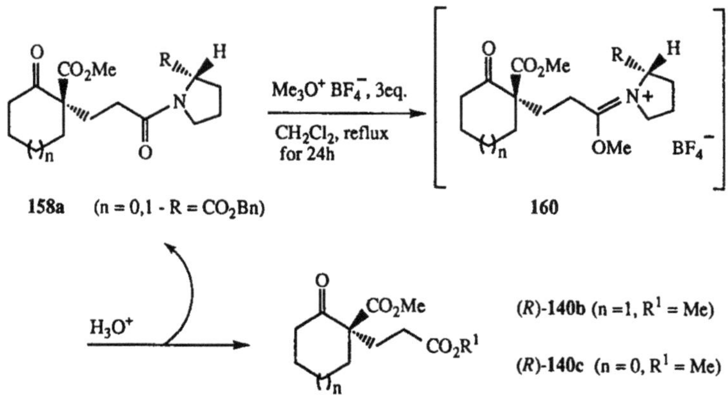

Scheme 71.

Reactions were performed using the optimized experimental conditions established for olefins **157**, that is, one molar equivalent of $TiCl_4$ in methylene chloride at 20 °C, and the results so obtained are summarized in Table 14.

The important observation to be made is that acrylates **161** behaved similarly to acrylamide **157** leading to adducts **162** and **163** with a high level of diastereoselection. From a synthetic point of view it is worthy of note that **161a,c** and **161b** yielded products of opposite configuration at the quaternary carbon center and that, as expected, removal of the chiral auxiliaries to give (*R*)- or (*S*)-**140b** (R^1 = Me) was easily accomplished (up to 88% yield) by treatment of adducts **162** and **163** with an excess of freshly prepared sodium methoxide in dry methanol.

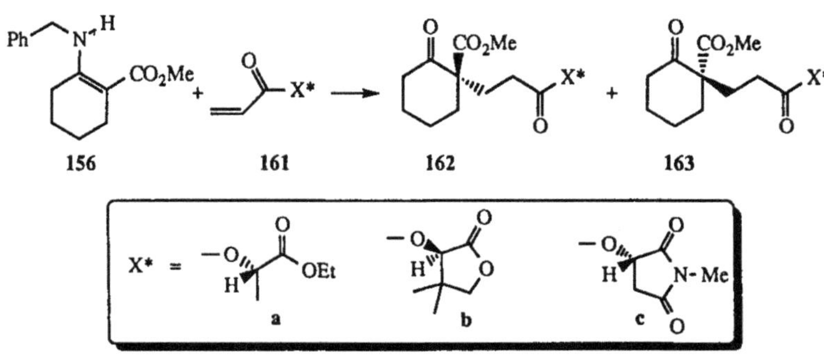

Scheme 72.

Table 14.

Entry	*Acrylate*	**162:163**	*% Yield*	*% ee*
1	**161a**	94/6	70	85 (*R*)
2	**161b**	8/92[a]	76	82 (*S*)
3	**161c**	97/3	75	95 (*R*)

Note: [a]4/96 after recrystallizations (2×) from AcOEt/petroleum ether.

The steric course of these addition reactions can be accounted for by assuming that:

1. In agreement with the pioneering work of Helmchen's group,[69] the electrophilic olefins **157** and **161**, in a *s-cis* conformation, form, with the bidentate Lewis acid a seven-membered ring chelate (e.g. **164**, Scheme 73) in which one of the diastereotopic faces of the reacting double bond is masked by a Ti–Cl bond. In other words, the chemical information borne by the stereogenic center of the chiral auxiliary is not directly responsible for the facial differentiation but is relayed to the vicinity of the reacting double bond by the intermediacy of the Lewis acid.
2. The reaction proceeds via a compact chair-like activated complex as depicted in the Newman projection **165** (the Lewis acid has been omitted for clarity) as previously established for the addition reactions of chiral 1-phenylethylamine-derived β-enaminoesters.

In full accord with such a mechanism it is important to note that the absolute configuration of the newly created quaternary carbon center is indicative of the

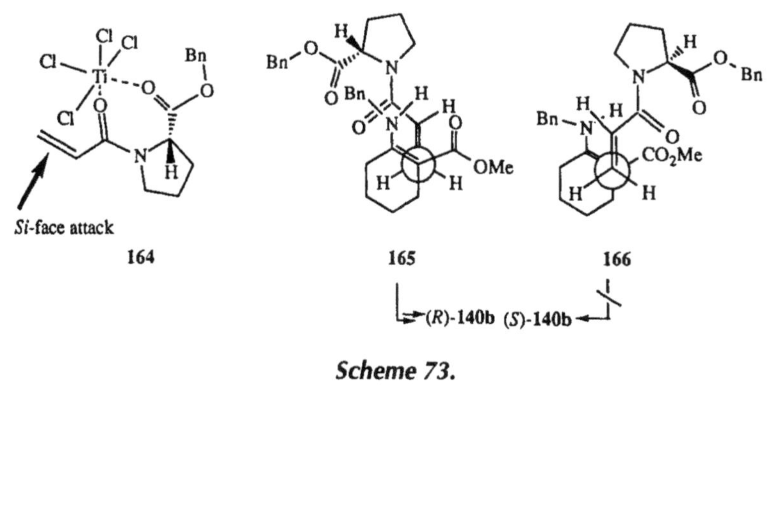

Scheme 73.

transition state topology. In fact, the boat-like arrangement **166** would have led to the establishment of the *S* configuration at that center rather than the *R* configuration, as experimentally observed.

V. CONCLUSION

This chapter has described asymmetric syntheses of α,α-disubstituted β-diketones and β-ketoesters. The problem in synthesizing these synthetically useful compounds from their α-monosubstituted precursors had been approached from two main angles, that is:

- use of a chiral catalyst (mainly in examples leading to α,α-disubstituted β-diketones)
- use of a chiral auxiliary (mainly in examples leading to α,α-disubstituted β-ketoesters)

The first approach proved to be moderately efficient and results were not commensurate with the efforts put into the enterprise. Indeed, enantioselectivity of 78% at best could be obtained in the conjugate addition reaction of methyl indane-1-one-2-carboxylate to MVK. Other examples involving structurally different β-diketones or β-ketoesters generally led to very poor results. Clearly, one of the most important and challenging future goals in this area will be the development of an efficient chiral catalyst and its use in addition reactions leading to adducts of synthetic interest.

The second approach was far more efficient, proceeding in high selectivity to deliver α,α-disubstituted β-ketoesters which are prone to further synthetically useful chemical derivations. In particular, chiral ketals and (*S*)-*tert*-butyl valine-derived β-enaminoesters, as chiral protected forms of α-monosubstituted β-ketoesters, allowed the obtention of extremely high enantioselectivities in alkylation reactions. As far as conjugate addition reactions are concerned, methods mainly developed in our laboratory and featuring the addition of chiral phenylethylamine-derived β-enaminoesters to achiral electrophilic olefins or achiral β-enaminoesters to chiral electrophilic olefins are the most general and efficient ones. Occasionally, good results were also obtained in case of conjugate addition reactions of α-monosubstituted β-ketoesters to certain electrophilic olefins. A simple and general method is still lacking, however.

It is hoped that the present article will give new impetus in this exciting area of asymmetric synthesis and will contribute in the development of new simple and efficient methodologies.

ACKNOWLEDGMENTS

I wish to express my sincere gratitude to Dr. H. Hammami and to K. Hervouet for carrying out part of the research reviewed in Sections IV.F and IV.G of this chapter.

ADDENDUM: ADDITIONAL EXAMPLES

This section provides the reader with a brief description of additional methods of preparation of α,α-disubstituted β-dicarbonyl compounds that do not follow the introductory Scheme 1.

α-Alkylation of Dianions Derived from Chiral α-Monosubstituted β-Hydroxy Esters

Fráter's group[70] has demonstrated that α-alkylation reaction of lithium dianions generated from α-monosubstituted β-hydroxy esters can be achieved with consistently good diastereoselectivity (> 90%). The resulting products are easily and efficiently transformed into α,α-dialkylated-β-keto esters by chromic oxidation. Because the starting chiral β-hydroxy esters are readily available by microbial reduction (baker's yeast) of the corresponding racemic β-keto esters with excellent control of the absolute configuration at C-3, this sequence of reactions, generally outlined in Scheme 74, represents an attractive method for the synthesis of chiral α,α-dialkylated-β-keto esters (Scheme 75).

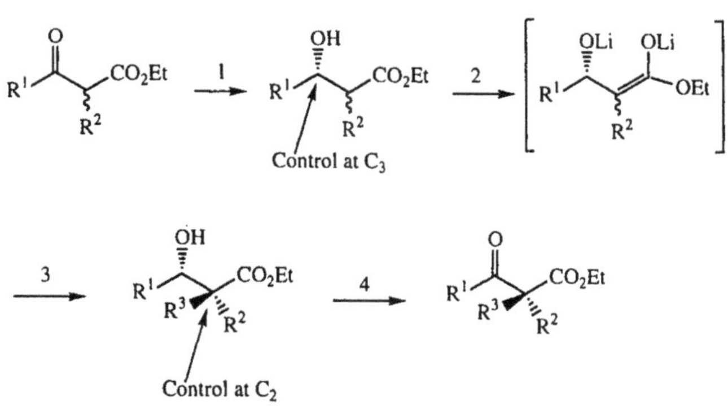

1: Baker's yeast-mediated reduction. 2: dianion generation (LDA, HMPA).
3: Alkylation (R^3X). 4: oxidation (Jones reagent)

Scheme 74.

Although the underlying causes of the good diastereoselectivity are still not completely clear at present, the rigid chelate **173** (Scheme 76) represents a useful model to account for and to predict the stereochemical course of the reaction.

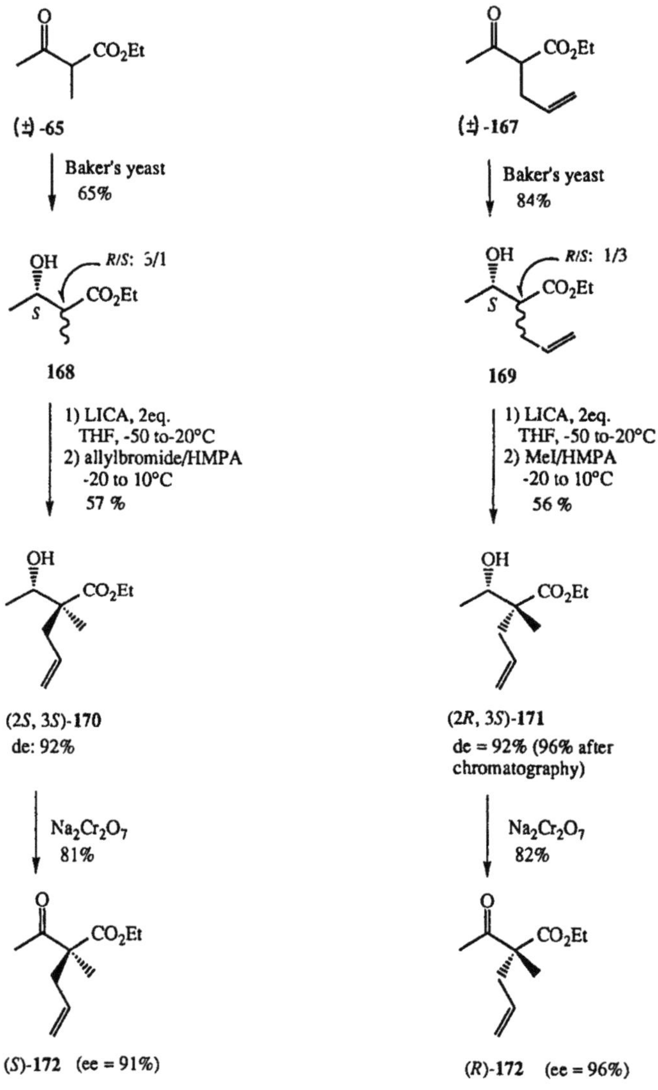

Scheme 75.

Scheme 76 also provides some other examples of chiral α,α-dialkylated-β-keto esters already prepared by Fráter's group.

It is also interesting to note that, in certain cases, baker's yeast-mediated reduction of α,α-disubstituted β-keto esters may be carried out with excellent diastereo- and enantioselectivities as demonstrated by the examples in Scheme 77. Chemical yields are, however, significantly lower than in the above transformation.

Improvement in the levels of diastereo- and enantioselectivity of the baker's yeast reduction as well as increase in the reaction velocity may sometimes be achieved by replacing the racemic α-monosubstituted β-keto ester to be reduced with a β-keto thioester. This has been successfully exploited by Sato et al.[71] to synthesize (–)-malyngolide **84** and (–)-frontalin **185** from the cyclic β-keto esters **183** and **184**, respectively (Scheme 78).

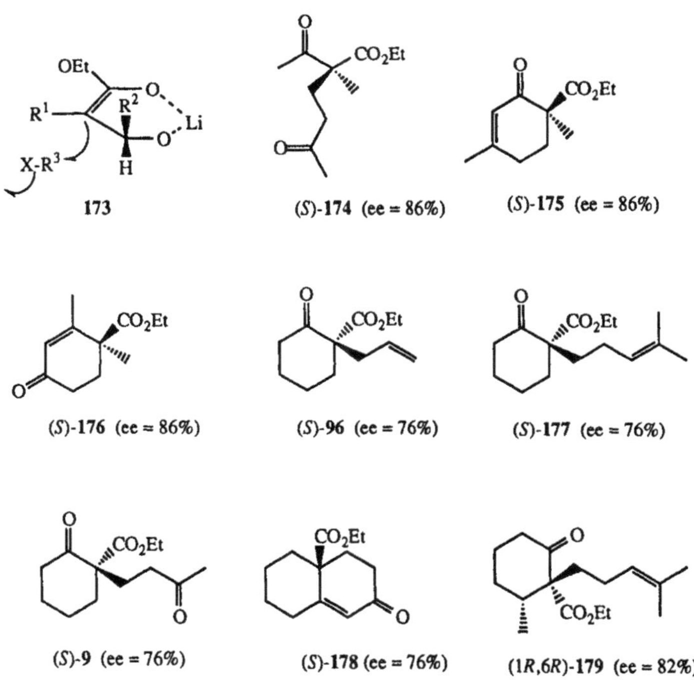

Scheme 76.

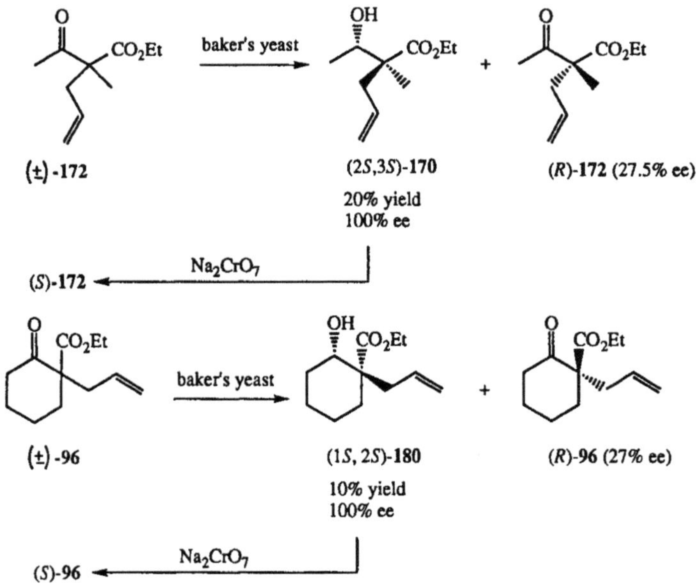

Scheme 77.

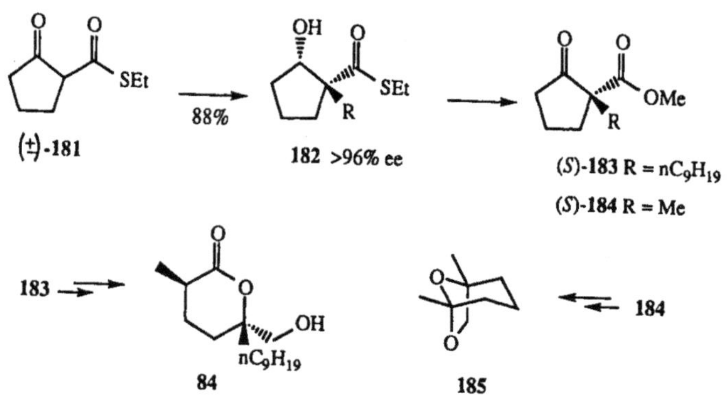

Scheme 78.

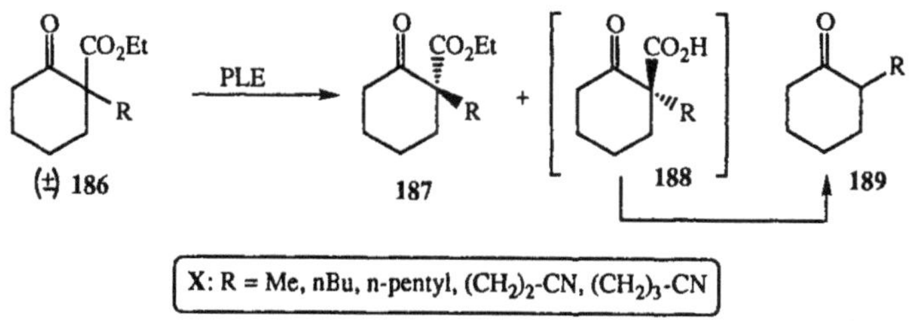

Scheme 79.

PLE-Catalyzed Resolution of α,α-Disubstituted β-Keto Esters

Recently, Westermann et al.[72] have reported that cyclic β-keto esters of the type **187** could be prepared in one step and with excellent enantiomeric purity (> 99%) by PLE-catalyzed hydrolysis of the racemic precursors **186** as shown in Scheme 79. The chemical yields are in the range of 40–88% based on 50% hydrolyzed **186**.

Compound **187** (R = CH_2–CH_2–CN) was utilized as a common intermediate for the syntheses of the spiroalkaloids (+)-nitramine **190** and (–)-isonitramine (unnatural enantiomer) **191** (Scheme 80).

Asymmetric Claisen-Type Acylation

An innovative asymmetric Claisen-type acylation employing (4*S*)-isopropyl-1,3-thiazolidine-2-thione (**193**) as a chiral leaving group has been reported by Nagao's group.[73] The method is particularly efficient for the preparation of chiral α,α-disubstituted β-keto esters **194** derived from methyl and *tert*-butyl-2-phenylpropionates **192** (Scheme 81).

The authors observed that the steric course of the reaction remained unaffected by modification of the enolate geometry and they tentatively proposed an

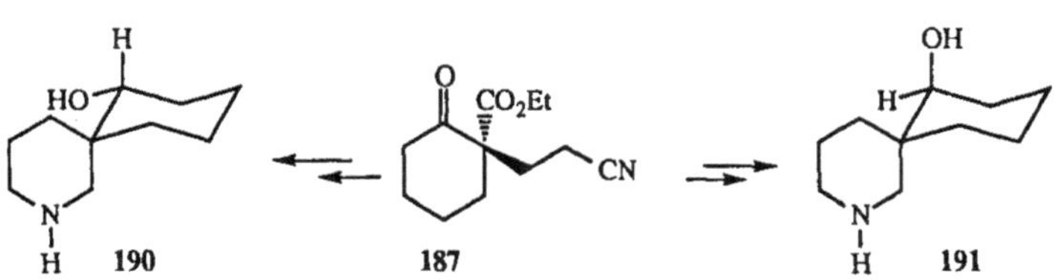

Scheme 80.

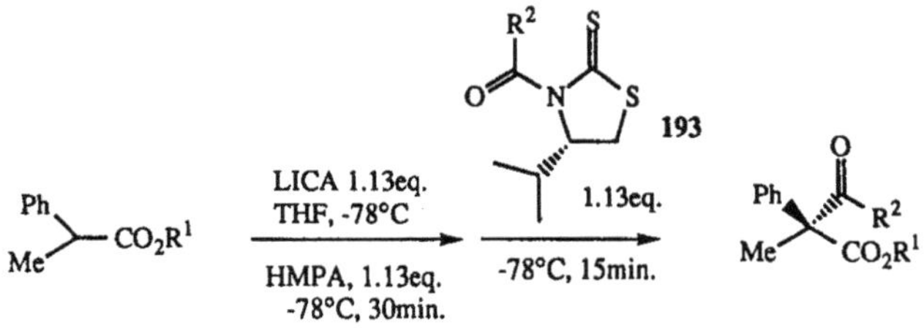

192 194

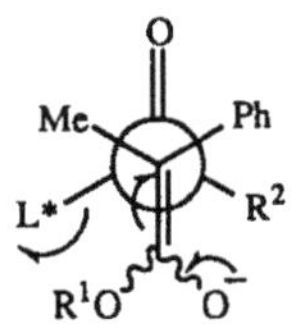

195

R^1	R^2	% yield	% ee
tert-But	Me	77	96
tert-But	nPr	75	95
Me	Me	57	94
Me	nPr	53	97

Scheme 81.

acyclic transition state model as depicted in **195** to account for the experimental results.

Carboxalkylation of Ketone Enolates

Hogeveen and Menge[74] have achieved the synthesis of the chiral α,α-disubstituted β-keto ester (*R*)-**199** by a tandem carboxylation–alkylation reaction of the

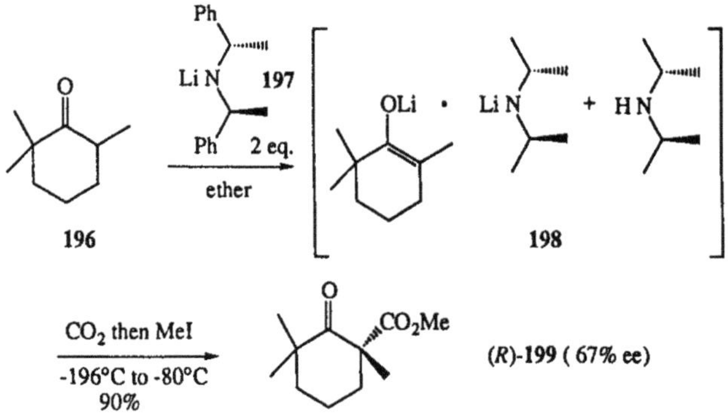

Scheme 82.

prochiral enolate **198** prepared in situ by deprotonation of the corresponding ketone **196** with a chiral amide base (Scheme 82). The reaction proceeded with 67% ee provided that 2 equiv of amide **197** was used and carbon dioxide was condensed onto the frozen solution of **198** in ether at −196 °C followed by slow warming at −80 °C. Although the reaction mechanism remains quite obscure, a mixed aggregate was assumed to be involved in the product-forming step.

In the context of their studies aimed at synthesizing enantiomerically pure Fredericamycin A, Braun's group[75] has reported the efficient diastereoselective carboxalkylation of ketone enolates **203** by treatment with (−)-menthyl chloroformate to produce regioisomeric β-keto esters **204** with virtually complete stereocontrol at the quaternary carbon center (absolute configuration not determined). The ketone enolates **203** were generated by quenching ketene **201** (liberated in situ by the action of BuLi on the BHT-ester **200**) with a mixture of anions **202** (Scheme 83).

Transformation of Diels–Alder Adducts

Koizumi's group[76] has exploited the potential of the Diels–Alder reaction to prepare the synthetically useful chiral α,α-disubstituted β-keto esters **209** and **210**. Cycloaddition of chiral α,β-unsaturated (*Z*)-sulfoxide **205** to cyclopentadi-

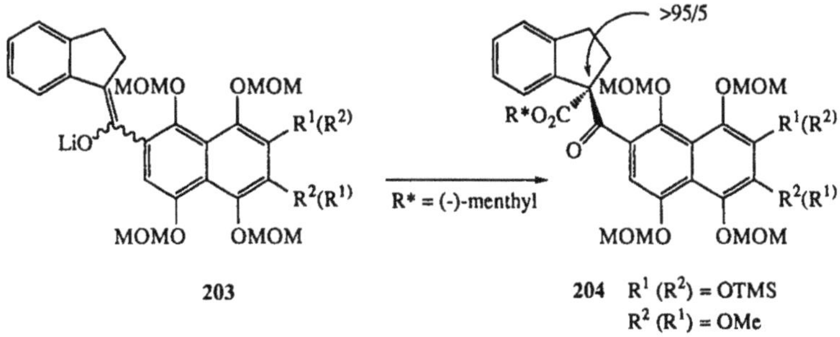

Scheme 83.

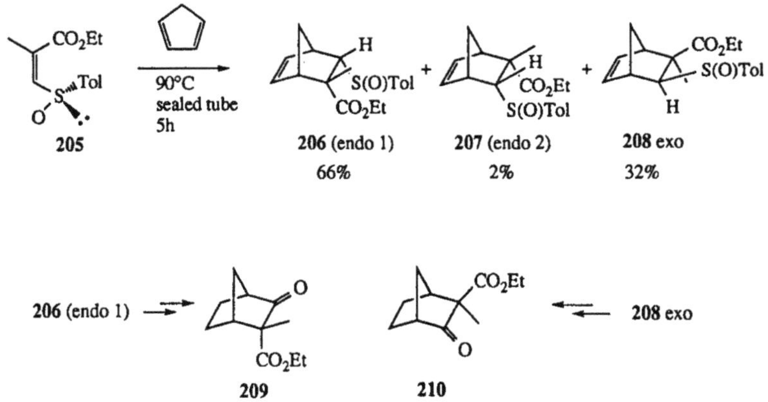

Scheme 84.

ene afforded *endo*-adducts **206** and **207** (de = 94%) along with *exo*-adduct **208** (de 100%). Adducts **206** and **208**, which could be separated by column chromatography, were transformed (three steps, 78% and 66% overall yields, respectively) into **209** and **210** (Scheme 84) which were subsequently used for the syntheses of β-santalene and β-santalol.

En route to their synthesis of (–)-bilobalide, Corey and Su[77] prepared the β-keto esters **212** and **213** by transformation of the Diels–Alder adduct **211** obtained by Lewis acid-catalyzed condensation of butadiene onto the (+)-menthol diester of fumaric acid (Scheme 85).

Robinson Annulation

The Robinson annulation is a powerful and well-known methodology to prepare cyclic enones and efficient asymmetric versions employing α-amino acids

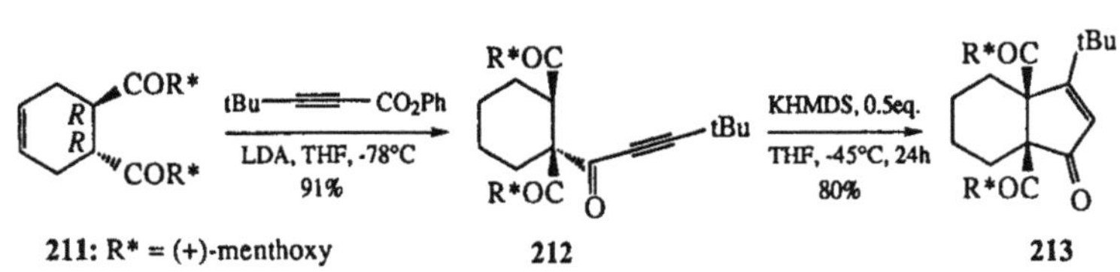

Scheme 85.

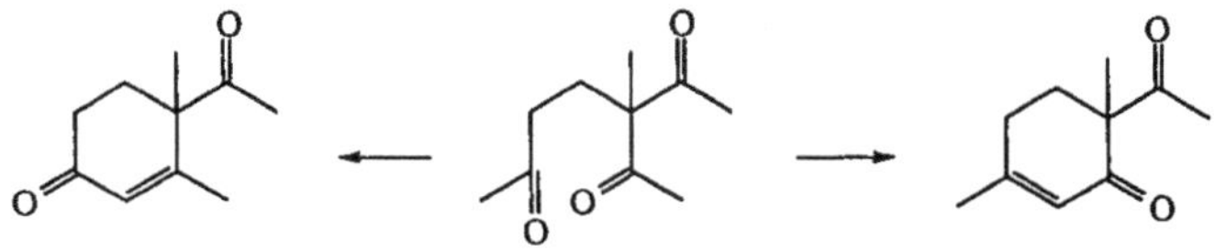
215 214 216

Scheme 86.

Table 15.

Entry	*α-Amino Acid*	*Experimental Conditions*	**215** *% yield*	**216** *% yield*	**216** *% ee*
		$MeOH-H_2O$ 100:8			
1	(*S*) phenylalanine	TsOH 0,25 equiv	19	38	56 (*R*)
	0,2 equiv	Reflux for 11 d			
2	(*S*) histidine	id	29	52	54 (*S*)
	0,05 equiv	Reflux for 8 d			
3	(*S*) phenylalanine	id	0	34	59 (*R*)
	1 equiv	Reflux for 4 d			

as chiral sources for the crucial intramolecular cyclization step have been devised.[78] Application of this methodology to the cyclization of the symmetrical triketone **214** (Scheme 86) has been explored by Terashima's group.[79] The regioselectivity of this intramolecular cyclization was found to be greatly influenced by the nature of the solvent. Thus, 1-5 diketone **215** is selectively produced under aprotic polar conditions whereas protic solvents (Table 15) favor the formation of the regioisomeric 1,3-diketone **216**. In the best circumstances (entry 3) **216** was the solely isolated annulation product albeit in low chemical yield and with moderate enantioselectivity. It is noteworthy that the steric course of the reaction is reversed if (*S*)-histidine is substituted for (*S*)-phenylalanine as the catalyst.

Synthesis of Chiral α,α-Disubstituted β-Aldehyde Esters and β-Keto Aldehyde

SAMP/RAMP Methodology

Enders's group[80] has also applied their efficient SAMP/RAMP methodology to prepare α,α-disubstituted β-aldehyde esters of the type **220** with excellent enantioselectivities and in good overall chemical yields. The reaction, portrayed in general terms in Scheme 87, was performed using a variety of aliphatic R groups.

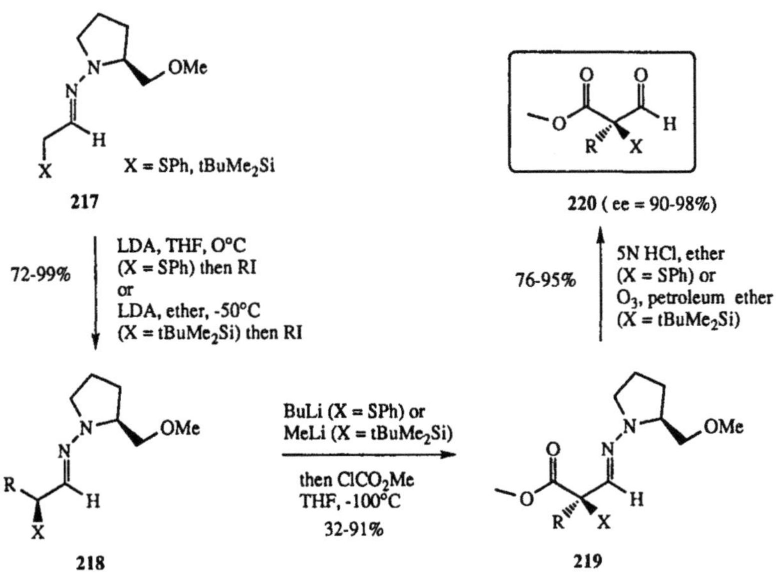

Scheme 87.

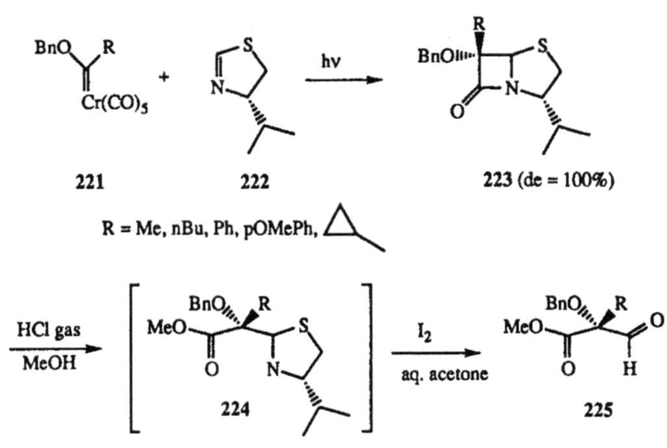

Scheme 88.

Photoaddition of Chromium–Alkoxycarbene Complexes to Optically Active Thiazolines

The photochemical reaction between alkoxycarbene complexes **221** and valine-derived thiazoline **222** was reported by Hegedus's group[81] to proceed with very high diastereoselectivity and in fair to good chemical yield to produce the β-lactam penam derivatives **223**. Cleavage of the penam system was accomplished in two steps to deliver the rather unstable α,α-disubstituted β-aldehyde esters **225** (Scheme 88).

Electrophilic α-Formylation of a Silylenolether

Scolastico's group[82] has reported one example of preparation of a α,α-disubstituted β-ketoaldehyde relying on the $BF_3 \cdot Et_2O$-mediated formylation reaction of silylenolether **226** with chiral (1*R*,2*S*)-nor-ephedrine-derived orthoamide **227**. After removal of the chiral auxiliary in the adduct **228**, β-ketoaldehyde **229** was isolated with good enantioselectivity (Scheme 89).

Although the exact mechanism of the reaction has not yet been elucidated, it presumably proceeds via a cationic intermediate **230** and through an acyclic transition state in which the carbon–carbon double bond (nucleophile) and the carbon–oxygen bond (electrophile) adopt an antiperiplanar disposition as depicted in **231**.

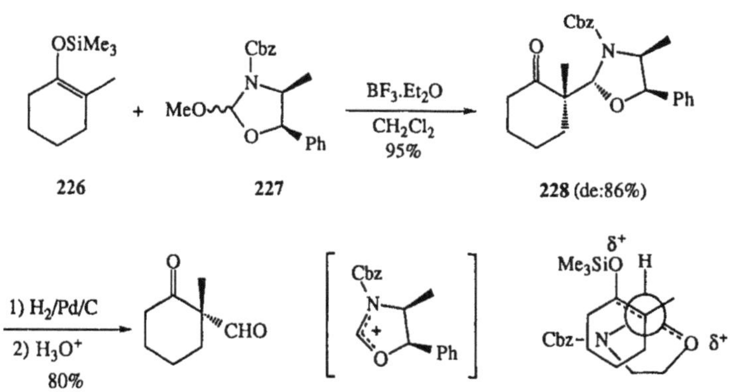

Scheme 89.

REFERENCES AND NOTES

1. This article is dedicated to the memory of my father.
2. For a recent review on the preparation of quaternary carbon centers, see: Fuji, K. *Chem. Rev.* **1993**, *93*, 2037–2066.
3. Långström, B.; Bergson, G. *Acta Chem. Scand.* **1973**, *27*, 3118–3119.
4. Wynberg, H.; Helder, R. *Tetrahedron Lett.* **1975**, 4057–4060.
5. Hermann, K.; Wynberg, H. *J. Org. Chem.* **1979**, *44*, 2238–2244.
6. Wynberg, H. *Top. Stereochem.* **1986**, *16*, 87–129.
7. Dijkstra, G.D.H; Kellogg, R.M.; Wynberg, H. *Recl. Trav. Chim. Pays-Bas* **1989**, *108*, 195–204.
8. Dijkstra, G.D.H; Kellogg, R.M.; Wynberg, H.; Svendsen, J.S.; Marko, I.; Sharpless, K.B. *J. Amer. Chem. Soc.* **1989**, *111*, 8069–8076.
9. A transition-state model has been offered by Sera's group (see ref. 13). Because it relies on the quinine "closed" conformation, its validity is quite doubtful (see text).
10. Hodge, P.; Khoshdel, E.; Waterhouse, J. *J. Chem. Soc., Perkin Trans. 1* **1983**, 2205–2209.
11. Hermann, K.; Wynberg, H. *Helv. Chim. Acta* **1977**, *60*, 2208–2212.
12. (a) Kobayashi, N.; Iwai, K. *J. Am. Chem. Soc.* **1978**, *100*, 7071–7072. (b) Kobayashi, N.; Iwai, K. *J. Polym. Sci. Polym. Chem. Ed.* **1980**, *18*, 923–932. (c) Kobayashi, N.: Iwai, K. *J. Polym. Sci. Polym. Lett. Ed.* **1982**, *20*, 85–90.
13. Sera, A.; Takagi, K.; Katayama, H.; Yamada, H.; Matsumoto, K. *J. Org. Chem.* **1988**, *53*, 1157–1161.
14. Inagaki, M.; Hiratake, J.; Yamamoto, Y.; Oda. *J. Bull. Chem. Soc. Jpn.* **1987**, *60*, 4121–4126.
15. Brünner, H.; Hammer, B. *Angew. Chem. Int. Ed. Engl.* **1984**, *23*, 312–313.
16. Botteghi, C.; Paganelli, S.; Schionato, A.; Boga, C; Fava, A. *J. Mol. Catal.* **1991**, *66*, 7–21.
17. Desimoni, G.; Quadrelli, P.; Righetti, P.P. *Tetrahedron* **1990**, *46*, 2927–2934.
18. Brünner, H.; Kraus, J. *J. Mol. Catal.* **1989**, *49*, 133–142.
19. (a) Cram, D.J.; Sogah, G.D.Y. *J. Chem. Soc., Chem. Comm.* **1981**, 625–628. (b) Cram, D.J.; Helgeson, R.C.; Peacock, S.C.; Kaplan, L.J.; Domeier, L.A.; Moreau, P.; Koga, K.; Mayer, J.M.; Mayer, J.M.; Chao, Y.; Siegel, M.G.; Hoffman, D.H.; Sogah, G.D.Y. *J. Org. Chem.* **1978**, *43*, 1930–1946.
20. Raguse, B.; Ridley, D.D. *Aust. J. Chem.* **1984**, *37*, 2059–2071.
21. Fiaud, J.C.; Hibon de Gournay, A.; Larcheveque, M.; Kagan, H.B. *J. Organomet. Chem.* **1978**, *154*, 175–185.
22. Hayashi, T.; Kanehira, K.; Tsuchiya, H.; Kumada, M. *J. Chem. Soc., Chem. Comm.* **1982**, 1162–1164.
23. Hayashi, T.; Kanehira, K.; Hagihara, T.; Kumada, M. *J. Org. Chem.* **1988**, *53*, 113–120.
24. Gamez, P.; Dunjic, B.; Fache, F.; Lemaire, M. *Tetrahedron: Asymmetry* **1995**, *6*, 1109–1116.
25. Keim, W.; Koehnes, A.; Roethel, T.; Enders, D. *J. Organomet. Chem.* **1990**, *382*, 295–301.
26. Fiaud, J.C. *Tetrahedron Lett.* **1975**, 3495–3496.
27. (a) Saigo, K.; Koda, H.; Nohira, H. *Bull. Chem. Soc. Jpn.* **1979**, *52*, 3119–3120. (b) Juliá, S.; Ginebreda, A.; Guixer, J.; Tomás, A. *Tetrahedron Lett.* **1980**, *21*, 3709–3712. (c) Colonna, S.; Annunziata, R. Juliá, S.; Guixer, J. *Afinidad* **1981**, 501–502. (d) Annunziata, R.; Cinquini, M.; Colonna, S. *J. Chem. Soc., Perkin Trans. 1* **1980**, 2422–2424. (e) Valli, V.L.K.; Sarma, G.V.M.; Choudary, B.M. *Indian J. Chem.* **1990**, *29B*, 481–482.
28. (a) Dolling, U.H.; Davis, P.; Grabowski, E.J.J. *J. Am. Chem. Soc.* **1984**, *106*, 446–447. (b) Hughes, D.L.; Dolling, U.H.; Ryan, K.M.; Schoenewaldt, E.F.; Grabowski, E.J.J. *J. Org. Chem.* **1987**, *52*, 4745–4752. (c) Masui, M.; Ando, A.; Shioiri, T. *Tetrahedron Lett.* **1988**, *29*,

2835–2838. (d) Nerinckx, W.; Vandewalle, M. *Tetrahedron: Asymmetry* **1990**, *1*, 265–276. (e) Lee, T.B.K.; Wong, G.S.K. *J. Org. Chem.* **1991**, *56*, 872–875.
29. Stork, G.; Saccomano, N.A. *Nouv. J. Chim.* **1986**, *10*, 677–679.
30. Stork, G.; Saccomano, N.A. *Tetrahedron Lett.* **1987**, *28*, 2087–2090.
31. Taber, D.F.; Mack, J.F.; Rheingold, A.L.; Geib, S.J. *J. Org. Chem.* **1989**, *54*, 3831–3836.
32. Yamada, F.; Kozikowski, A.P.; Reddy, E.R.; Pang, Y-P.; Miller, J.H.; McKinney, M. *J. Am. Chem. Soc.* **1991**, *113*, 4695–4696.
33. Ito, I.; Ueda, T.; Asano, T.; Oda, N. *Chem. Pharm. Bull.* **1978**, *26*, 1328–1332.
34. Kunisch, F.; Hobert, K.; Welzel, P. *Tetrahedron Lett.* **1985**, *26*, 5433–5436.
35. (a) Kobayashi, M.; Umemura, K.; Watanabe, N.; Matsuyama, H. *Chem. Lett.* **1985**, 1067–1070. (b) Kobayashi, M.; Koyabu, K.; Shimizu, T.; Umemura, K.; Matsuyama, H. *Chem. Lett.* **1986**, 2117–2120. (c) Umemura, K.; Matsuyama, H.; Watanabe, N.; Kobayashi, M.; Kamigata, N. *J. Org. Chem.* **1989**, *54*, 2374–2383. (d) Umemura, K.; Matsuyama, H.; Kobayashi, M.; Kamigata, N. *Bull. Chem. Soc. Jpn.* **1989**, *62*, 3026-3028.
36. Differding, E.; Lang, R.W. *Tetrahedron Lett.* **1988**, *29*, 6087–6090.
37. Davies, F.A.; Zhou, P.; Murphy, C.K. *Tetrahedron Lett.* **1993**, *34*, 3971–3974.
38. Taber, D.F.; Amedio, J.C, Jr.; Raman, K. *J. Org. Chem.* **1988**, *53*, 2984–2990.
39. (a) Kato, K.; Suemune, H.; Sakai, K. *Tetrahedron Lett.* **1992**, *33*, 247–250. (b) Sakai, K.; Suemune, H. *Tetrahedron: Asymmetry* **1993**, *4*, 2109–2118 (a review).
40. Kato, K.; Suemune, H.; Sakai, K. *Tetrahedron Lett.* **1994**, *35*, 3103–3104.
41. Chitkul, B.; Pinyopronpanich, Y.; Thebtaranonth, C.; Thebtaranonth, Y. *Tetrahedron Lett.* **1994**, *35*, 1099–1102.
42. Kato, K.; Suemune, H.; Sakai, K. *Tetrahedron Lett.* **1992**, *33*, 3481–3482.
43. Iwaoka, T.; Murohashi, T.; Sato, M.; Kaneko, C. *Tetrahedron: Asymmetry* **1992**, *3*, 1025–1028.
44. Hodgson, A.; Marshall, J.; Hallett, P.; Gallagher, T. *J. Chem. Soc., Perkin Trans 1* **1992**, 2169–2174.
45. Hiroi, K.; Abe, J. *Tetrahedron Lett.* **1990**, *31*, 3623–3626.
46. (a) Tomioka, K.; Ando, K.; Takemasa, Y.; Koga, K. *J. Am. Chem. Soc.* **1984**, *106*, 2718–2719. (b) Tomioka, K.; Ando, K.; Takemasa, Y.; Koga, K. *Tetrahedron Lett.* **1984**, *25*, 5677–5680. (c) Ando, K.; Takemasa, Y.; Tomioka, K.; Koga, K. *Tetrahedron* **1993**, *49*, 1579–1588.
47. Shilay, V.; Alyev, A.; Grishina, G. *Bull. Soc. Chim. Belg.* **1992**, *101*, 751–752.
48. Georg, G.I.; Guan, X.; Kant, J. *Bioorg. Med. Chem. Lett.* **1991**, *1*, 125–128.
49. Georg, G.I.; Guan, X.; Kant, J. *Tetrahedron Lett.* **1988**, *29*, 403–406.
50. Huang, P.Q.; Zhou, W.S. *Tetrahedron: Asymmetry* **1991**, *2*, 875–878.
51. Tomioka, K.; Ando, K.; Yasuda K.; Koga, K. *Tetrahedron Lett.* **1986**, *27*, 715–716.
52. Tomioka, K.; Yasuda K.; Koga, K. *J. Chem. Soc., Chem. Soc.* **1987**, 1345–1346.
53. (a) Tomioka, K.; Seo, W.; Ando, K.; Koga, K. *Tetrahedron Lett.* **1987**, *28*, 6637–6640. (b) Ando, K.; Seo, W.; Tomioka, K.; Koga, K. *Tetrahedron* **1994**, *50*, 13081–13088.
54. Lee, J.; Oya, S.; Snyder, J.K. *Tetrahedron Lett.* **1991**, *32*, 5899–5902.
55. (a) Pfau, M.; Revial, G.; Guingant, A.; d'Angelo, J. *J. Am. Chem. Soc.* **1985**, *107*, 273–274. (b) d'Angelo, J.; Desmaele, D.; Dumas, D.; Guingant, A. *Tetrahedron: Asymmetry* **1992**, *3*, 459–505. (c) d'Angelo, J.; Cave, C.; Desmaele, D.; Dumas, D. *Trends in Org. Chem.* **1993**, *4*, 555–616.
56. Guingant, A.; Hammami, H. *Tetrahedron: Asymmetry* **1991**, *2*, 411–414.
57. (a) Brunner, H.; Kraus, J.; Lautenschlager, H-J. *Monatshefte für Chemie* **1988**, *119*, 1161–1167. (b) The ee value for **9**, as deduced from the work of Wynberg,[5] is overestimated (ref 56).
58. Guingant, A. *Tetrahedron: Asymmetry* **1991**, *2*, 415–418.

59. Guingant, A.; Hammami, H. *Tetrahedron: Asymmetry* **1993**, *4*, 25–26. In this paper, β-enaminoester **136b** was erroneously represented by its ethyl ester analog.
60. Pinheiro, S.; Guingant, A.; Desmaele, D.; d'Angelo, J. *Tetrahedron: Asymmetry* **1992**, *3*, 1003–1006.
61. Felk, A.; Revial, G.; Viossat, B.; Lemoine, P.; Pfau, M. *Tetrahedron: Asymmetry* **1994**, *5*, 1459–1462.
62. For a review on the use of allylic $A^{(1,3)}$ strain in stereoselective transformations, see: Hoffmann, R.W. *Chem. Rev.* **1989**, *89*, 1841–1860.
63. (a) d'Angelo, J.; Revial, G.; Guingant, A.; Riche, C.; Chiaroni, A. *Tetrahedron Lett.* **1989**, *30*, 2645–2648. (b) Chiaroni, A.; Riche, C.; Guingant, A.; d'Angelo, J. *J. Acta Cryst. C49* **1993**, 1406–1409
64. Sevin, A.; Tortajada, J.; Pfau, M. *J. Org. Chem.* **1986**, *51*, 2671–267. Sevin, A.; Masure, D., Giessner-Prettre, C.; Pfau, M. *Helv. Chim. Acta* **1990**, *73*, 552–573.
65. Seebach, D.; Golinski, J. *Helv. Chim. Acta* **1981**, *64*, 1413–1423.
66. (a) Cavé, C.; Daley, V.; d'Angelo, J.; Guingant, A. *Tetrahedron: Asymmetry* **1995**, *6*, 79–82. (b) d'Angelo, J.; Guingant, A.; Riche, C.; Chiaroni, A. *Tetrahedron Lett.* **1988**, *29*, 2667–2670. (c) Ambroise, L.; Desmaele, D.; Mahuteau, J.; d'Angelo, J. *Tetrahedron Lett.* **1994**, *35*, 9705–9709.
67. Hervouet, K.; Guingant, A. *Tetrahedron: Asymmetry* **1996**, *7*, 421–424.
68. Hervouet, K.; Guingant, A. *Tetrahedron: Asymmetry* **1996**, *7*, 425–426.
69. Poll, T.; Metter, J.O.; Helmchen, G. *Angew. Chem. Int. Ed. Engl.* **1985**, *24*, 112–114.
70. (a) Fráter, G. *Helv. Chim. Acta* **1979**, *62*, 2825–2828. Fráter, G. *Helv. Chim. Acta* **1980**, *63*, 1383–1390. Fráter, G. *Tetrahedron Lett.* **1981**, *22*, 425–428. (b) Fráter, G.; Müller, U.; Gunther, W. *Tetrahedron* **1984**, *40*, 1269–1277. (c) Fráter, G.; Günther, W.; Müller, U. *Helv. Chim. Acta* **1989**, *72*, 1846–1851. See also, Nagumo, S.; Suemune, H.; Sakai, K. *Tetrahedron Lett.* **1991**, *32*, 5585–5588.
71. Sato, T.; Maeno, H.; Noro, T.; Fujisawa, T. *Chem. Lett.* **1988**, 1739–1742.
72. Westermann, B.; Große Scharmann, H.; Kortmann, I. *Tetrahedron: Asymmetry* **1993**, *4*, 2119–2122. See also, Hölscher, P.; Knölker, H-J.; Winterfeldt, E. *Tetrahedron Lett.* **1990**, *31*, 2705–2706.
73. Nagao, Y.; Hagiwara, Y.; Tohjo, T.; Hasegawa, Y.; Ochiai, M.; Shiro, M. *J. Org. Chem.* **1988**, *53*, 5986–5988.
74. Hogeveen, H.; Menge, W.M.P.B. *Tetrahedron Lett.* **1986**, *27*, 2767–2770. See also: Cain, C.M.; Cousins, R.P.C.; Coumbarides, G.; Simpkins, N.S. *Tetrahedron* **1990**, *46*, 523–544.
75. Trypke, W.; Steigel, A.; Braun, M. *Synlett* **1992**, 827–829.
76. Arai, Y.; Yamamoto, M.; Koizumi, T. *Chem. Lett.* **1986**, 1225–1228. Arai, Y.; Yamamoto, M.; Koizumi, T. *Bull. Chem. Soc. Jpn.* **1988**, *61*, 467–473.
77. Corey, E.J.; Su, W-G. *Tetrahedron Lett.* **1988**, *29*, 3423–3426.
78. For a review on the mechanism of proline-catalyzed intramolecular aldol reaction, see: Agami, C. *Bull. Soc. Chem. Fr.* **1988**, 499–507.
79. Terashima, S.; Sato, S.; Koga, K. *Tetrahedron Lett.* **1979**, *36*, 3469–3472.
80. Enders, D.; Zamponi, A.; Raabe, G. *Synlett* **1992**, 897–900.
81. Thompson, D.K.; Suzuki, N.; Hegedus, L.S.; Satoh, Y. *J. Org. Chem.* **1992**, *57*, 1461–1467.
82. Palazzi, C.; Poli, G.; Scolastico, C.; Villa, R. *Tetrahedron Lett.* **1990**, *31*, 4223–4226.

PREPARATION OF CHIRAL FERROCENES BY ASYMMETRIC SYNTHESIS OR BY KINETIC RESOLUTION

H. B. Kagan and O. Riant

Advances in Asymmetric Synthesis
Volume 2, pages 189–235

ISBN 1-55938-797-1

I. INTRODUCTION

Ferrocene was discovered in 1951[1] and its unusual structure **1** (Scheme 1) was established in 1952.[2] The aromatic character of this molecule inspired chemists to prepare ferrocene derivatives by electrophilic substitution. In that way many monosubstituted and polysubstituted ferrocenes were synthesized (usually as mixtures such as **2**→**3** which have to be separated). It soon became apparent that some substitution pattern, e.g. **4**, **5**, or **6** (Scheme 2), must give rise to racemic mixtures, because each molecule is a chiral species. The first chiral ferrocene to be resolved was compound **7** in 1959.[3] The name *planar chirality* was proposed by Cahn, Ingold, and Prelog (CIP) to characterize the chirality associated with a distribution of substituents which breaks a symmetry plane of compounds such paracyclophanes, ferrocene, or benzenechromiumtricarbonyl (benchrotrene). The CIP rules give the *R*-configuration to **7**, the observation being from the iron

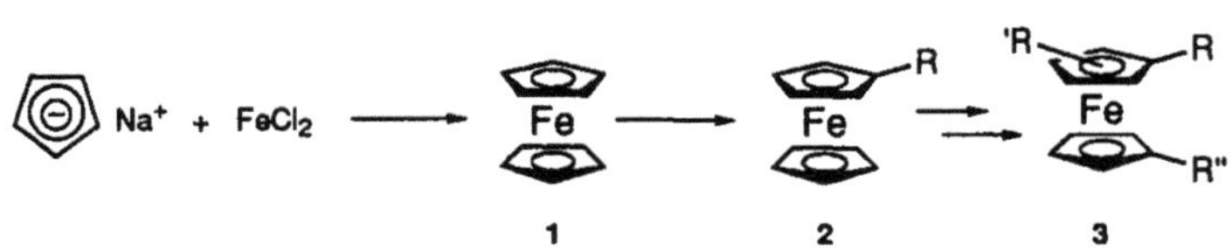

Scheme 1.

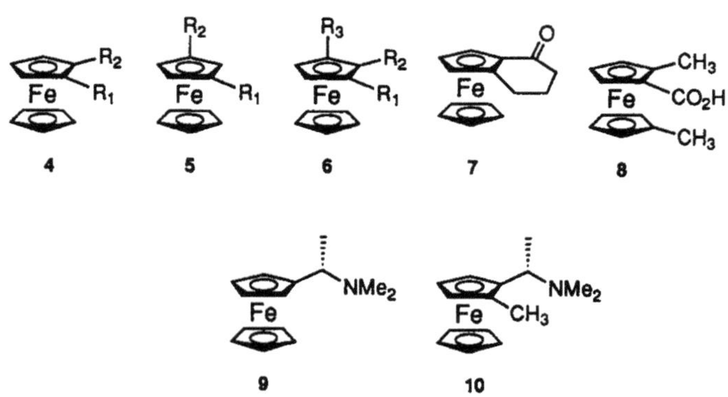

Scheme 2.

showing a clockwise distribution of the substituents of decreasing priority on the ring. Schlögl proposed an alternate nomenclature where the molecule is viewed from above and where the clockwise or anticlockwise sequence of groups of decreasing priority determines the *R* or *S* configuration.[4] A revision of CIP rules has been proposed for chiral ferrocenes with a substitution pattern such as in **3**.[5] We will use here the Schlögl nomenclature. The first resolution of a trisubstituted ferrocene, namely **8**, was reported in 1962.[6] Since that time stereochemistry of ferrocenes with planar chirality became a rich and active area.[7,8]

Another class of chiral ferrocenes involves ferrocenes having a *lateral chiral unit*, e.g. amine **9**, which is easily resolvable.[9] And a last class of compounds has both *planar chirality* and a *lateral chirality* as amine **10**.[10]

Chiral ferrocenes are of interest in several areas. The main use is in *asymmetric catalysis* as chiral ligands of transition-metal complexes.[11] Some examples are indicated in Scheme 3. Applications have been found in asymmetric hydrogenation,[12] asymmetric hydrosilylation,[13] asymmetric aldolization,[14] asymmetric coupling reactions,[15] asymmetric Michael additions,[16] and asymmetric addition of organozincs to aldehydes.[17]

Chiral ferrocenes have also been used as modular units in *material science*, e.g. in ferroelectric liquid crystals or for nonlinear optics.[18] Finally chiral ferrocenes could be of importance as a structural unit of products with biological or biochemical activity.[19–21] We intend in this review to present the various ways to prepare chiral ferrocenes, focusing on asymmetric synthesis or kinetic resolution. This chapter considers both the chemical and biochemical approaches.

II. ASYMMETRIC SYNTHESIS OF FERROCENES WITH CHIRAL CENTERS

A. Chemical Approach

Chiral Ferrocenyl Alcohols by Asymmetric Reduction and Asymmetric Hydroboration

The success of *N,N*-dimethyl-(ferrocenylethyl)amine (Ugi's amine) **9** in asymmetric synthesis has driven chemists to search for new methods to synthesize this molecule and related analogs. Thus, access to chiral ferrocenyl alcohols such as **11** (Scheme 4) should provide a large number of chiral synthons through activation of the alcoholic function and stereospecific nucleophilic substitution.[22] One possibility is the asymmetric reduction of ferrocenyl ketones. Encapsulation of

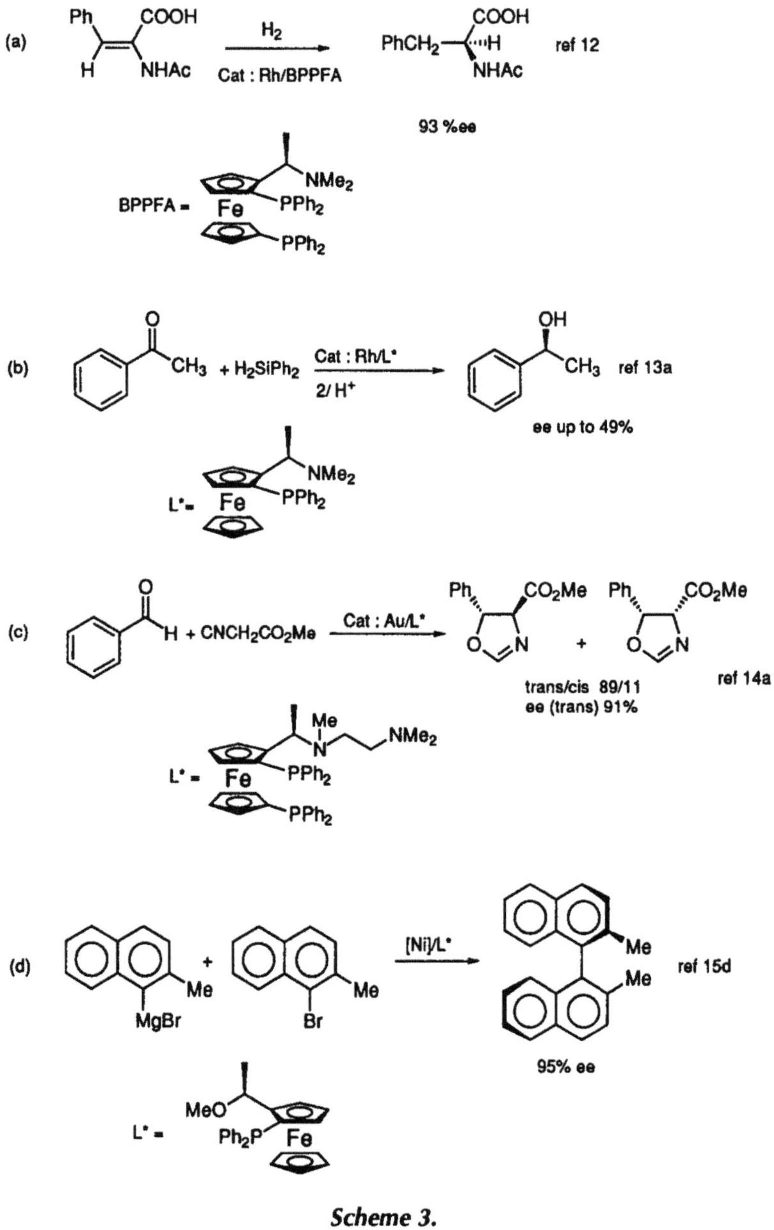
(a)
Ph
COOH
H
NHAc
H2
Cat : Rh/BPPFA
PhCH2
COOH
H
NHAc
ref 12
93 %ee
BPPFA =
NMe2
Fe
PPh2
PPh2
(b)
O
CH3
+ H2SiPh2
Cat : Rh/L*
2/ H+
OH
CH3
ref 13a
ee up to 49%
L* =
NMe2
Fe
PPh2
(c)
O
H + CNCH2CO2Me
Cat : Au/L*
Ph
CO2Me
O
N
+
Ph
CO2Me
O
N
trans/cis 89/11
ee (trans) 91%
ref 14a
L* =
Me
N
NMe2
Fe
PPh2
PPh2
(d)
Me
MgBr
+
Me
Br
[Ni]/L*
Me
Me
ref 15d
95% ee
L* =
MeO
Ph2P
Fe

Scheme 3.

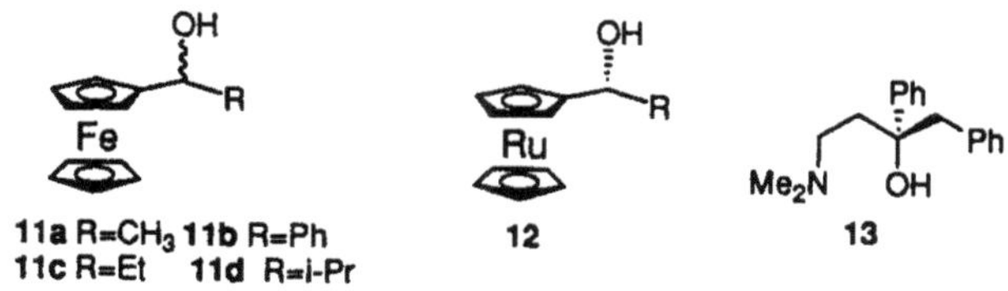

Scheme 4.

acetyl ferrocene or benzoyl ferrocene in β-cyclodextrins gave a chiral complex which, on reduction by sodium borohydride in aqueous solution and in the presence of potassium chloride, followed by extraction, furnished alcohols (–)-**11a** and (–)-**11b** with 40 and 84% enantiomeric excesses, respectively.[23]

Sokolov et al. have also reported asymmetric reduction of benzoylmetallocenes into carbinols (*R*)-**11b** and **12** using lithium aluminum hydride modified with Darvon's alcohol **13**.[24] Although here reduction of benzoyl ferrocene occurs with a rather modest enantioselectivity (53% ee), a much better result is obtained in the case of the ruthenocene analog (83% ee). Another method using borane as reducing agent and a chiral oxazaborolidine borane as a catalyst (Scheme 5) proved to be one of the most efficient method for the preparation of various enantiopure ferrocenyl alcohols.[25]

Conversion of vinyl ferrocene into optically active alcohol **11a** (Scheme 6) through a rhodium-catalyzed asymmetric hydroboration/oxidation sequence has been independently reported by two groups. In both cases catecholborane is used as the hydroborating agent and binap as the chiral ligand but the enantiomeric excesses of the alcohol **11a** remained modest.[26]

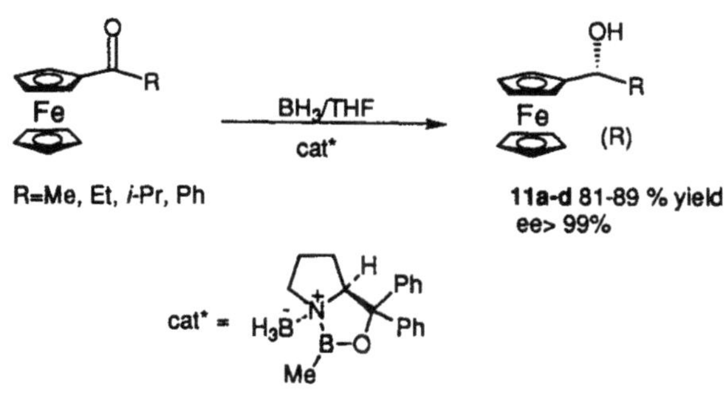

Scheme 5.

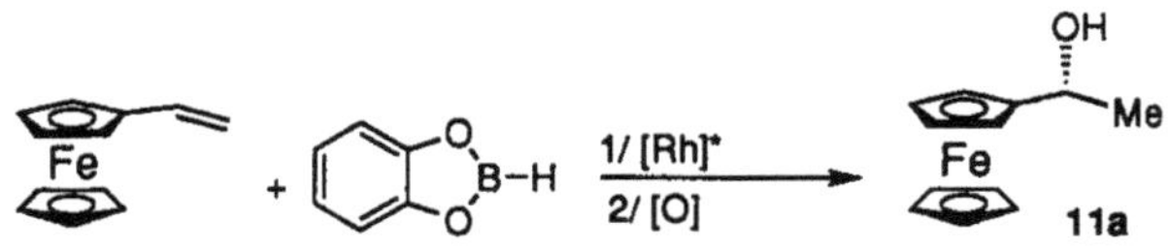

Ito *et al.*[26a] : $[Rh(COD)_2]BF_4$/(R)-(binap), THF, -30°C, 58% ee (R)
Zhang *et al.*[26b] : $[RhCl(COD)]_2$/(2S)-binap, DME, -78°C, 18.4% ee (R)

Scheme 6.

Chiral Ferrocenyl Alcohols by Asymmetric Alkylation

Asymmetric alkylation of ferrocene carboxaldehyde can also give entry to chiral ferrocenyl alcohol as demonstrated by Noyori et al. (Scheme 7a). Thus, reaction of ferrocene aldehyde **14** and dimethylzinc catalyzed by chiral amino alcohol (–)-DAIB **15** furnished alcohol (*S*)-**11a** with a 60% yield and 81% ee.[27]

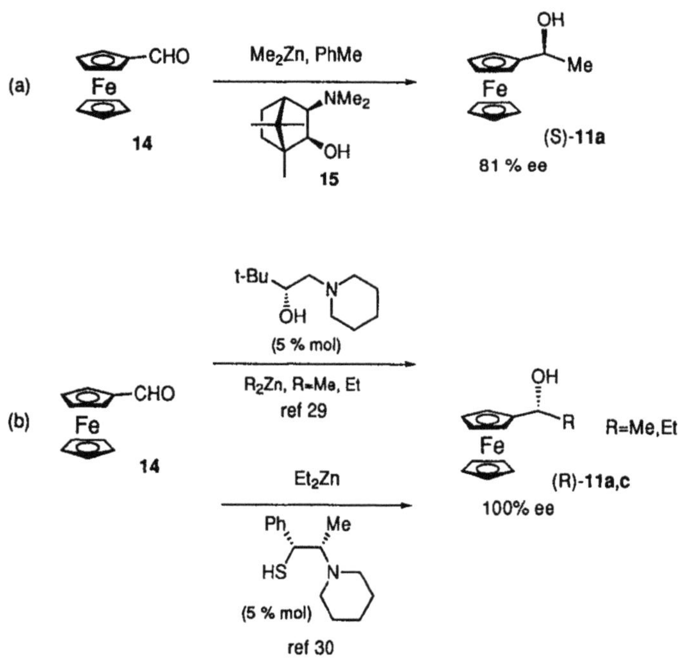

Scheme 7.

A more recent publication by Hayashi et al. described the reactions of dimethyl and diethylzinc on ferrocene and ruthenocene aldehydes, catalyzed by a chiral amino alcohol (Scheme 7b).[28] Optimized conditions permitted the isolation of the chiral carbinols in excellent yields and high enantiomeric excesses (> 90–95%).

Further conversion of the alcohols to the corresponding *N,N*-dimethylamines was reported in view of the synthesis of chiral ferrocenyl and ruthenocenyl phosphines.[29] A chiral aminothiol was also an excellent catalyst for the formation of **11e** (98% ee, Scheme 7b).[30] The reaction of diisopropylzinc on ferrocene carboxaldehyde **14** and 1,1′-ferrocene dicarboxaldehyde **16** using dibutyl norephedrine (DBNE) as a catalyst furnished the corresponding alcohol (*S*)-**11d** and diol (*S,S*)-**17** with high enantioselectivities[31] (Scheme 8).

Chiral Ferrocenyl Amino Acids by Asymmetric Hydrogenation

Asymmetric synthesis of chiral ferrocenyl amino acid derivatives **19**[32] and **22**[33] (Schemes 9 and 10) have been reported. Thus, Heck reaction of a protected dehydroalanine derivative with 1,1′-diiodoferrocene furnished the bis-dehydroamino acid derivative **18** which is then subjected to asymmetric hydrogenation using [Rh(*R*)–prophos(NBD)]ClO_4 as a chiral catalyst to give **19** as a (–)/(+)/*meso*:5/57/38 mixture (23.5% de and 84% ee). Deprotection of the amino and carboxy groups and crystallization finally gave the bis-amino acid (+)-**20** incor-

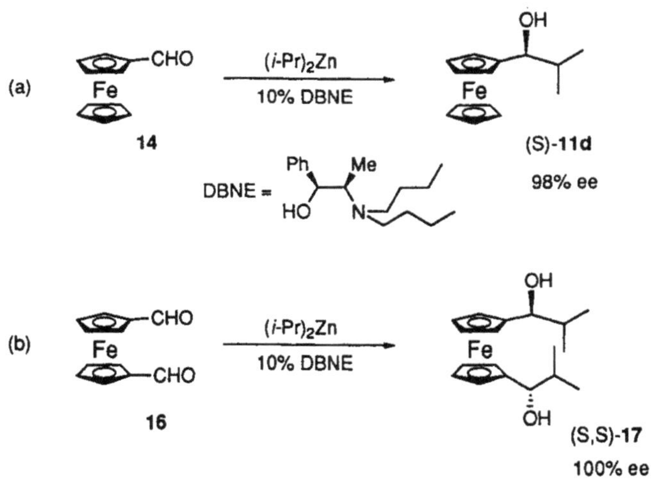

Scheme 8.

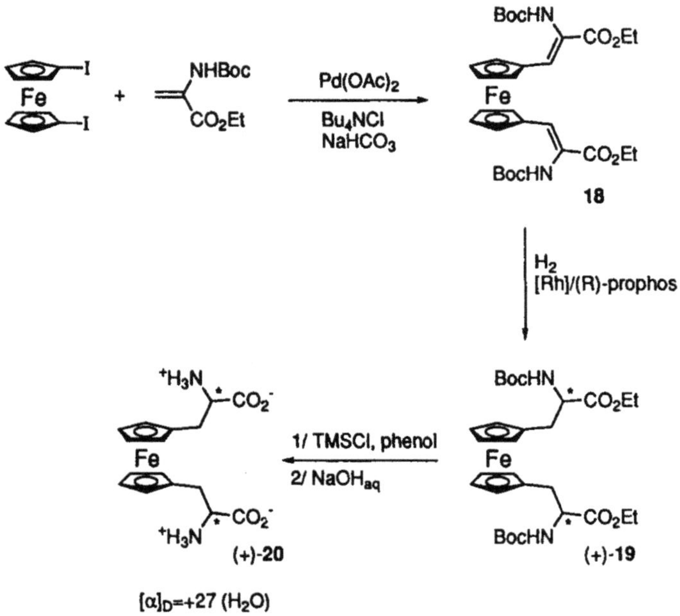

Scheme 9.

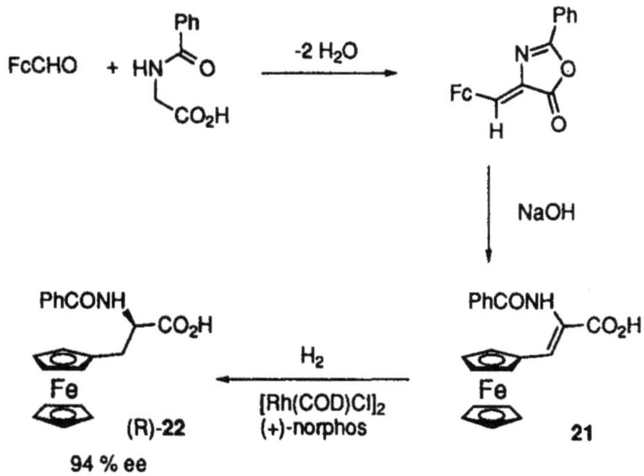

Scheme 10.

porating a rigid backbone between the two amino acid residues. Similarly, protected ferrocenyl alanine **22** is obtained with 94% ee from hydrogenation of alkene **21** using a norphos–rhodium complex as a catalyst. In this case, the absolute configuration of the N-protected amino acid was determined by X-ray analysis of **22**.

Chiral Ferrocenyl Sulfoxides

Chiral ferrocenyl sulfoxides **26a-b** bearing the sulfur atom in the α-position of the ferrocenyl moiety have been recently prepared via an asymmetric synthesis.[34] *tert*-Butyl ferrrocenylsulfoxide **26a** has been prepared by two different methods (Scheme 11). Opening of the cyclic sulfite **23** by *tert*-butylmagnesium bromide gave sulfinate **24** which is then reacted with 2 equiv of ferrocenyl lithium to deliver enantiomerically pure **26a**. Alternatively, sulfoxide **26a** can be prepared in 90% ee starting from sulfide **25** using a chiral oxidation (Scheme 11a).

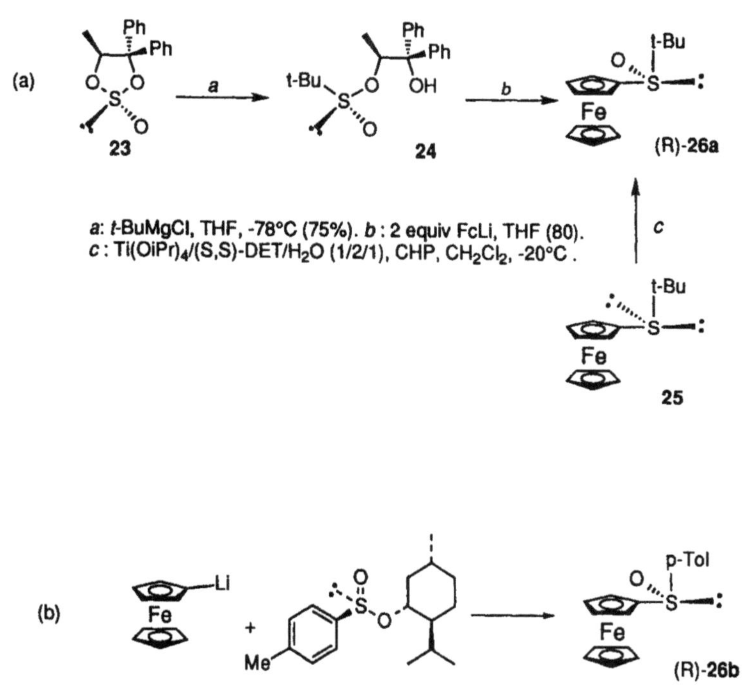

Scheme 11.

Ferrocenyl *p*-tolylsulfoxide **26b** has also been prepared starting from commercially available menthyl *p*-tolylsulfinate via the well-known Andersen method (Scheme 11b). However, in the initial report partial racemization (81% ee) occurred during the synthesis of the sulfoxide when ferrocenyllithium was prepared from ferrocene and *tert*-butyllithium. It was reported later that racemization could be avoided by a careful control of the reaction conditions and if the ferrocenyllithium was prepared by transmetallation of tris *n*-butylstannyl ferrocene.[35]

The asymmetric oxidation of ferrocenyl *tert*-butyl sulfide (**25**) has been recently optimized to 95% ee.[36] Thus, application of optimized conditions for the oxidation of some ferrocenyl sulfides **25** Fc–S–R (R = Ph, *p*-Tol, *t*-Bu, Me, *i*-Pr) gave high ee for the corresponding sulfoxides **26** (ee = 84→99%), allowing the preparation of various enantiomerically pure ferrocenyl sulfoxides (Scheme 12). Lower enantiomeric excesses were also observed in some cases (R = *n*-Oct, *p*-Cl-C_6H_4, *n*-Bu; ee = 2–45%).

During the course of these studies on asymmetric oxidation of prochiral sulfides into sulfoxides, a phenomenon of enantiomeric enrichment by flash chromatography on silica gel of some sulfoxides like ferrocenyl phenyl sulfoxide and ferrocenyl methyl sulfoxide has also been discovered.[37]

Diastereoselective Reactions

A straightforward synthesis of the acetate **29** from inexpensive commercially available starting materials has been reported by Pyne et al.[38] (Scheme 13a). Reduction of the imine **27** (readily prepared from acetyl ferrocene and α-phenylethylamine) by sodium borohydride led to amine **28** as a 9:1 mixture of diastereomers. The major diastereomer (*R,R*)-**28** was isolated in 73% yield after a single recrystallization and can be further transformed into the acetate **29** by reaction with formaldehyde and acetic acid in acetonitrile. An earlier example was related

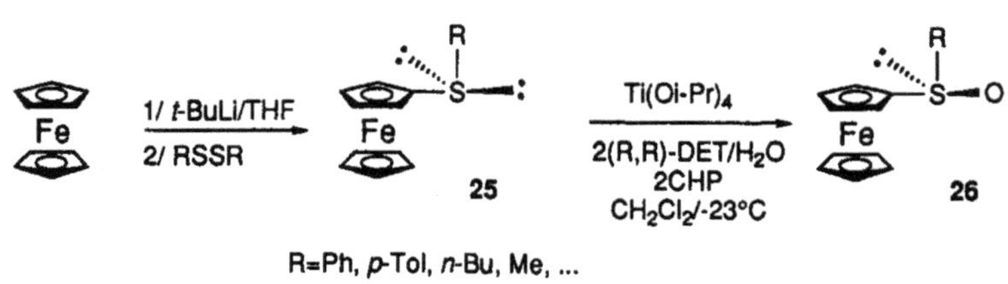

Scheme 12.

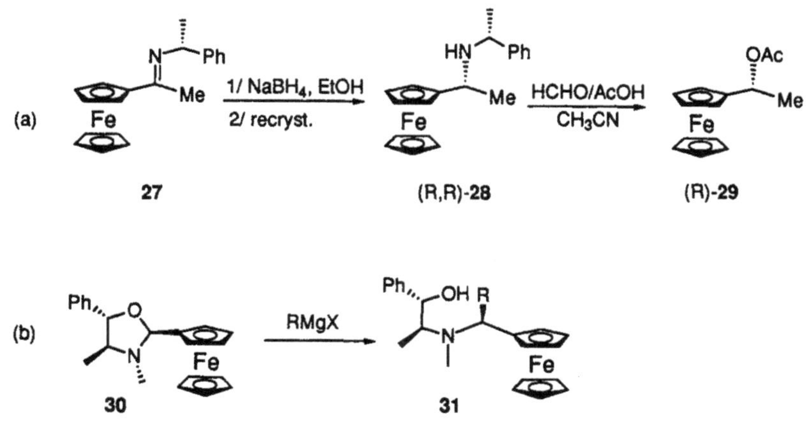

Scheme 13.

to the diastereoselective opening of chiral ferrocenyl-derived oxazolidines (Scheme 13b).[39] Thus, reaction of oxazolidine **30** with various organomagnesium reagents led to amino alcohols **31** in excellent yields, but unfortunately, the diastereoselectivities obtained in this reaction were not reported.

Finally, a set of diastereoisomeric chiral ferrocenyl amines bearing a permanent chiral group, such as **32a-d**, have been prepared (Scheme 14)[40] starting from chiral terpenes derivatives (menthone, α-pinene) by asymmetric synthesis. One example is described in Scheme 15. Reaction of an enantiopure bicyclic aldehyde

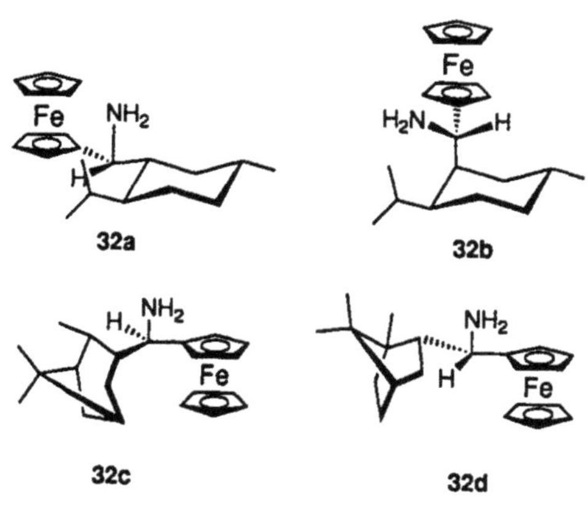

Scheme 14.

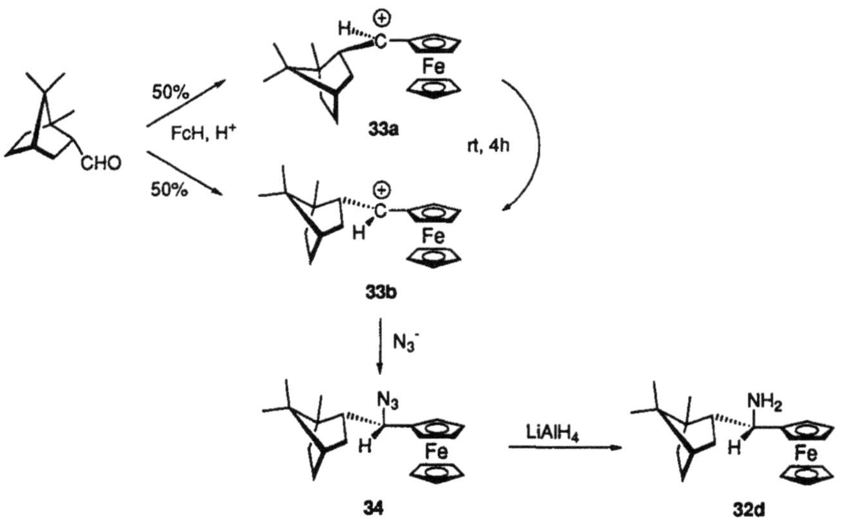

Scheme 15.

(from α-pinene) with ferrocene in acidic medium gave a 1:1 mixture of diastereoisomeric carbocations **33a-b**. Upon standing at room temperature, isomerization occurred and the thermodynamically more stable carbocation **33b** remained as the only product. Trapping of the cation by azide ion and subsequent reduction of **34** yielded the enantiomerically pure amine **32d**. Some applications including the

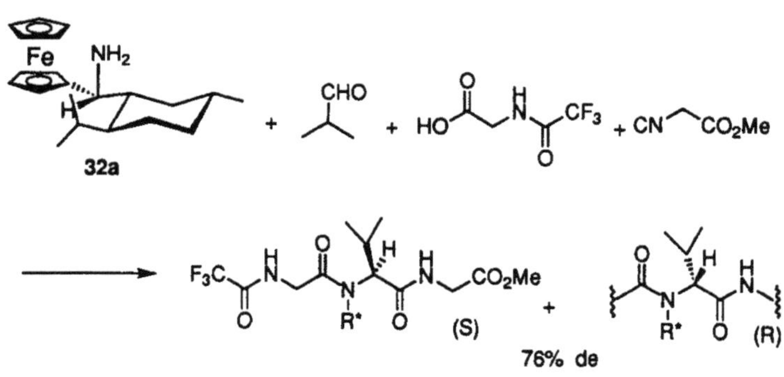

Scheme 16.

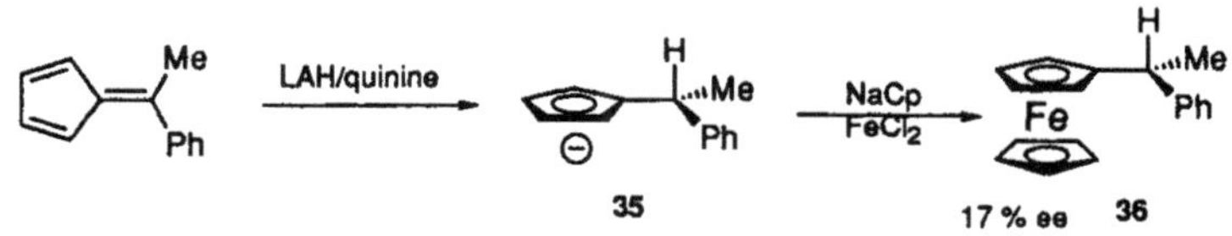

Scheme 17.

preparation of chiral phosphine ligands for asymmetric catalysis[41] and the four-component condensation[42] (Scheme 16) have been described for those amines.

Construction of the ferrocenyl moiety using a chiral functionalized cyclopentadienyl group is another possibility for the preparation of chiral ferrocenes. The cyclopentadienyl could in turn be prepared from a prochiral fulvene by asymmetric synthesis. The first and only example of such a method was reported in 1976 by Leblanc and Moise[43] (Scheme 17). Asymmetric reduction of methyl phenylfulvene with a LAH/quinine complex gave the cyclopentadienide ligand **35**

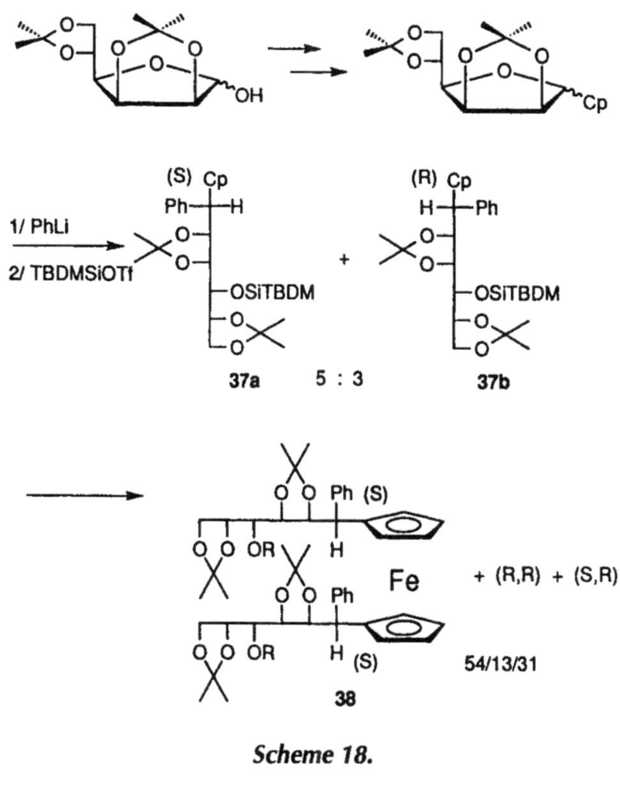

Scheme 18.

which is directly reacted with sodium cyclopentadienide and ferrous chloride to yield the chiral ferrocene **36** with a 17% ee.

Another approach related to this methodology and involving the synthesis of a fulvene bearing a permanent or removable chiral auxiliary has been recently described.[44,45] Thus, cyclopentadienyl mannitol derivatives such as **37** were chosen by Vasella et al.[44] as precursors of the chiral ferrocenes (Scheme 18). Deprotonation of the Cp ligands and subsequent reaction with iron(II) chloride yielded the 1,1′-disubstituted ferrocenes **38** as a mixture of diastereomers. A similar approach was described by Togni et al. starting with amino acids as the chiral building blocks[45] (Scheme 19). Reaction of fulvene **39** with methyl lithium and iron(II) chloride gave the 1,1′-disubstituted ferrocene **40** with a ds up to 87%. Using Cp*Fe(acac) as a precursor, the amine **41** could be isolated after removal of the chiral auxiliary. This hindered analog of Ugi's amine **9** was also trans-

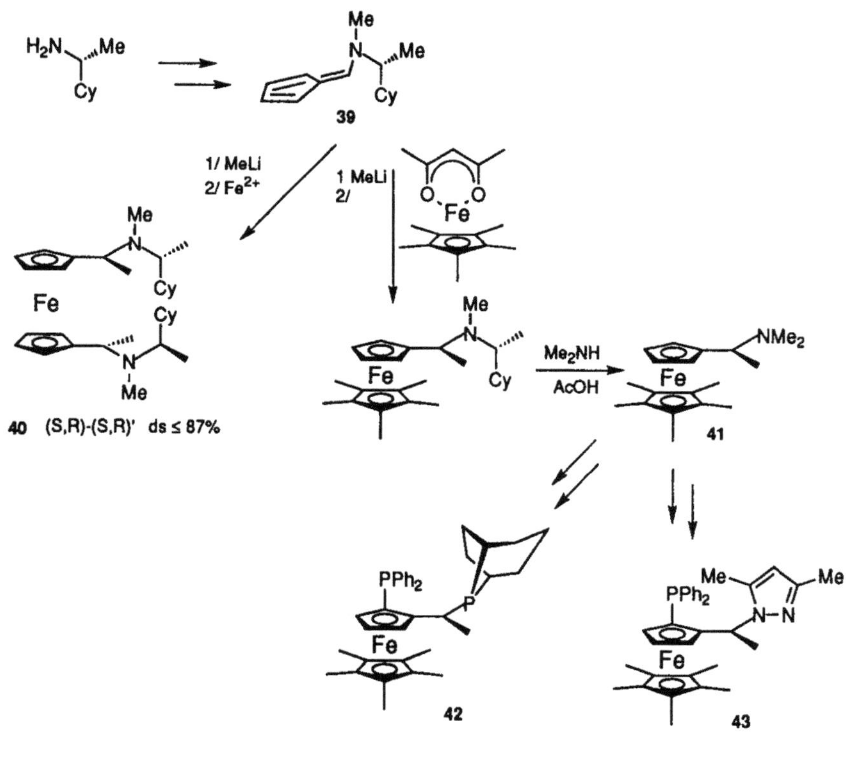

Scheme 19.

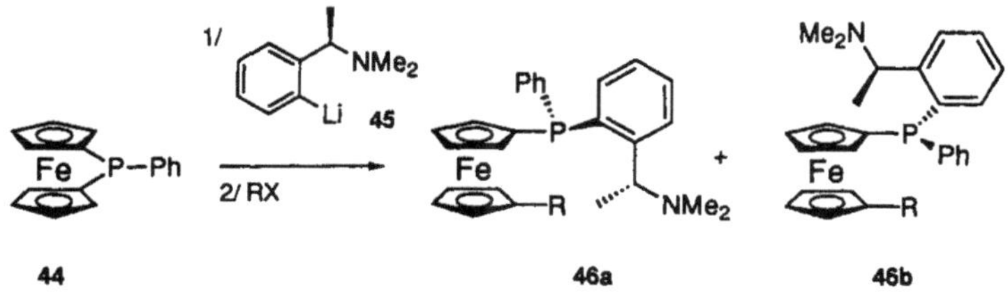

Scheme 20.

formed into various ligands such as **42–43** bearing both planar and lateral chirality.

Only one example is known of a ferrocenylphosphine bearing chirality at phosphorus.[46] Diastereoselective opening of the ferrocenophane **44** with the ortholithio derivative of (*R*)-*N,N*-dimethyl-α-phenylethylamine (**45**) followed by electrophilic trapping gave a mixture of the two diastereoisomeric phosphines **46a-b** (Scheme 20). The configuration at phosphorus was determined by X-ray analysis after separation of the diastereomers.

B. Biochemical Methods

Various ferrocenyl ketones have been transformed into the corresponding chiral alcohols by microbial reduction (Scheme 21). Reduction of ferrocenyltrifluoromethyl ketone (**47**)[47] and ferrocenophane-1-one (**48**)[48] by baker's yeast occurred in good yields but the enantioselectivities were not given. Stereospecific reduction with baker's yeast of ferrocene-1-D-aldehyde **49** into hydroxy-1-D-methylferrocene has also been described.[49] More recently, Yamasaki and

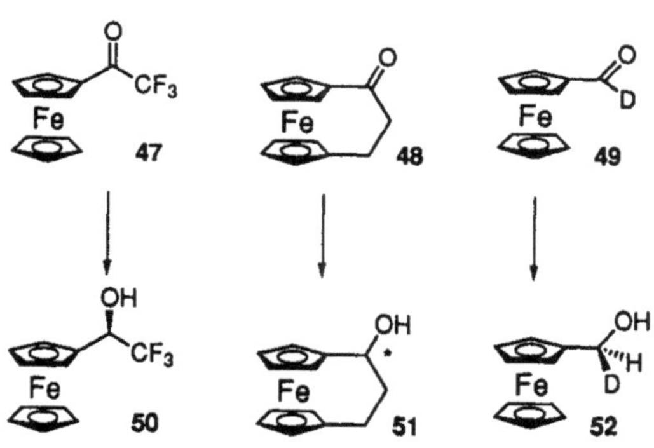

Scheme 21.

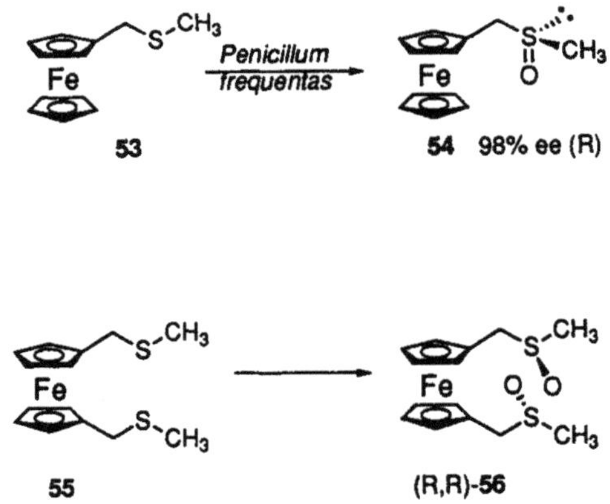

Scheme 22.

Hosono[50] reduced acetylferrocene into carbinol **11a** with high ee (83–96%) using various species of yeasts and bacterias.

So far, only a few cases of asymmetric oxidation using biochemical methods are known.[51] Asymmetric oxidation of prochiral ferrocenyl sulfides has been studied using various microbial and yeasts cultures (Scheme 22) and either the (*R*)-(–) (ee up to 98%) or (*S*)-(+) (ee up to 69%) methylsulfinylmethyl)ferrocene (**54**) could be selectively produced. The C_2-symmetric bissulfoxide (*R*,*R*)-**56** [97% ee isolated as a 3:4 (*R*,*R*)/(*R*,*S*) mixture] could also be prepared starting from the disulfide **55**.[52]

III. ASYMMETRIC SYNTHESIS OF FERROCENES WITH PLANAR CHIRALITY

A. Introduction

The creation of planar chirality starting from an achiral ferrocenyl compound needs the regioselective introduction of a group in an achiral ferrocene. The functionalization must remove all the S_n symmetry elements of the starting material. A good way to avoid creating a mixture of regioisomers is to start from a monosubstituted ferrocene having an *ortho*-directing group for metallation. Such a process is schematized in Scheme 23. In order to control the creation of planar chirality in a given absolute configuration it is necessary to use an external chirality, such as a chiral base. This process is called enantioselective functionalization because it directly gives the desired enantiomer. Another possibility is to com-

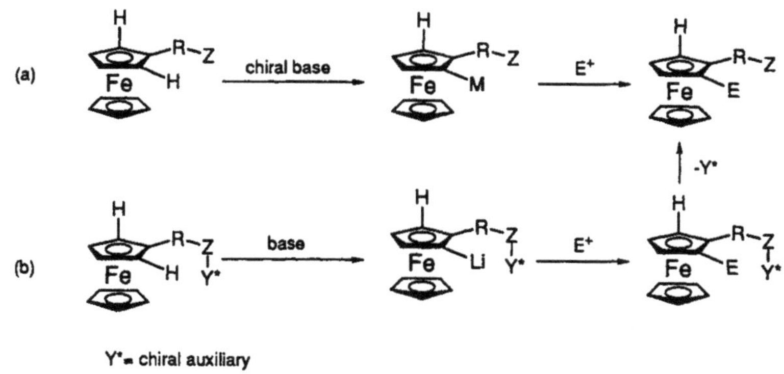

Scheme 23.

bine temporarily a chiral auxiliary to the *ortho*-directing group. The reaction will be called diastereoselective functionalization since a mixture of diastereomers is the initial product, before the removal of the chiral auxiliary. Both routes are known and are described below.

B. Enantioselective Functionalization

Chemical Methods

The first enantioselective deprotonation of a monosubstituted ferrocene has been carried out by Nozaki et al.[53] These authors used 2.5 mol.equiv of the combination, (–)-sparteine / *n*-BuLi = 1:1, for the dilithiation of isopropylferrocene (Scheme 24) into **57**. Further treatment by $ClSiMe_3$ led to isolation of **58a** in

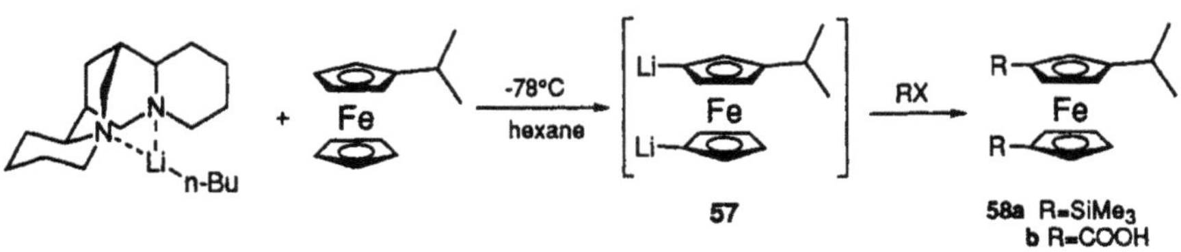

Scheme 24.

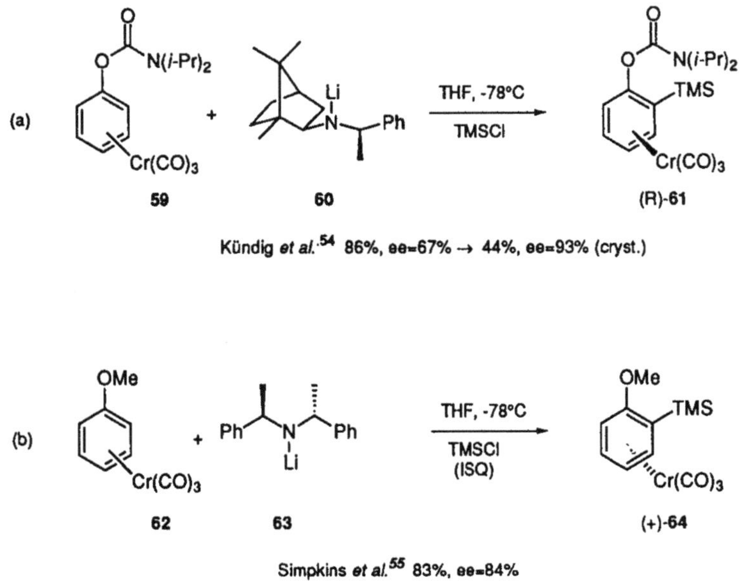

Scheme 25.

good yield with a weak optical activity. Quenching of **57** by CO_2 gave the dicarboxylic acid **58b** with 3% ee.

A similar approach for the enantioselective ortholithiation of arenechromium–tricarbonyl complexes using a chiral base has been recently reported.[54,55] Two examples of this methodology are described in Scheme 25. Deprotonation of both complexes **59** or **62** with chiral lithium amide **60** or **63** followed by electrophilic quenching with chlorotrimethylsilane (stepwise procedure or in situ quenching (ISQ) procedure) give the *ortho*-substituted complexes with good regio- and enantioselectivity. It has been shown that the regioselectivity (*ortho* vs. *meta* + *para*) as well as the enantioselectivity was highly dependent on the direct-

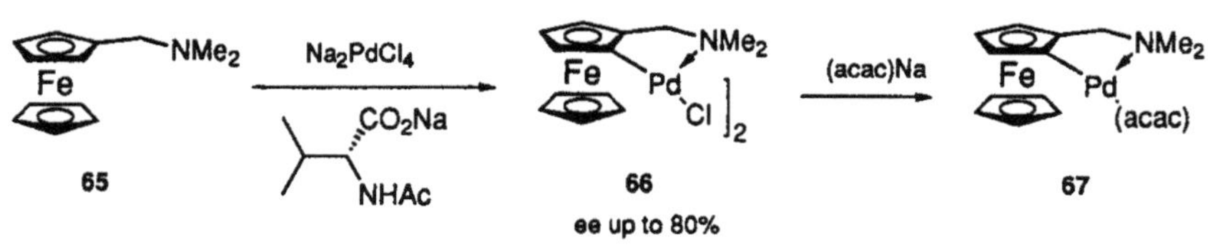

Scheme 26.

ing group (alkoxy, acyloxy, acetal, chloro, fluoro, carboxamide) present on the aromatic ring of the starting chromium complex. This concept of combining the effect of an *ortho*-directing group and a chiral lithium base remains to be studied in the case of ferrocene.

Another approach using the *ortho*-directing group methodology was devised by Sokolov and Troitskaya.[56] The tertiary amine **65** (Scheme 26) was orthopalladated by Na_2PdCl_4 in methanol in the presence of 1 mol equiv of the sodium salt of a chiral acid. The best results were obtained with the sodium salt of *N*-acetyl valine. The palladium complexes **66** or **67** were prepared with enantiomeric excesses up to 60–68% (ee up to 80% after crystallization). This approach was used to synthesize optically active analogs of prostaglandins[19a] (Scheme 27). Complex **66** has been transformed into various functionalized chiral ferrocenes (Scheme 28) and can be used either as a nucleophile (Heck reaction or carbonylation[57]) or as an electrophile (reaction with lithium diphenylphosphide giving a chiral aminophosphine **67**[58]).

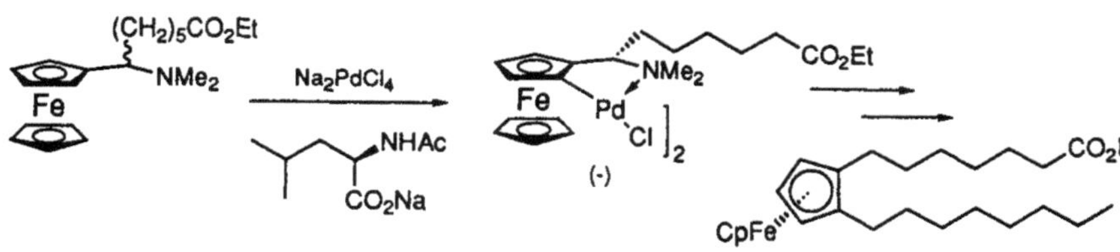

Scheme 27.

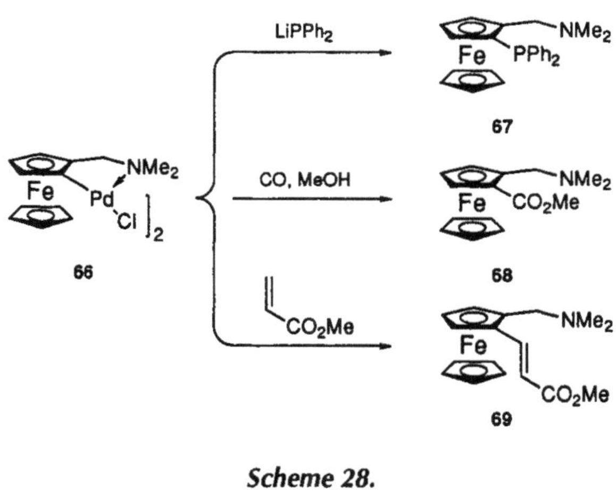

Scheme 28.

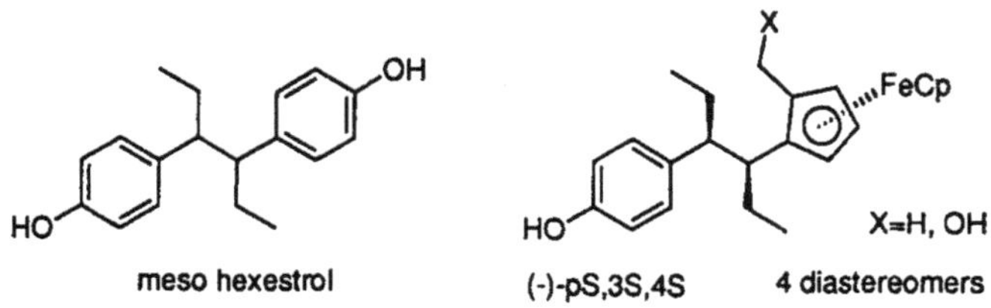

Scheme 29.

More recently, some chiral ferrocenyl analogs of hexestrol and estradiol have been prepared (Scheme 29) starting from complex **66** and their relative binding affinity toward estradiol receptors were explored.[59] The palladium complex **66** has also been used for desymmetrization of a *meso*-cymantrene diol through a carbonylation reaction with de up to 35% (Scheme 30).[60]

Biochemical Methods

Nicolosi et al. recently investigated the use of lipase in organic medium for the asymmetric synthesis of 1,2-disubstituted ferrocenes.[61] Interesting results were obtained in the monoacetylation by vinyl acetate of diol **70** (Scheme 31). The absolute configurations of the products were assigned by a chemical correlation to **72**, further transformed into the known 1-formyl-2-hydroxymethyl ferrocene

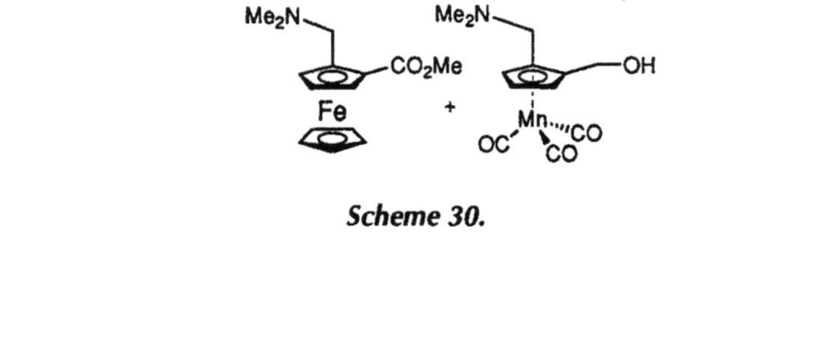

Scheme 30.

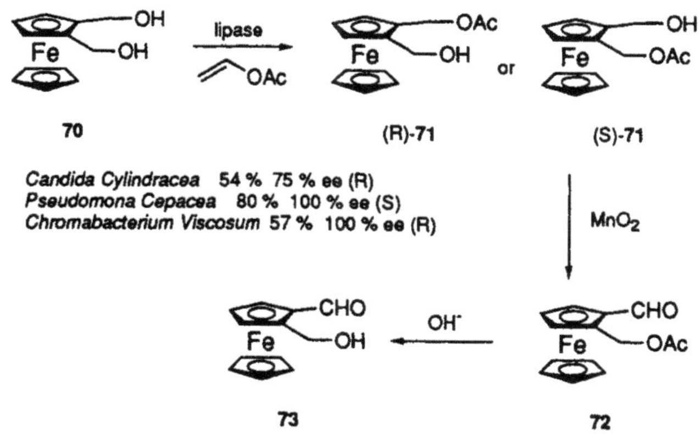

Scheme 31.

73. By this route many 1,2-disubstituted ferrocenes should be available with high enantiomeric excess.

Biooxidations are an additional way to prepare ferrocenes with planar chirality. The asymmetric synthesis of monosulfoxides (R_{Fc},S_S)-**75a** and (R_{Fc},R_S)-**75b** (as a 4/1 mixture) from bissulfide **74** has been achieved by some microorganisms (Scheme 32).[52] Although the oxidation occurred with a rather modest enantioselection on sulfur, it was shown that the selection between enantiotopic side chain was high and led to a single planar chirality for both sulfoxides.

The monooxidation of diol **70** catalyzed by horse liver alcohol deshydrogenase (HLADH) provided hydroxyaldehyde (+)-**76a** with good enantioselectivity (Scheme 33a), while the reduction of the corresponding bisaldehyde **77** gave the enantiomer of the hydroxyaldehyde (−)-**76b** (Scheme 33b).[62,63] Those reactions were studied in the case of the 1,3-disubstituted diol **78** and bisaldehyde **80**, thus providing one of the very rare examples of asymmetric synthesis of 1,3-disubstituted ferrocenyl complexes (Scheme 34). Unfortunately, in those cases, the enantioselectivities for the monooxidation and reduction were low (≤ 10% ee).

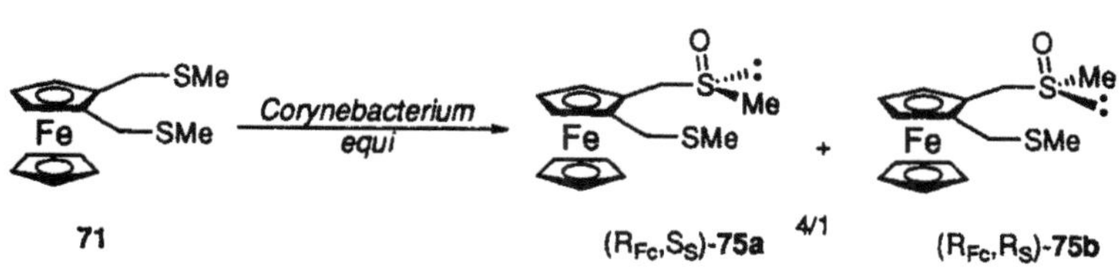

Scheme 32.

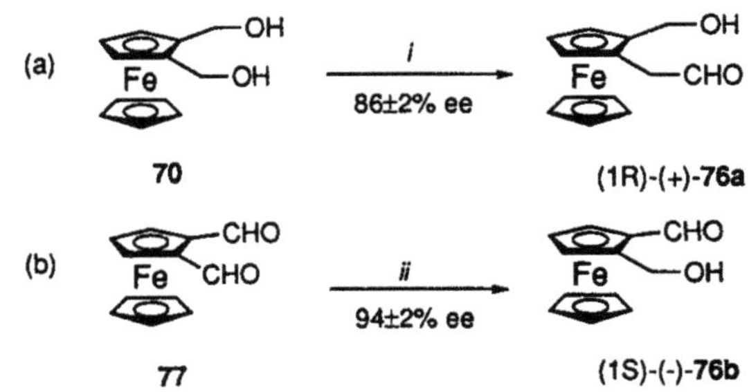

i: HLADH, NAD, 2-oxoglutarate, NH_4^+, glutamate deshydrogenase
ii: HLADH, NADH, ethanol

Scheme 33.

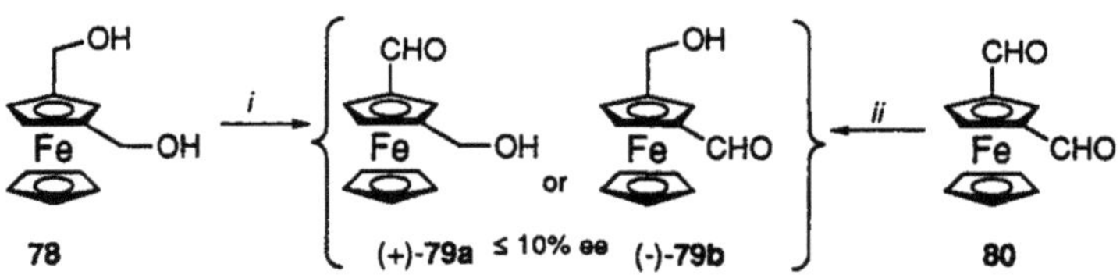

i: HLADH, NAD, 2-oxoglutarate, NH_4^+, glutamate deshydrogenase
ii: HLADH, NADH, ethanol

Scheme 34.

In conclusion biochemical methods are very promising for the enantioselective synthesis of ferrocenes with planar chirality.

C. Diastereoselective Functionalization

Ortholithiation

Tertiary Amines as Directing Groups. Ortholithiation of chiral ferrocenylamine **81** (Scheme 35) was investigated by Nozaki et al. in 1969.[64] These authors assumed that almost exclusive formation of ferrocenyllithium derivative **82** had occurred, which was subsequently transformed through electrophilic trap-

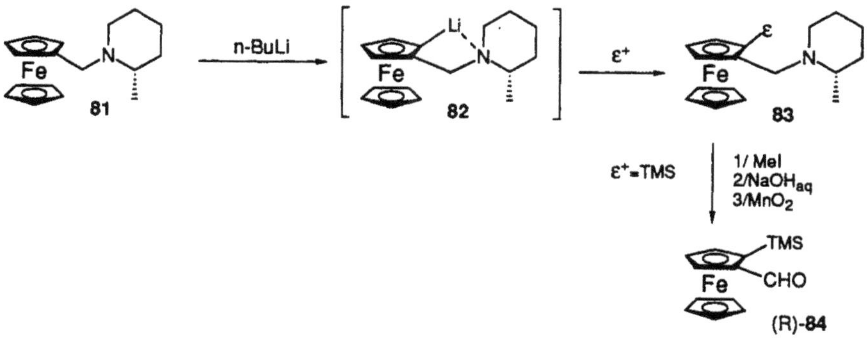

Scheme 35.

ping into **83**. The chiral amine auxiliary could then be easily removed by quaternization of the amine followed by hydrolysis of the quaternary ammonium salt. Oxidation of the resulting alcohol provided the chiral aldehyde **84** which was believed to be enantiomerically pure. However, Ugi et al. demonstrated that **84** was not enantiopure as expected from a fully diastereoselective deprotonation of **81**, but was of 67% ee.[65] However, it should be noted that, to our knowledge, this reaction is the very first example of a diastereoselective metallation on a metallocene ring using a temporary chiral auxiliary.

Ugi et al. devised a very useful method (Scheme 36) to produce ferrocenes with planar chirality.[66] The authors started from the α-ferrocenyl tertiary amine (*R*)-**9**, easily prepared by resolution of the racemic amine with tartaric acid.[9b] Metallation with *n*-butyllithium followed by treatment with an electrophile produced compounds **86** as a diastereomeric mixture (96:4, i.e. a 92% de) (only the

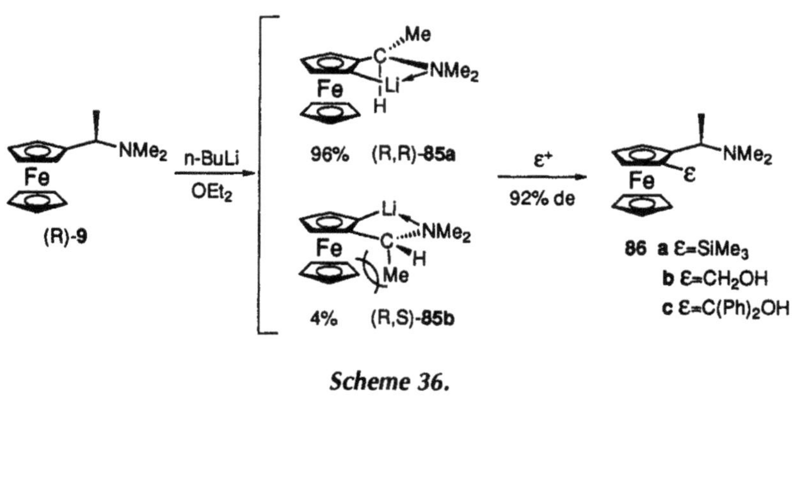

Scheme 36.

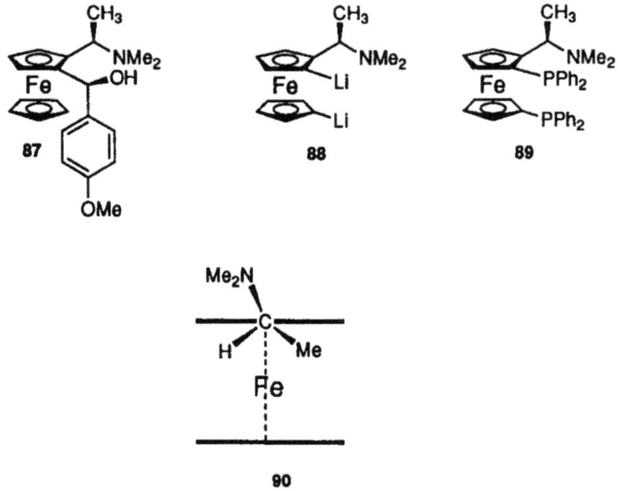

Scheme 37.

major diastereoisomer is represented in Scheme 36). The stereochemistry of the major product **86** was firmly established through a single crystal X-ray structure determination of **87**.[67] The favored conformation of ferrocenyllithium **85a** was interpreted by a preferred transition state where the C–methyl group stays away from the upper ring of the ferrocene moiety.

This interpretation was questioned by Cullen et al.[68] who studied the conformational analysis of the *N,N*-dimethyl aminoethyl group connected to various ferrocene frameworks (Scheme 37). The X-ray structure determination of the TMEDA adduct of dilithiocompound **88**[69] and ^{1}H NMR study of **89** (using NOE) gave indications that in solution conformation **90** is preferred, where the C-methyl group lies between the two rings while $-NMe_2$ is outside the rings. The dilithio compound **88** is formed from the etheral solution of monolithio derivative **85** by addition of a second molar equivalent of *n*-butyllithium and a molar equivalent of TMEDA. It opens the route to many trisubstituted ferrocenes with planar chirality.

The Ugi method has been widely exploited for the preparation of chiral ligands for asymmetric catalysis, mainly mono- and diphosphines. Some examples are given in Scheme 38. For some additional examples on ferrocene based chiral ligands, see refs. 11–17.

Sulfoxides as Directing Groups. Sulfoxides easily coordinate to alkali earth, and are known as good *ortho*-directing groups in aromatic chemistry. Recently the sulfoxides prepared by asymmetric synthesis such as **26a** and **26b** (Schemes

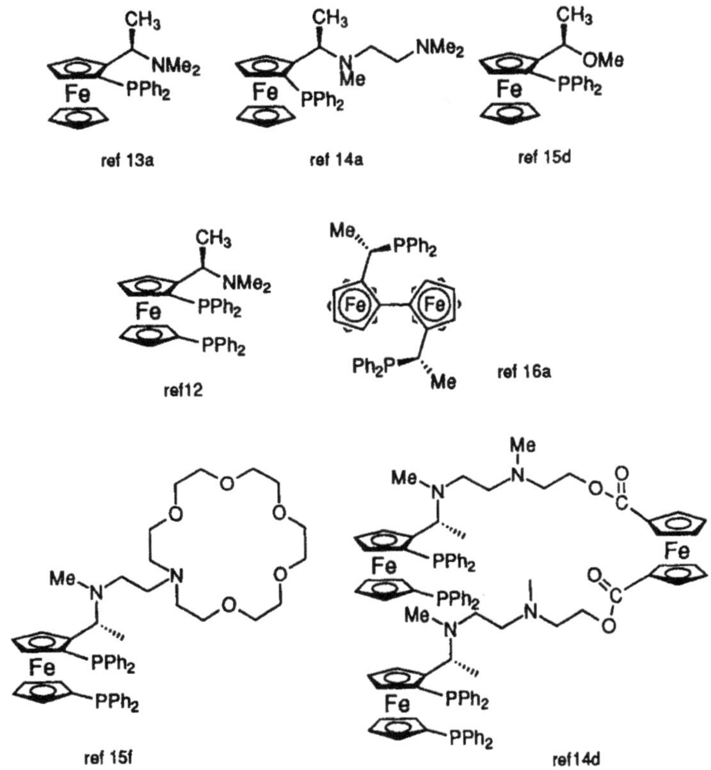

Scheme 38.

10, 11) were transformed into 1,2-disubstituted ferrocenes with a high diastereoselectivity.[34] Some examples are presented in Schemes 39 and 40. Metallation of **26a** into **91a-b** occurred with high diastereoselectivity, giving after electrophilic trapping the sulfoxides **92** with de = 96% (Scheme 39). The stereochemistry was established by a single crystal X-ray structure on **92** (R_1 = Me). Oxidation into sulfone **93** and further metallation allowed functionalization of the second *ortho*-position,[34,70] yielding enantiopure 1,2,3-trisubstituted ferrocene complex **94**. A second lithiation with an excess of *n*-butyl lithium followed by electrophilic trapping with chlorodiphenylphosphine gave the 1,2,1′-trisubstituted diphosphine **95**. Sulfoxide **26b**, easily prepared by the Andersen method, also gave a fully diastereoselective ortholithiation (Scheme 40), but in this case LDA is the preferred base.[34]

Chiral Acetal as Directing Group. A general asymmetric synthesis of ferrocenes carboxaldehydes with planar chirality **101** was recently published by the

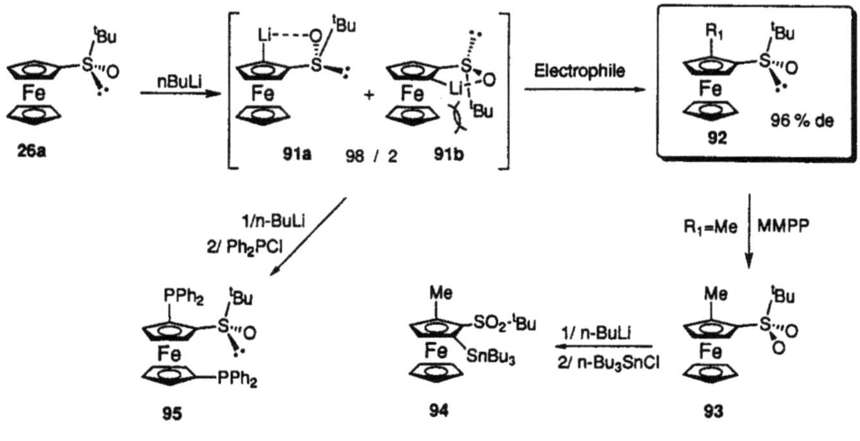

Scheme 39.

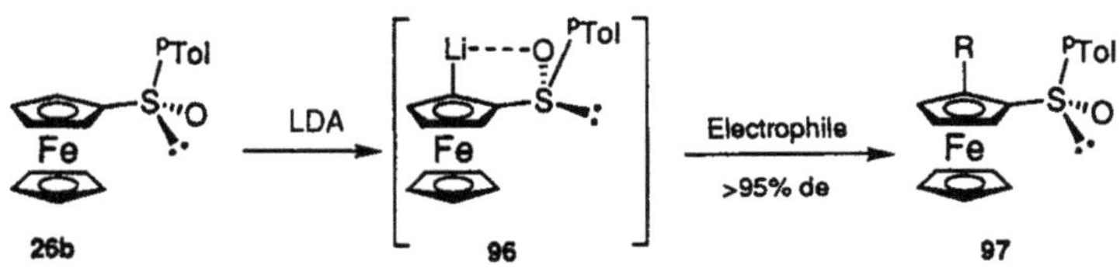

Scheme 40.

authors of this present review.[71] This method is based on the transformation of commercially available ferrocene carboxaldehyde into chiral acetal **98** (Scheme 41). The ortholithiation in diethyl ether with *t*-BuLi is fully diastereoselective, as judged by the very high diastereomeric excess of the *ortho*-substituted acetals **100** (R = $SiMe_3$, PPh_2, CO_2Me, etc.). Acid-catalyzed hydrolysis allowed the recovery of enantiopure aldehydes **101** in good yields. The stereochemistry of the orthodeprotonation was established by chemical correlation. The specific rotation of **101** (R = $SiMe_3$ or R = CO_2Me) was compared to the previous data of those compounds of known absolute configuration.[72,73] The comparison gave the absolute configuration as indicated in **100**, and confirmed that the ee's were close to 100%.

From aldehydes **101** it was possible to prepare many ferrocenes with planar chirality by chemical manipulations of the aldehydic function (oxidation, reduction, pinacolization) or by transformation of the R group. A few examples of

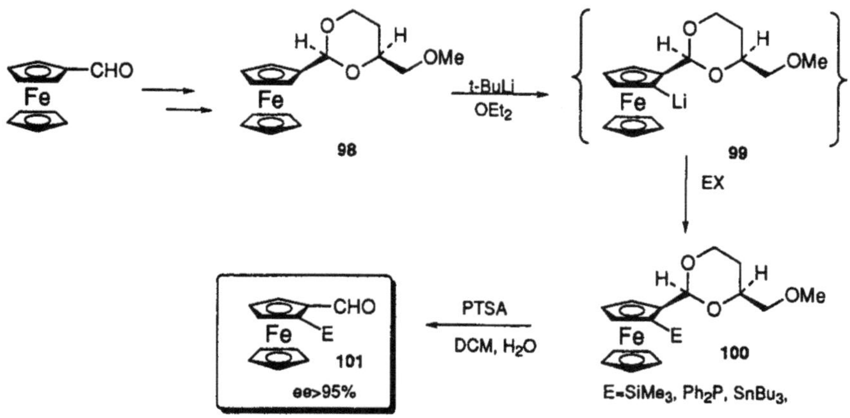

Scheme 41.

transformation of the key iodoaldehyde (*S*)-**102** are depicted in Scheme 42. Palladium coupling reactions with C–C bond formation (Heck reaction, Suzuki reaction, etc.) are easily performed with the iodoaldehyde **102** to give an entry to a wide range of enantiopure *ortho*-alkenyl, alkynyl, and aryl ferrocene carboxaldehydes **103**–**105**.[74]

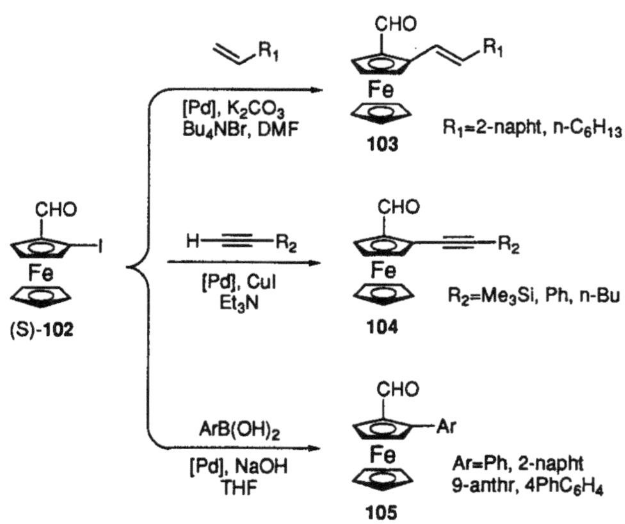

Scheme 42.

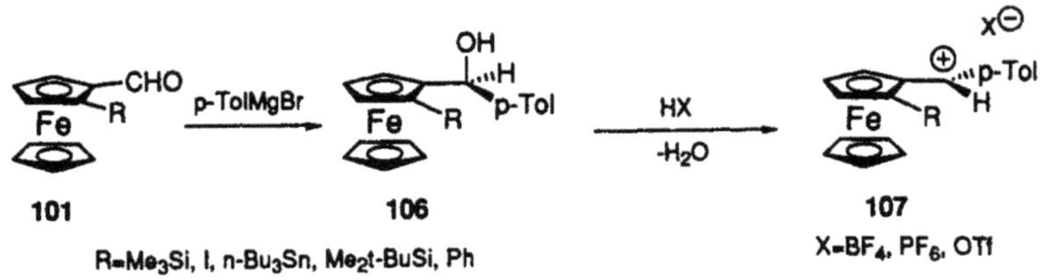

Scheme 43.

A first application of this methodology for the preparation of new chiral catalysts has also been recently described (Scheme 43).[75] Reaction of various *ortho*-functionalized ferrocene aldehydes **101** with *p*-tolylmagnesium bromide gave the carbinols **106** which are in turn transformed to the ferrocenyl-stabilized carbenium salts **107** after treatment with strong acids and precipitation from diethyl ether. Those salts which present both planar chirality and a stereochemically stable chiral carbenium center were isolated and characterized by spectroscopic methods. Their potential as chiral Lewis acid catalyst has also been investigated.[75]

Oxazolines as Directing Groups. The orthodeprotonation of an achiral ferrocenyl oxazoline has been reported by Ebertz et al. in 1982.[76] The asymmetric modification of this reaction has recently been simultaneously described by Richards,[77] Sammakia,[78] and Uemura.[79] Those oxazolines **108** are easily prepared starting from a chiral amino alcohol and either ferrocene carboxylic acid or cyanoferrocene (Scheme 44). It has been shown that the diastereoisomeric excesses for **109a-b** strongly relied upon the deprotonation system and the R

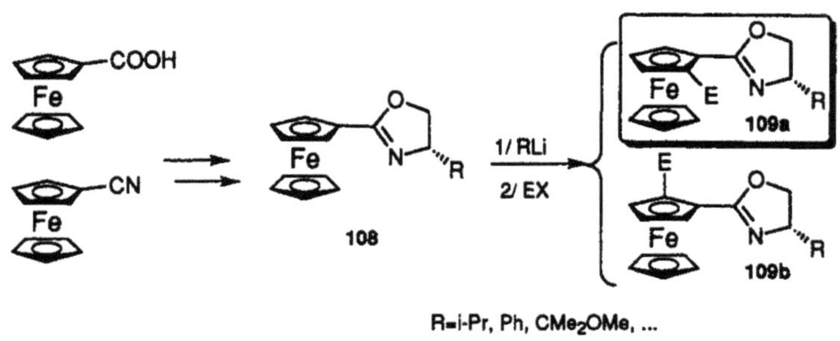

Scheme 44.

Table 1. Oxazolines as Directing Group

Entry	*R*	*Conditions*[a]	*E*	*Yield*	*de or ratio*	*Ref.*
1	*i*-Pr	A	$-PPh_2$	64	2.5:1	77
2	*t*-Bu	B	$-SiMe_3$	b	36:1	78
3	*i*-Pr	B	$-SiMe_3$	b	8:1	78
4	Ph	B	$-SiMe_3$	b	6:1	78
5	CMe_2OMe	B	$-SiMe_3$	b	19:1	78
6	*i*-Pr	C	$-PPh_2$	77	95%	79
7	*t*-Bu	C	$-PPh_2$	51	84%	79
8	Ph	C	$-PPh_2$	55	>99%	79

Notes: [a]Conditions for deprotonation are as follows: (A) *n*-BuLi, diethyl ether, rt (B) 1.2 equiv *s*-BuLi in THF, −78°C, 2 h, then 0°C, 5 min (C) 1.1 equiv of *s*-BuLi in diethyl ether, −78°C, 2 h.
[b]Reported yields are in the 83–96% range in the case of R = *t*-Bu.

group on the chiral center of the oxazoline heterocycle. In Table 1 are shown some representative results.

Depending on the reaction conditions, the best R groups are *t*-Bu, Ph, and *i*-Pr. Temporary protection of one site by a TMS group followed by another ortholithiation and electrophilic capture and further elimination of the TMS group allowed Richards et al. to synthesize the diastereoisomeric phosphine **112** (Scheme 45).[77] Uemura et al. used phosphine **113** (Scheme 46)[79] as a chiral ligand in the rhodium(I)-catalyzed asymmetric hydrosilylation of acetophenone, albeit with modest enantioselection[79] (60% ee). The sense of asymmetric induction was proved

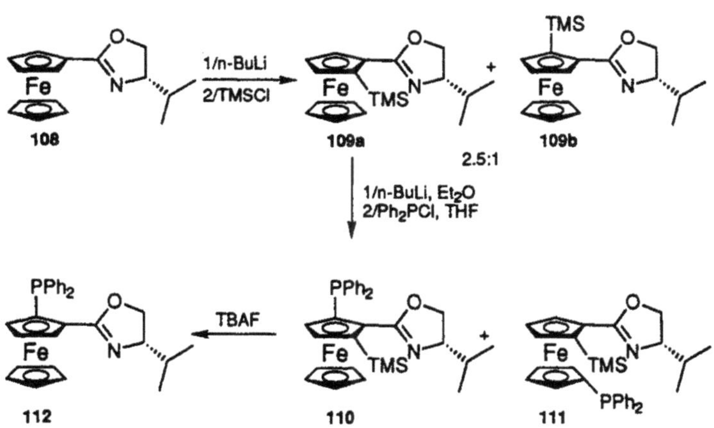

Scheme 45.

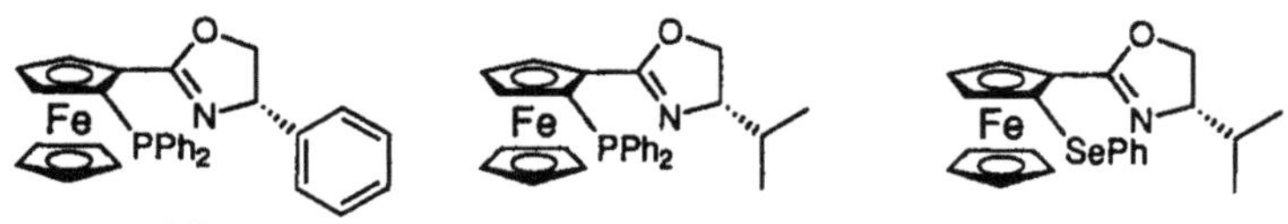

Scheme 46.

in two cases by X-ray crystallography analysis of the *ortho*-functionalized ferrocenyl oxazolines **114**[77] and **115**.[79] An important feature lies in the work of Sammakia et al. who showed that the functionalized oxazolines could be reduced to the corresponding aldehydes (Scheme 47).[78] Thus, oxazoline **116** was alkylated with iodomethane in nitromethane and after reduction and acid-catalyzed hydrolysis of the resulting oxazolidine 1-formyl-2-methyl-ferrocene (*R*)-**117** was isolated. Further transformation to the methyl ketone (*R*)-**118** allowed chemical correlation and attribution of the sense of asymmetric induction for the orthometallation reaction.

Other Metallocenes. Although the scope of this review is limited to chiral ferrocenyl complexes, it should be noted that the concept of diastereoselective metallation with the use of a removable chiral auxiliary has been previously described for other metallocenes, namely cyclopentadienyl–manganesetricarbonyl (cymantrene) and arenechromium–tricarbonyl (benchrotrene) complexes. The very first example was reported by Sokolov et al. for the regio- and diastereoselective lithiation of a chiral cymantrene oxazolidine **119** prepared from cymantrene carboxaldehyde and ephedrine (Scheme 48a).[80] Although the reaction yielded a mixture of *ortho*- and *meta*-substituted complexes, the diastereoselectivity for the orthodeprotonation proved to be quite high (85–90%). Moreover,

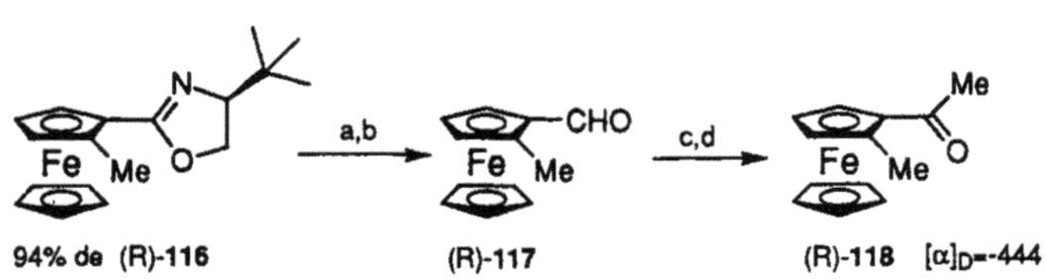

Scheme 47.

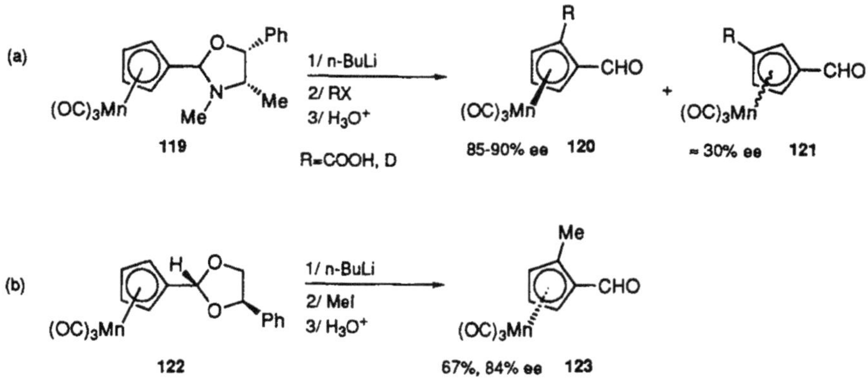

Scheme 48.

a better regioselectivity (with a 84% de) was recently published for the metallation of a chiral acetal **122** derived from phenylethanediol (Scheme 48b).[81] The orthometallation of some chiral 2-cymantreneoxazolines has also been realized, but without diastereoselectivity.[82]

Asymmetric synthesis of arenechromium–tricarbonyl complexes through diastereoselective deprotonation is also known in the case of chiral acetals and aminal derivatives of benzaldehyde such as **124**[83] and **127**[84] (Scheme 49) and acetophenonechromium–tricarbonyl derivatives such as **130**[85] (Scheme 50). It was shown in all cases that the yields and diastereoselectivities were highly dependent of the reaction conditions (base, solvent, and temperature). However, those methods proved to be a straightforward pathway for the production of enantiopure *ortho*-substituted benchrotrene aldehydes and methyl ketone derivatives, and offer an elegant and efficient alternative to the classical resolution.

In the case of acetal derivatives, deprotection of the acetal group for recovery of the chiral aldehydes or ketones proved somewhat difficult or even impossible in some cases. As a comparison, the aminal derivatives could be easily deprotected to the corresponding aldehydes using mild conditions. It should also be noted that the benchrotrene **132** equivalent of Ugi's amine **9** has been used for the creation of planar chirality via diastereoselective ortholithiation[86] (Scheme 51). *ortho*-Functionalized complexes **133** have found application in asymmetric synthesis[87] and for the preparation of new chiral ligands. Diastereoselective orthometallation of racemic benchrotrene sulfoxide **134** has also been recently described.[88]

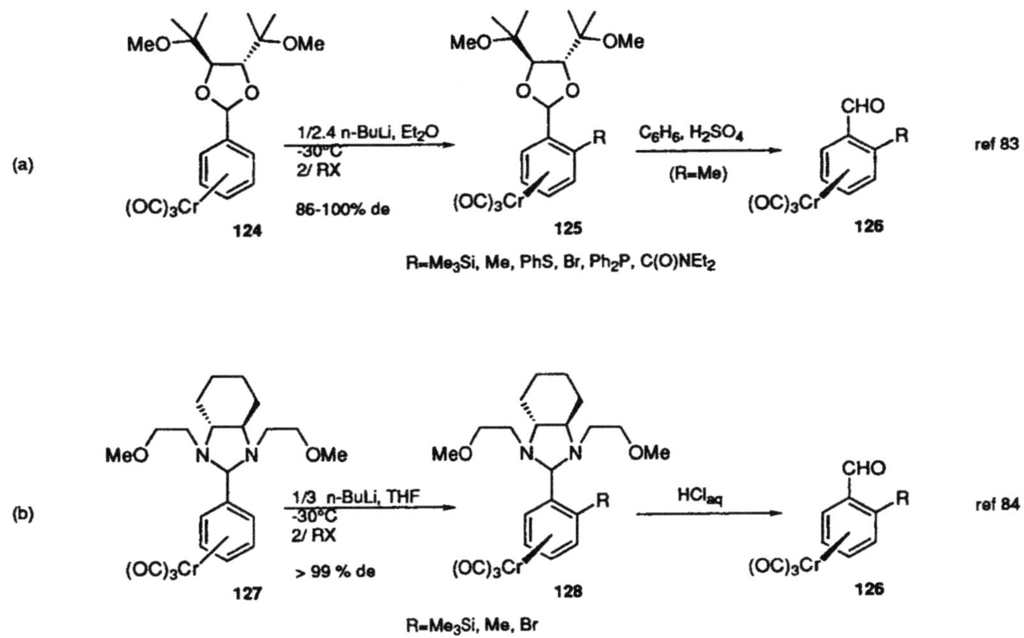
(a)
MeO
OMe
O
O
(OC)3Cr
124
1/2.4 n-BuLi, Et2O
-30°C
2/ RX
86-100% de
R
125
C6H6, H2SO4
(R=Me)
CHO
126
ref 83
R=Me3Si, Me, PhS, Br, Ph2P, C(O)NEt2
(b)
N
127
1/3 n-BuLi, THF
-30°C
2/ RX
> 99 % de
128
HClaq
ref 84
R=Me3Si, Me, Br

Scheme 49.

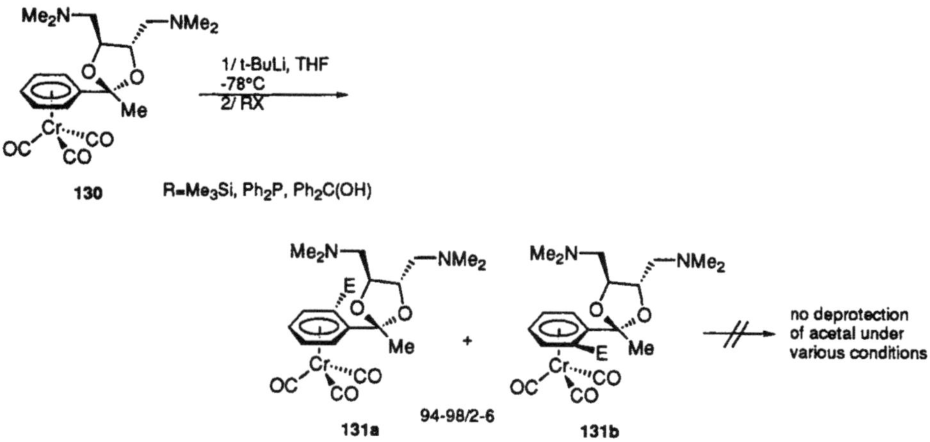

Scheme 50.

Orthopalladation

Orthopalladation of ferrocenylamines has been studied extensively by Sokolov et al.[89] As opposed to ortholithiation, it is here possible to isolate the primary product, a palladium complex with a σ-bond, as described in Section III.B (Scheme 26). The procedure involves sodium tetrachloropalladate and a sodium salt of a carboxylic acid. In the case of the palladation of **9** (Scheme 52), sodium acetate was used as a cocatalyst, as already described for [(*N,N*-dimethyl-amino)methyl]-ferrocene.[90] Complex **135** was obtained and transformed into **136** which is a mixture of diastereomers (85:15). The asymmetric induction for ortho-palladation (70% de) is therefore inferior to that observed for lithiation (92% de), but is in the same direction. The absolute configuration was assigned

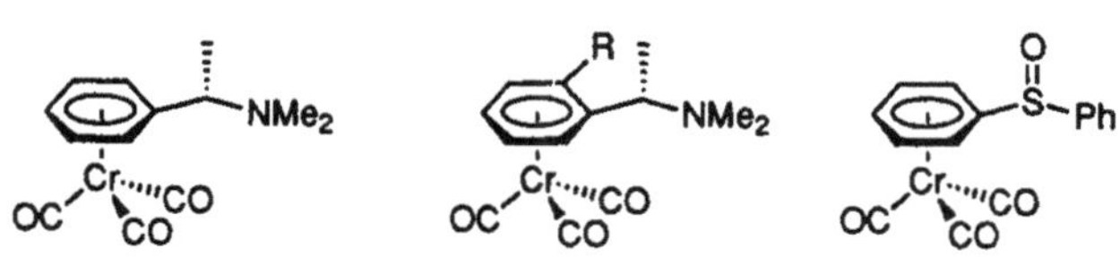

Scheme 51.

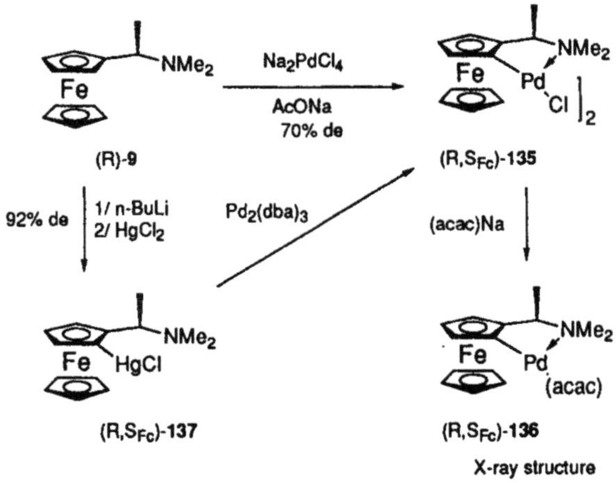

Scheme 52.

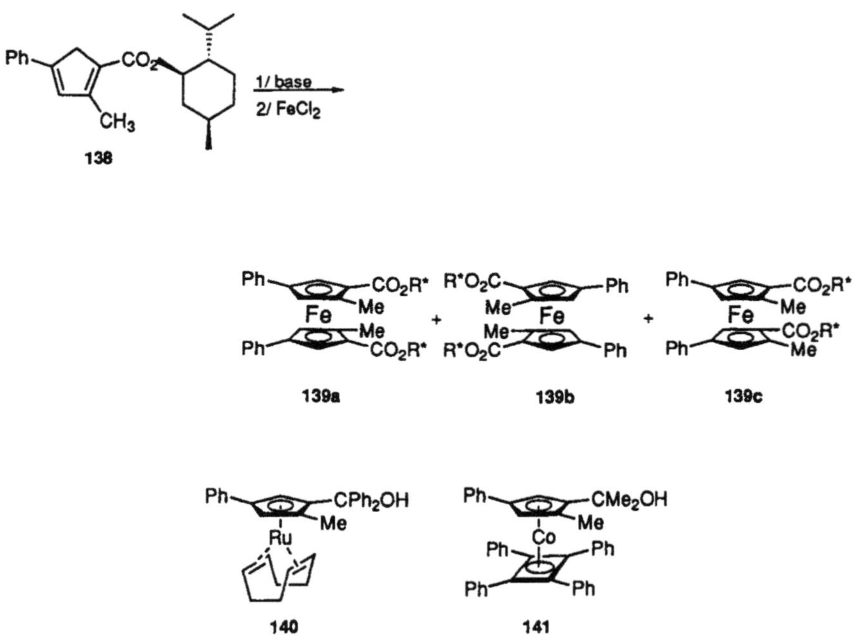

Scheme 53.

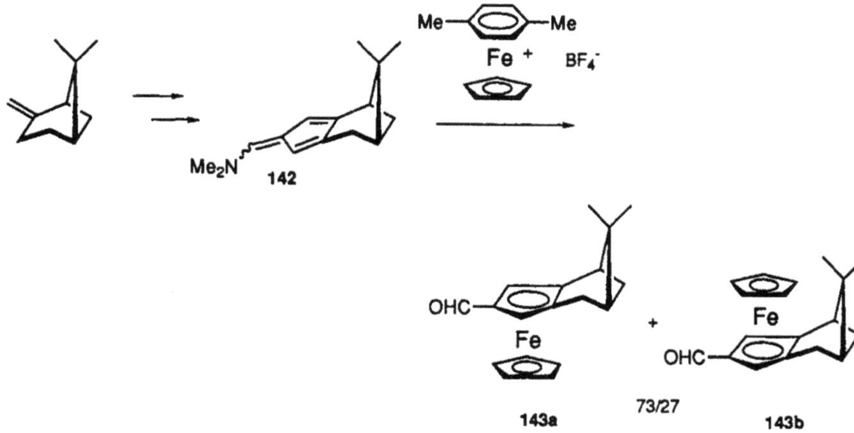

Scheme 54.

by chemical correlation through the mercury complex **137** and independently by a X-ray structure analysis of **136**.[91] Kinetic resolution of racemic amines **9**[92] and 1′-dimethylamino [3]ferrocenophane[93] has also been obtained for the orthopalladation reaction using salts of chiral amino acids.

D. Diastereoselective Formation of the Ferrocene Moiety

An unusual asymmetric synthesis of chiral ferrocenes is the construction of the ferrocenyl system from chiral sodium cyclopentadienide and $FeCl_2$ (Scheme 53). Takahashi et al. described the syntheses of hexa-substituted ferrocenes **139a-c** from menthyl ester **138**.[94] The mixture of the three diastereomers (in relative proportions 2:3:5) was prepared in 67% yield. The product distribution indicates that some asymmetric induction had been realized. One can calculate 20% ee for the chiral ferrocenes **139a-b**.

This method has also been used for the preparation of new ruthenium **140**[94] and cobalt **141**[95] complexes bearing planar chirality. An earlier approach was depicted by Paquette et al.[96] who used a chiral fulvene **142** prepared from β-pinene as a precursor to the ferrocene complex (Scheme 54). Ligand exchange with a cationic $[(Cp)Fe(p\text{-xylidene})]^+$ salt gave two diastereosiomeric ferrocene aldehydes **143a-b** in a 73/27 ratio.

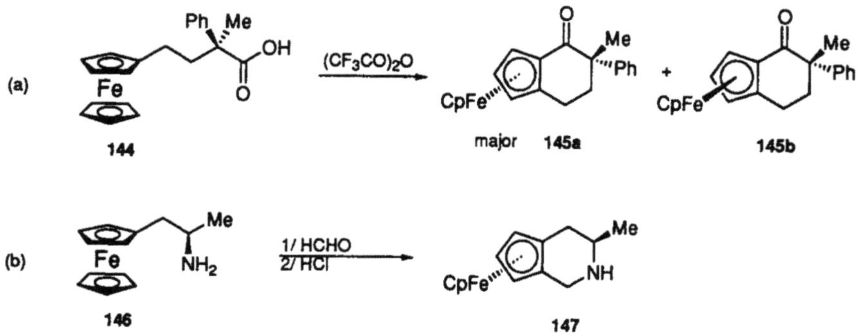

Scheme 55.

E. Miscellaneous Reactions

Intramolecular ring closure on a ferrocene ring with creation of planar chirality has been investigated. Des Abbayes and Dabard described the intramolecular ring closure of a chiral ferrocene butanoic acid **144** by trifluoroacetic anhydride into ketones **145a-b** with modest diastereoselectivity (Scheme 55a).[97] An asymmetric Pictet–Spengler-like cyclization was also reported starting from chiral ferrocenylamine **146** (Scheme 55b). Upon treatment of **146** with formaldehyde and cyclization of the resulting imine under acidic conditions, the planar chiral secondary amine **147** was isolated as a single diastereomer.[98]

Troitskaya et al. prepared some chiral allenes starting from planar chiral ferrocenes (Scheme 56).[99] *ortho*-Substituted ferrocenyl aldehydes **148** were transformed into propargylic ferrocenes **149** in two steps. Complexes **149** were in turn isomerized into the chiral allenes **150** with diastereoisomeric excesses up to 38%.

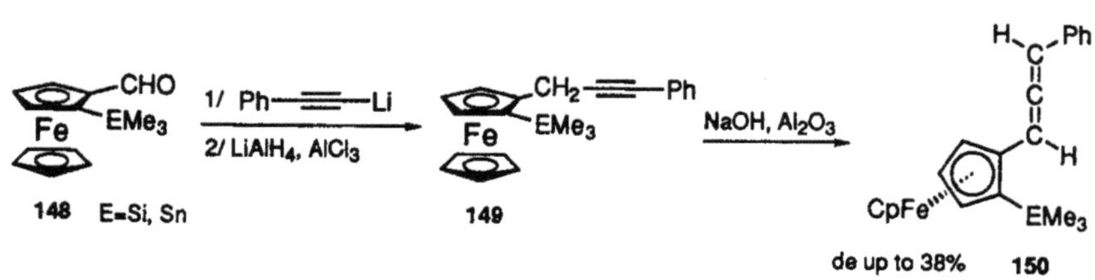

Scheme 56.

IV. KINETIC RESOLUTION OF FERROCENES

A. Lateral Chirality

Enzymatic resolution in organic medium of racemic α-hydroxymethylferrocene (**11a**) (Schemes 57, 58) provided an interesting alternative for the preparation of *N,N*-dimethylaminoethylferrocene (**9**). This has first been described by Wong et al. for a transesterification reaction of the racemic alcohol (±)-**11a** with vinyl propionate using lipases from porcine pancreas (PPL) as a catalyst (Scheme 57).[100] Optimized results (84% ee) were obtained for the alcohol (*S*)-**11a** at 60% conversion. Another resolution of this alcohol has been realized later using lipases of *Pseudomonas fluorescens* (PFL), with 92% ee at 56% conversion (Scheme 58).[101] Efficient separation of the remaining alcohol and ester was obtained by direct reaction of the crude reaction mixture with dimethylamine in methanol. The corresponding acetate was then cleanly converted to the amine **9**

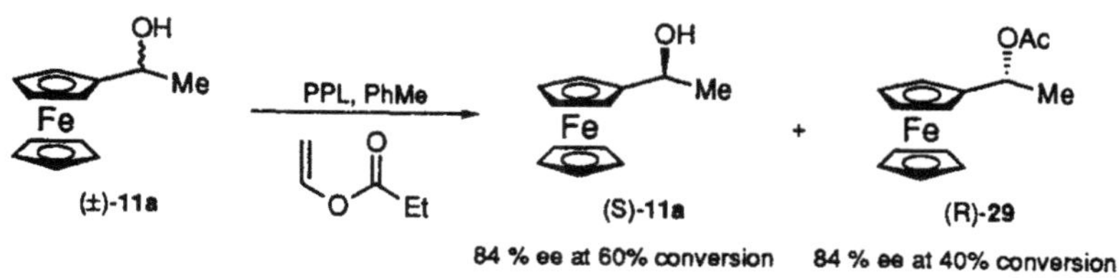

Scheme 57.

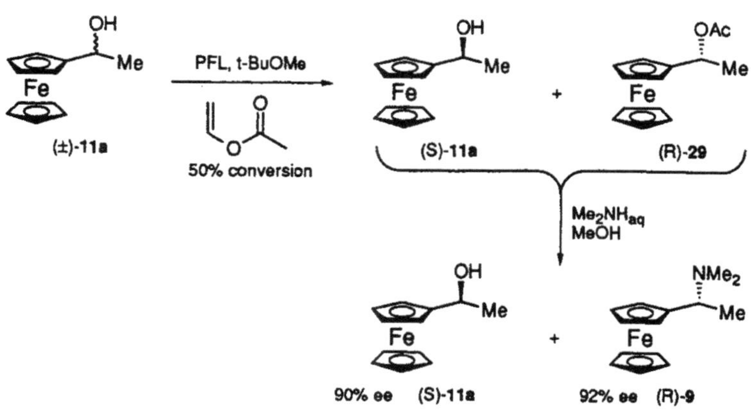

Scheme 58.

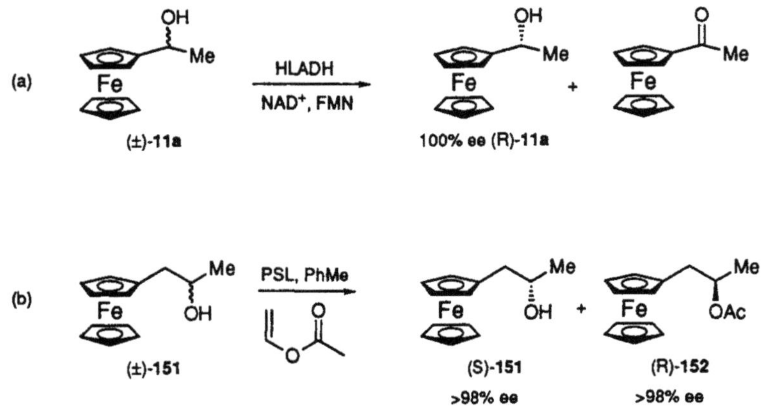

Scheme 59.

with retention of configuration, **9** being isolated from the alcohol **11a** by simple acid/base extraction.

Biochemical oxidation of racemic alcohols such as **11a** also provides an entry to optically active ferrocenylcarbinols (Scheme 59).[63,102] Thus, use of the combination of horse liver alcohol dehydrogenase (HLADH) and cofactors NAD^+ and FMN yielded 1-hydroxyethylferrocene (*R*)-**11a** in 44% yield and 100% ee (Scheme 59a). This reaction has also been studied in the case of the ruthenocene and osmocene analogs with very good enantioselectivities for each case (92% ee). Only one example of resolution has so far been reported for a ferrocene com-

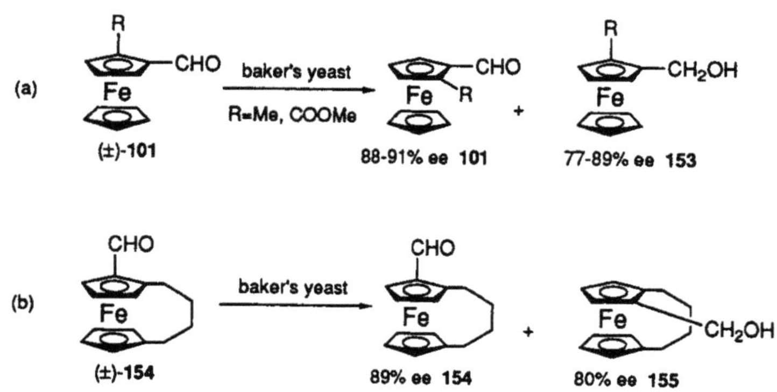

Scheme 60.

plex bearing chirality on a β-position (Scheme 59b).[103] Thus, it has been shown that *Pseudomonas* lipases selectively acylated one enantiomer of alcohol **151** (ee > 98%) and selectively hydrolyzed one enantiomer of acetate **152** (ee > 98%).

B. Planar Chirality

As for lateral chirality, the only case of kinetic resolution of a racemic ferrocenyl compound bearing planar chirality deals with biochemical methods. Baker's yeast has thus been shown to selectively reduce a racemic mixture of aldehydes **101** (**154**) to yield the enantiomerically enriched alcohols **153** (**155**) (ee = 77–89%) and the remaining aldehydes **101a-b** (ee = 88–91%) (Scheme 60).[104] Similarly, asymmetric oxidation of racemic alcohol (±)-**153** by HLADH/NAD/2-oxoglutarate/NH_4^+/glutamate dehydrogenase gave the aldehyde **101** with modest enantioselectivity (64% at 50% conversion) (Scheme 61a).[63] However, low enantioselectivity (ee = 18% at 50% conversion) was observed for the alcohol **153** upon reduction of the racemic aldehyde (±)-**101** using HLADH/NADH/EtOH

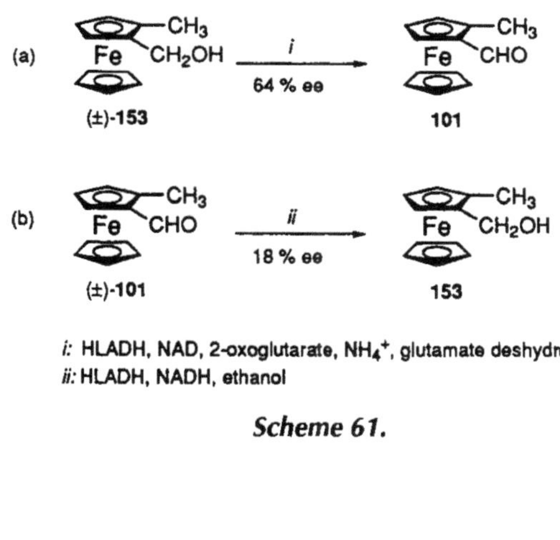

i: HLADH, NAD, 2-oxoglutarate, NH_4^+, glutamate deshydrogenase
ii: HLADH, NADH, ethanol

Scheme 61.

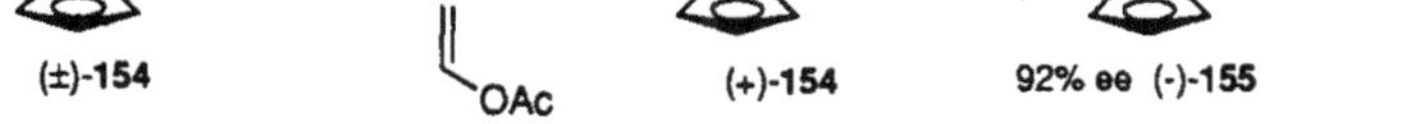

CCL= *Candida cylindracea* lipase

Scheme 62.

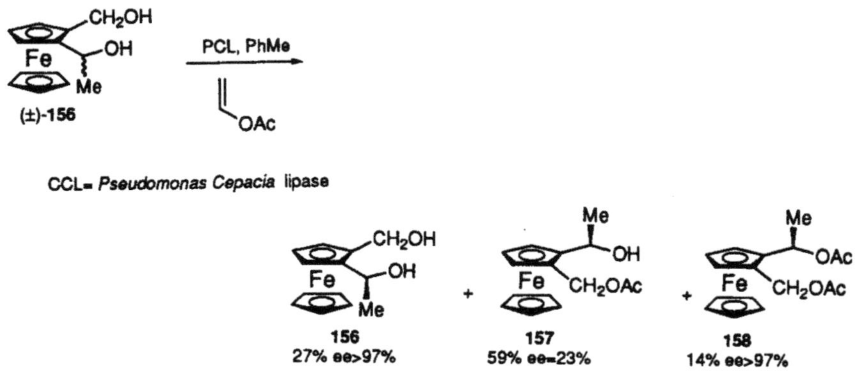

Scheme 63.

(Scheme 61b). It has to be noted that this reaction gave full enantioselectivity in the case of the cymantrene and benchrotrene analogs.[105]

Lipase-catalyzed transesterification of racemic 1-hydroxymethyl-2-dimethylaminomethylferrocene [(±)-**154**] furnished the acetate (–)-**155** with ee up to 92% (Scheme 62).[106] Enzymatic resolution of ferrocenyl complexes bearing both central and planar chirality have finally been investigated (Scheme 63).[107] Enzymatic acylation of the racemate (±)-**156** with vinyl acetate using immobilized *Pseudomonas cepacia* lipase gave both the recovered diol **156** and the diacetate **158** with ee > 97%. The monoacetate **157** could also be isolated at lower conversion with a good enantiomeric excess (ee = 79%).

V. CONCLUSION

Many chiral ferrocenes are now available in high enantiomeric excesses and are finding useful applications. Besides classical resolution, biochemical resolution is a very efficient way to prepare various types of enantiopure ferrocenes. Enantioselective reactions readily introduce a lateral chirality but are still inefficient (apart from some biochemical reactions) for the large-scale synthesis of ferrocene with planar chirality. In the recent past, general highly diastereoselective reactions using the chiral *ortho*-directing group methodology were developed and allowed access to a wide variety of ferrocenes with planar chirality.

This account was focused on the ferrocene family, but the main strategies presented here have been applied to other metallocenes. Chiral metallocenes on the whole is a very rich area in stereochemistry which is finding more and more

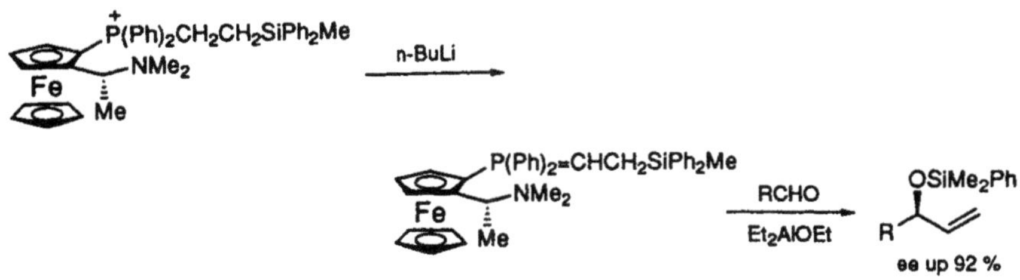

Scheme 64.

applications in stereoselective organic synthesis, either in asymmetric catalysis or in material sciences.

Apart from asymmetric catalysis, some ferrocenyl complexes have been used as chiral auxiliaries in stoichiometric asymmetric synthesis. The first developments in this area concern ferrocenes with a lateral chirality (those complexes have been extensively used in the asymmetric four-component condensation by Ugi et al. (see Scheme 16) and have been recently reviewed.[108] The use of planar chiral ferrocenes as chiral auxiliaries in asymmetric synthesis is a recent field of investigation and few examples have been reported. Enantioselective vinylation of aldehydes by a chiral ferrocenyl β-silylphosphorus ylide was the first example reported in 1991 with high enantioselectivities (Scheme 64).[109] More recently, some chiral ferrocenyl selenium reagents have been prepared and used in some asymmetric reactions, including methoxyselenation of alkenes (Scheme 65)[110] and [2,3]sigmatropic rearrangements (Scheme 66).[111]

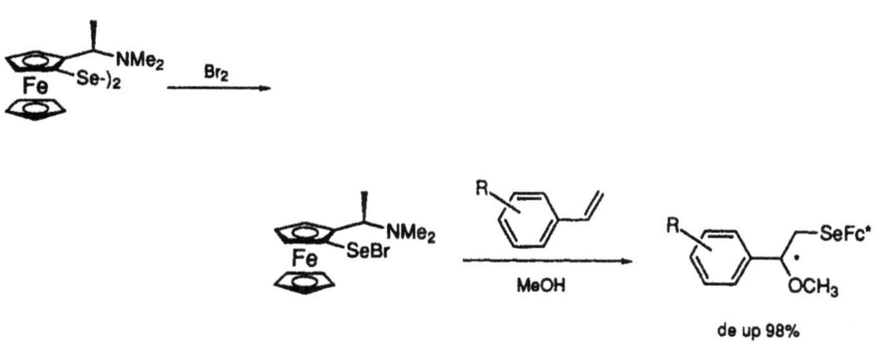

Scheme 65.

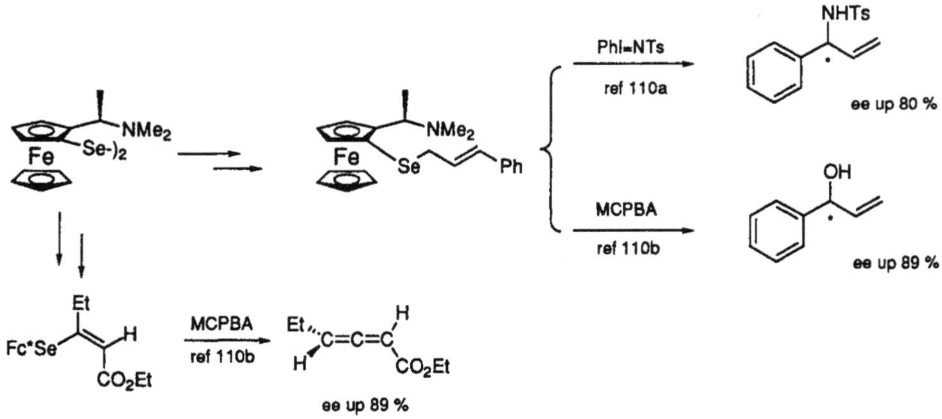

Scheme 66.

NOTE ADDED IN PROOF

After the completion of this manuscript, the first example of enantioselective orthodeprotonation of a ferrocenylphosphine oxide was reported by Price and Simpkins.[112] Similarly, enantioselective deprotonation/silylation of some 1,2-dimethoxybenzenechromium–tricarbonyl complexes was also described by Schmalz and Schellhaas.[113] Enantioselective ortholithiation of *N,N*-diisopropyl ferrocene carboxamide with a *n*-BuLi-sparteine combination was performed by Snieckus et al.[114] Enantioselectivities in the range of 81–98% were reported for various electrophiles. Another example was later published by Uemura et al.[115] for the orthometallation of *N,N*-dimethylaminoferrocene with up to 80% ee. In this later case, *N,N,N′,N′*-tetramethyl-1(*R*),2(*R*)-cyclohexanediamine was the most efficient among the chiral ligands tested. Ferrocenylalkylamines were prepared by the SAMP–hydrazone methodology.[116] Asymmetric alkylation of ferrocene carboxaldehyde and 1,1′-bisformylferrocene with dialkylzinc using a planary chiral ferrocenyl aminoalcohol was described by Watanabe.[117] In a recent publication Knochel et al.[118] applied the CBS–reduction method on 1,1′-ferrocenyl diketones to prepare new C_2-symmetrical ferrocenyl diols. Finally, the synthesis of C_2-symmetrical tetra-substituted ferrocenes by an ortholithiation–electrophilic quenching sequence of chiral 1,1′-bis(oxazolinyl)ferrocenes was realized independently by two groups.[119]

REFERENCES AND NOTES

1. Kealy, T.J.; Pauson, P.J. *Nature* **1951**, *168*, 1039.
2. Wilkinson, G.; Rosenblum, M.; Whiting, M.C.; Woodward, R.B. *J. Am. Chem. Soc.* **1952**, *74*, 2125.
3. Thomson, J.B. *Tetrahedron Lett.* **1959**, *1*, 26.
4. Schlögl, K. *J. Organometal. Chem.* **1986**, *300*, 219.
5. (a) Cahn, R.S.; Ingold, C.; Prelog, V. *Angew. Chem. Int. Ed. Engl.* **1966**, *4*, 385. (b) Prelog, V.; Helmchen, G. *Angew. Chem. Int. Ed. Engl.* **1982**, *21*, 567.
6. Westman, L.; Rinehart Jr., K.L. *Acta Chim. Scand.* **1962**, *16*, 1199.
7. For some reviews on ferrocene chirality see refs. 4, 8.
8. (a) Schlögl, K. *Topics in Stereochem.* **1967**, *1*, 39. (b) *Ferrocenes*; Hayashi, T.; Togni, A., Eds.; VCH: Weinheim, 1995, Chap. 4.
9. (a) Marquarding, D.; Hoffmann, P.; Heitzer, H.; Ugi, I. *J. Am. Chem. Soc.* **1970**, *92*, 1969. (b) Gokel, G.W.; Ugi, I.K. *J. Chem. Ed.*, **1972**, *49*, 294.
10. Marquarding, D.; Klusacek, H.; Gokel, G.; Hoffmann, P.; Ugi, I. *J. Am. Chem. Soc.* **1970**, *92*, 5389.
11. (a) Hayashi, T.; Mise, T.; Fukushima, M.; Kagotani, M.; Nagashima, N.; Hamada, Y.; Matsumoto, A.; Kawakami, S.; Konishi, M.; M.; Yamamoto, K.; Kumada, M. *Bull. Chem. Soc. Jpn.* **1980**, *53*, 1138. (b) Hayashi, T.; Kumada, M. *Acc. Chem. Res.* **1982**, *15*, 395. For a recent review, see, Chap. 2 in ref. 8b.
12. Hayashi, T.; Mise, T.; Mitashi, S.; Yamamoto, K.; Kumada, M. *Tetrahedron Lett.* **1976**, *17*, 1133.
13. (a) Hayashi, T.; Yamamoto, K.; Kumada, M. *Tetrahedron Lett.* **1974**, *15*, 405. (b) Hayashi, T.; Matsumoto, Y.; Morikawa, I.; Ito, Y. *Tetrahedron: Asymmetry* **1990**, *1*, 151.
14. (a) Ito, Y.; Sawamura, M.; Hayashi, T. *J. Am. Chem. Soc.* **1986**, *108*, 6405. (b) Togni, A.; Pastor, S.D. *J. Org. Chem.* **1990**, *55*, 1649. (c) Sawamura, M.; Hamashima, H.; Ito, Y. *J. Org. Chem.* **1990**, *55*, 5935. (d) Pastor, S.D.; Togni, A. *Helv. Chim. Acta* **1991**, *74*, 905.
15. (a) Hayashi, T.; Tajika, M.; Tamao, K.; Kumaeda, M. *J. Am. Chem. Soc.* **1976**, *98*, 3718. (b) Hayashi, T.; Kanehira, K.; Hagihara, T.; Kumada, M. *J. Org. Chem.* **1988**, *53*, 113. (c) Ororafer, M.G.; Wara, D.L.; Brubaker Jr., C.H. *Organometallics* **1988**, *7*, 1504. (d) Hayashi, T.; Hayashizaki, K.; Kiyoi, T.; Ito, Y. *J. Am. Chem. Soc.* **1988**, *110*, 8153. (e) Naiini, A.A.; Lai, C.K.; Waed, D.L.; Brubaker Jr., C.H. *J. Organometal. Chem.* **1990**, *390*, 73. (f) Sawamura, M.; Nagata, H.; Sakamoto, H.; Ito, Y. *J. Am. Chem. Soc.* **1992**, *114*, 2586.
16. (a) Sawamura, M.; Hamashima, H.; Ito, Y. *Tetrahedron: Asymmetry* **1991**, *2*, 593. (b) Sawamura, M.; Hamashima, H.; Ito, Y. *J. Am. Chem. Soc.* **1992**, *114*, 8295. (c) Sawamura, M.; Hamashima, H.; Ito, Y. *Tetrahedron* **1994**, *50*, 4439.
17. (a) For a leading reference, see Watanabe, M.; Araki, S.; Butsukan, Y.; Uemura, M. *J. Org. Chem.* **1991**, *56*, 2218. (b) For a recent review, see Chap. 3 in ref. 8b.
18. (a) Chap. 9 in ref. 8b. (b) Long, N.J. *Angew. Chem., Int. Ed. Engl.* **1995**, *34*, 21.
19. (a) Sokolov, V.I.; Troitskaya, L.L.; Khrushchova, N.S. *J. Organometal. Chem.* **1983**, *250*, 439. (b) Ryabov, A.D. *Angew. Chem. Int. Ed. Engl.* **1991**, *30*, 931.
20. (a) The sodium salt of *o*-ferrocenylbenzoic acid (under the trade name 'Ferrocenon') has been used for many years in Russia as a drug against iron deficiency anemia.
21. Ferricinium salts have been extensively studied for their antineoplastic activity. For some review, see (a) Köpf-Maier, P.; Köpf, H. *Chem. Rev.* **1987**, *87*, 1137. (b) Neuse, E.W. *Makromol. Symp.* **1994**, *80*, 111.
22. Gokel, G.W.; Marquarding, D.; Ugi, I. *J. Org. Chem.* **1972**, *37*, 3052.

23. (a) Fornasier, R.; Reniero, F.; Scrimin, P., Tonellato, U. *J. Org. Chem.* **1985**, *50*, 3209. (b) Sokolov, V.I.; Bondavera, V.L. *Izv. Akad. Nauk. SSSR, Ser. Khim.* **1987**, 460. (c) Kawajiri, Y.; Motohashi, N. *J. Chem. Soc., Chem. Commun.* **1989**, 1336.
24. Nefedova, M.N.; Mamedyarova, I.A.; Petrovski, P.P., Sokolov, V.I. *J. Organometal. Chem.* **1992**, *425*, 125.
25. Wright, J.; Frambes, L.; Reeves, P. *J. Organomet. Chem.* **1994**, *476*, 215.
26. (a) Hayashi, T.; Matsumoto, Y.; Ito, Y. *Tetrahedron: Asymmetry* **1991**, *2*, 601. (b) Zhang, J.; Lou, B.; Guo, G., Dai, L. *J. Org. Chem.* **1991**, *56*, 1670.
27. Noyori, R.; Suga, S.; Kawai, K.; Okada, S.; Kitamura, M.; Oguni, N.; Hayashi, M.; Kaneto, T.; Matsuda, Y. *J. Organometal. Chem.* **1990**, *382*, 19.
28. Matsumoto, Y.; Ohno, A.; Lu, S.-J.; Hayashi, T.; Oguni, N.; Hayashi, M. *Tetrahedron Asymmetry* **1993**, *4*, 1763.
29. Hayashi, T.; Ohno, A.; Lu, S.-J.; Matsumoto, Y.; Fukuyo, E.; Yanagi, K. *J. Am. Chem. Soc.* **1994**, *116*, 4221.
30. Kang, J.; Lee, J.W.; Kim, J.I. *J. Chem. Soc., Chem. Commun.* **1994**, 2009.
31. Soai, K.; Hayase, T.; Takai, K.; Sugiyama, T. *J. Org. Chem.* **1994**, *59*, 7908.
32. Carlström, A.-S.; Frejd, T. *J. Org. Chem.* **1990**, *55*, 4175.
33. Brunner, H.; König, W.; Nuber, B. *Tetrahedron: Asymmetry* **1993**, *4*, 699.
34. Rebiere, F.; Riant, O.; Ricard, L.; Kagan, H.B. *Ang. Chem. Int. Ed. Engl.* **1993**, *32*, 568.
35. Guillaneux, D.; Kagan, H.B. *J. Org. Chem.* **1995**, *60*, 2502.
36. Diter, P.; Samuel, O.; Taudien, S.; Kagan, H.B. *Tetrahedron: Asymmetry* **1994**, *5*, 549.
37. Diter, P.; Taudien, S.; Samuel, O.; Kagan, H.B. *J. Org. Chem.* **1994**, *59*, 370.
38. Davis, D.M.; Kane-Maguire, L.A.P.; Pyne, S.G. *J. Chem. Soc., Chem. Commun.* **1990**, 888.
39. Ugi, I.; Aigner, H.; Beiger, B.; Ben-Efraim, D.; Bughard, H.; Bukall, P.; Erberle, G.; Eckert, H.; Marquarding, D.; Rhen, D.; Urban, R.; Wackerte, L.; Von Zynchlinsky, H. "New Methods for Peptid Synthesis with Organometallic Reagents and Isocyanides," *Pept. Proc. Eur. Pept. Symp. 14th*, 1976, 159–181. *Chem. Abst. 88*, 191359y.
40. (a) Gieren, A.; Kaerlein, C.-P.; Hubner, T.; Herrmann, R.; Sigmüller, F.; Ugi, I. *Tetrahedron* **1986**, *42*, 427. (b) Sigmüller, F.; Herrmann, R.; Ugi, I. *Tetrahedron* **1986**, *42*, 5931. (c) Ketter, A.; Herrmann, R. *J. Organometal. Chem.* **1990**, *386*, 241. (c) Siglmüller, F.; Herrmann, R.; Ugi, I. *Liebigs Ann. Chem.* **1989**, 623.
41. Deus, N.; Robles, D.; Herrmann, R. *J. Organometal. Chem.* **1990**, *386*, 253.
42. Demharter, A.; Ugi, I. *J. Prakt. Chem.* **1993**, *335*, 244. For earlier works on asymmetric four-component condensation using chiral ferrocenyl amines, see: Marquarding, D.; Hoffmann, P.; Heitzer, H.; Ugi, I. *J. Am. Chem. Soc.* **1970**, *92*, 1969.
43. Leblanc, J.C.; Moise, C. *J. Organometal. Chem.* **1976**, *126*, 65.
44. VedsØ, P.; Chauvin, R.; Li, Z.; Bernet, B.; Vasella, A. *Helv. Chim. Acta* **1994**, *77*, 1631.
45. Abbenhuis, H.C.L.; Burckhardt, U.; Gramlich, V.; Togni, A.; Albinati, A.; Müller, B. *Organometallics* **1994**, *13*, 4481.
46. Butler, I.R.; Cullen, W.R.; Rettig, S.J.; White, A.S.C. *J. Organometal. Chem.*, **1995**, *492*, 157.
47. (a) Sokolov, V.I.; Troitskaya, L.L.; Rozhkova, T.I. *Gazz. Chim. Ital.* **1987**, *117*, 525. (b) Yamazaki, Y.; Kobayashi, H. *Tetrahedron: Asymmetry* **1993**, *4*, 1287.
48. Ratajczak, A.; Misterkiewicz, B. *Chem. Ind. (London)* **1976**, 902.
49. Sokolov, V.I.; Troitskaya, L.L.; Reutov, O.A. *Dokl. Akad. Nauk. SSSR* **1977**, *237*, 1376.
50. Yamazaki, Y.; Hosono, K. *Agric. Biol. Chem.* **1988**, *52*, 3239.
51. Yamazaki, Y.; Hosono, K. *Biotechnol. Letters* **1989**, *II*, 627.
52. Yamazaki, Y.; Hosono, K. *Agric. Biol. Chem.* **1990**, *54*, 2183.
53. Aratani, T.; Gonda, T.; Nozaki, H. *Tetrahedron* **1970**, *26*, 5453.

54. Kündig, E.P.; Quattropani, A. *Tetrahedron Lett.* **1994**, *35*, 3497.
55. Price, D.A.; Simpkins, N.S.; McLeod, A.M.; Watts, A.P. *J. Org. Chem.* **1994**, *59*, 1961.
56. Sokolov, V.I.; Troitskaya, L.L. *Chimia* **1978**, *32*, 122.
57. Sokolov, V.I.; Troitskaya, L.L.; Reutov, O.A. *J. Organometal. Chem.* **1979**, *182*, 537.
58. Sokolov, V.I.; Troitskaya, L.L.; Reutov, O.A. *J. Organometal. Chem.*, **1980**, *202*, C58.
59. (a) Gruselle, M.; Malezieux, B.; Troitskaya, L.L.; Sokolov, V.I.; Epstein, L.M.; Shubina, Y.S.; Vaisserman, *Organometallics* **1994**, *13*, 200. (b) Gruselle, M.; Malezieux, B.; Sokolov, V.I.; Troitskaya, L.L. *Inorg. Chim. Acta* **1994**, *222*, 51.
60. Troitskaya, L.L.; Bulygina, L.A.; Sokolov, V.I. *Russ. Chem. Bull.* **1993**, *42*, 1724. See also Troitskaya, L.L.; Sokolov, V.I. *J. Organometal. Chem.* **1985**, *285*, 389.
61. Nicolosi, G.; Morrone, R.; Patti, A.; Piatelli, M. *Tetrahedron: Asymmetry* **1992**, *3*, 753.
62. Yamazaki, Y.; Hosono, K. *Tetrahedron Lett.* **1988**, *29*, 5769.
63. Yamazaki, Y.; Uebayasi, M.; Hosono, K. *Eur. J. Biochem.* **1989**, *184*, 671.
64. (a) Aratani, T.; Gonda, T.; Nozaki, H. *Tetrahedron Lett.* **1969**, *10*, 2265. (b) Aratani, T.; Gonda, T.; Nozaki, H. *Tetrahedron* **1970**, *26*, 5453.
65. Gokel, G.; Hoffman, P.; Kleimann, H.; Klusacek, H.; Marquarding, D.; Ugi, I. *Tetrahedron Lett.* **1970**, *11*, 1771.
66. (a) Marquarding, D.; Klusacek, H., Gokel, G.; Hoffmann, P.; Ugi, I. *J. Am. Chem. Soc.* **1970**, *92*, 5389. (b) *Angew. Chem. Int. Ed. Engl.* **1970**, *9*, 371.
67. Battelle, L.F.; Bau, R.B.; Gokel, G.W.; Oyakawa, R.T.; Ugi, I.K. *J. Am. Chem. Soc.* **1973**, *95*, 482.
68. Butler, I.R.; Cullen, W.R.; Herring, F.G.; Jagannathan, N.R. *Can. J. Chem.* **1986**, *64*, 667.
69. Butler, I.R.; Cullen, W.R.; Reglinski, J.; Retlig, J.J. *J. Organometal. Chem.* **1983**, *249*, 183.
70. Guillaneux, D. Ph.D. Thesis, Paris Sud, Orsay, 1995.
71. Riant, O.; Samuel, O.; Kagan, H.B. *J. Am. Chem. Soc.* **1993**, *115*, 5835. This method has recently been applied in the benchrotrene serie. See: Uemura, M.; Daimon, A.; Hayashi, Y. *J. Chem. Soc., Chem. Commun.* **1995**, 1943.
72. Rapic, V.; Schlögl, K.; Steinitz, B. *Monatsch. Chem.* **1977**, *108*, 767.
73. Gokel, K.; Hoffmann, P.; Kleimann, H.; Klusacek, H.; Marquarding, D.; Ugi, I. *Tetrahedron Lett.* **1970**, *11*, 1771.
74. Riant, O.; Flessner, T.; Kagan, H.B., manuscript in preparation.
75. Taudien, S.; Riant, O.; Kagan, H.B. *Tetrahedron Lett.* **1995**, *36*, 3513.
76. Schmitt, G.; Klein, P.; Ebertz, W. *J. Organometal. Chem.* **1982**, *234*, 63.
77. Richards, C.J.; Damalidis, T.; Hibbs, D.E.; Hursthouse, M.B. *Synlett* **1995**, 74.
78. (a) Sammakia, T.; Latham, H.A.; Schaad, D.R. *J. Org. Chem.* **1995**, *60*, 10. (b) Sammakia, T.; Latham, H.A.; Schaad, D.R. *J. Org. Chem.* **1995**, *60*, 6002. (c) Sammakia, T.; Latham, H. *J. Org. Chem.* **1996**, *61*, 1629.
79. Nishibayashi, Y.; Uemura, S. *Synlett* **1995**, 79.
80. (a) Loim, N.M.; Kondratenko, M.A.; Sokolov, V.I. *Dokl. Akad. Nauk. SSSR* **1991**, *320*, 1123. (b) Loim, N.M.; Kondratenko, M.A.; Sokolov, V.I.; Turdybekov, K.M.; Struchnov, Y.T. *J. Organometal. Chem. USSR (Engl.)* **1992**, *5*, 575.
81. Loim, N.M.; Kondratenko, M.A.; Sokolov, V.I. *J. Org. Chem.* **1994**, *59*, 7485.
82. Loim, N.M.; Kondratenko, M.A.; Sokolov, V.I. *Russ. Chem. Bull.* **1993**, *42*, 166.
83. (a) Kondo, Y.; Green, J.R.; Ho, J. *J. Org. Chem.* **1991**, *56*, 7199. (b) Kondo, Y.; Green, J.R.; Ho, J. *J. Org. Chem.* **1993**, *58*, 6182.
84. Alexakis, A.; Kanger, T.; Mangeney, P.; Rose-Munch, F.; Perrotey, A.; Rose, E. *Tetrahedron: Asymmetry* **1995**, *6*, 47.
85. Aubé, ; Herppert, J.A.; Milligan, M.L.; Smith, M.J.; Zenk, P. *J. Org. Chem.* **1992**, *57*, 3563.

86. (a) Blagg, Davies, S.G.; Goodfellow, C.L.; Sutton, H.K. *J. Chem. Soc., Perkin Trans. 1* **1987**, 1805. (b) Heppert, J.A.; Aubé, J.; Thomas-Miller, M.E.; Milligan, M.L.; Tagusagawa, F. *Organometallics* **1990**, *9*, 727. (c) Uemura, M.; Mikaye, R.; Nakayama, K.; Shiro, M.; Hayashi, Y. *J. Org. Chem.* **1993**, *58*, 1238.
87. For a recent application, see: Christian, P.W.N. Gibson (nee Thomas), S.E.; Gil, R.; Jones, C.V.; Marcos, C.F.; Prechtl, F.; Wierzchleyski, A.T. *Rec. Trav. Chim. Pays-Bas* **1995**, *114*, 195 and references therein.
88. Davies, S.G.; Loveridge, T.; Clough, J.M. *J. Chem. Soc., Chem. Commun.* **1995**, 817.
89. (a) Troitskaya, L.L.; Sokolov, V.I.; Reutov, O.A. *Dokl. Akad. Nauk SSSR* **1977**, *236*, 371. (b) Sokolov, V.I.; Troitskaya, L.L.; Reutov, O.A.; *J. Organometal. Chem.* **1980**, *202*, C58.
90. Gacent, J.C.; Shaw, B.L. *J. Organometal. Chem.* **1975**, *102*, 511.
91. Kuz'min, L.G.; Struchkov, Y.T.; Troitskaya, L.L.; Sokolov, V.I.; Reutov, O.A. *Izv. Akad. Nauk SSSR, Ser. Khim* **1979**, 1528.
92. Troitskaya, L.L.; Bulygina, L.A.; Sokolov, V.I. *Russ. Chem. Bull.* **1994**, *43*, 1253.
93. Sokolov, V.I.; Troitskaya, L.L.; Gautheron, B.; Tainturier, G. *J. Organometal. Chem.* **1982**, *235*, 369.
94. Uno, M.; Ando, K.; Komatsuzaki, N.; Takahashi, S. *J. Chem. Soc., Chem. Commun.* **1992**, 964.
95. Uno, M.; Ando, K; Komatsuzaki, N.; Tsuda, T.; Tanaka, T.; Sawada, M.; Takahashi, S. *J. Organometal. Chem.* **1994**, *473*, 303.
96. Paquette, L.A.; Gubelchuk, M.; McLaughlin, M.L. *J. Org. Chem.* **1987**, *52*, 4732.
97. Des Abbayes, H.; Dabard, R. *Tetrahedron* **1975**, *31*, 2111.
98. (a) Ratajczak, A.; Zmuda, H. *Bull. Acad. Pol., Ser. Sci. Chim.* **1974**, *22*, 261. (b) Ratajczak, A.; Zmuda, H. *Bull. Acad. Pol., Ser. Sci. Chim.* **1977**, *25*, 35.
99. Troitskaya, L.L.; Bulygina, L.A.; Sokolov, V.I. *Russ. Chem. Bull.* **1994**, *43*, 1050.
100. Wang, Y.-F.; Lalonde, J.J.; Momongan, M.; Bergbreiter, D.E.; Wong, C.-H. *J. Am. Chem. Soc.* **1988**, *110*, 7200.
101. Boaz, N.W. *Tetrahedron Lett.* **1989**, *30*, 2061.
102. Yamazaki, Y.; Hosono, K. *Tetrahedron Lett.* **1989**, *30*, 5313.
103. Kim, M.-J.; Cho, H.; Choi, Y.K. *J. Chem. Soc., Perkin Trans. 1* **1991**, 2270.
104. (a) Izumi, T.; Hino, T.; Shoji, K.; Sasaki, K.; Kasahara, A. *Chem. Ind. (London)* **1989**, 457. (b) Izumi, T.; Murakami, S.; Kasahara, A. *Chem. Ind. (London)* **1990**, 79.
105. Yamazaki, Y.; Uebayasi, M.; Someya, I-I.; Hosono, K. *Agric. Biol. Chem.* **1990**, *54*, 1781.
106. Nicolosi, G.; Patti, A.; Morrone, R.; Piattelli, M. *Tetrahedron: Asymmetry* **1994**, *5*, 1275.
107. Nicolosi, G.; Patti, A.; Piattelli, M. *J. Org. Chem.*, **1994**, *59*, 251.
108. See Chap. 4 in ref. 8b.
109. Iio, H.; Fujii, A.; Ishii, M.; Tokoroyama, T. *J. Chem. Soc., Chem. Commun.* **1991**, 1390.
110. Fukuzawa, S.-I.; Kasugahara, Y.; Uemura, S. *Tetrahedron Lett.* **1994**, *35*, 9403 and references therein.
111. (a) Nishibayashi, Y.; Chiba, T.; Ohe, K.; Uemura, S. *J. Chem. Soc., Chem. Commun.* **1995**, 1243. (b) Nishibayashi, Y.; Singh, J.D.; Fukuzawa, S.-I.; Uemura, S. *J. Org. Chem.* **1995**, *60*, 4114.
112. Price, D.; Simpkin, N.S. *Tetrahedron Lett.* **1995**, *36*, 6135.
113. Schmalz, H.-G.; Schellhaas, K. *Tetrahedron Lett.* **1995**, *36*, 5515.
114. Tsukazaki, M.; Tinkl, M.; Roglans, A.; Chapell, B.J.; Taylor, N.J.; Snieckus, V. *J. Am. Chem. Soc.* **1995**, *118*, 685.
115. Nishibayashi, Y.; Arikawa, Y.; Ohe, K.; Uemura, S. *J. Org. Chem.* **1996**, *61*, 1172.
116. Enders, D.; Lochtman, R.; Raabe, G. *Synlett* **1996**, 126.

117. (a) Watanabe, M. *Synlett* **1995**, 1050. (b) Watanabe, M. *Tetrahedron Lett.* **1995**, *36*, 8991.
118. Schwink, L.; Knochel, P. *Tetrahedron Lett.* **1996**, *37*, 25. See also the following reference for the preparation of chiral ferrocene using optically active α-hydroxycyclopentadienes as building blocks: Schwink, L.; Vettel, S.; Knochel, P. *Organometallics* **1995**, *14*, 5000.
119. (a) Park, J.; Lee, S.; Ahn, K.H.; Cho, C.-W. *Tetrahedron Lett.* **1995**, *36*, 7263. (b) Zhang, W.; Adashi, Y.; Hirao, T.; Ikeda, I. *Tetrahedron: Asymmetry* **1996**, *7*, 451. (c) Zhang, W.; Hirao, T.; Ikeda, I. *Tetrahedron Lett.* **1996**, *37*, 4545.

BIOCATALYSIS AS A NEW POWERFUL TOOL FOR THE SYNTHESIS OF ENANTIOMERICALLY PURE CHIRAL BUILDING BLOCKS

Enzo Santaniello and Patrizia Ferraboschi

Advances in Asymmetric Synthesis
Volume 2, pages 237–283

ISBN 1-55938-797-1

I. INTRODUCTION

For the asymmetric synthesis of compounds containing stereogenic centers, organic chemists can rely upon an extraordinarily vast arsenal of chiral building blocks and a multitude of methods for their preparation and elaboration. Mother Nature offers a great variety of opportunities within the so-called *chiral pool*[1] and has furnished biological auxiliaries that are capable to create stereogenic centers or discriminate stereoisomers. In fact, cells of microorganisms, plants, and animals containing enzymes and coenzymes essential for their vital cycle can be used *in toto* to carry out reactions on substrates different from the physiological ones. A purified enzyme constitutes a more sophisticated biocatalytic system for the same purpose, and all these biocatalysts have been already introduced into the practice of organic chemists for the purpose of asymmetric syntheses.[2]

The synthesis of enantiomerically pure compounds can be divided in two big chapters: creation of stereogenic centers with formation of a single stereoisomer; or, given one or more stereogenic centers, separation of each stereoisomer present in a mixture. From a practical point of view, both approaches present advantages and disadvantages and the solution of the synthetic problem may dictate the choice of the most suitable biocatalyst.

A classical example of creation of a stereogenic center with formation of a single stereoisomer is the reduction of a ketone to the corresponding alcohol. When the biocatalyst used is a microorganism or a purified dehydrogenase, usually only one enantiomer should be formed. Several reductions follow a common stereochemical outcome, and for many microorganisms and dehydrogenases this pattern can be rationalized with the aid of the so-called *Prelog's rule*.[3] If the carbonyl is bound to two groups of different size (L, large, and S, small), the alcohol biocatalytically obtained will have the configuration indicated in Scheme 1.

Scheme 1.

II. REDUCTION OF CARBONYL COMPOUNDS WITH BAKER'S YEAST

The above rule is also valid for many bioreductions realized by means of *Saccharomyces cerevisiae* (baker's yeast, BY) a common microorganism that can be used by organic chemists not trained in microbiology and that can be considered as a biological reagent for asymmetric synthesis.[4] According to the expectations of the Prelog's rule, ethyl acetoacetate (**1**) is reduced to nearly optically pure ethyl (*S*)-3-hydroxybutyrate (**2**) (Scheme 2),[5] a structurally simple but important chiral synthon.[6] The experimental procedure of this biocatalytic methodology has been accurately described in *Organic Syntheses.*[7]

We experienced this simple and effective microbiological approach, when we prepared gram quantities of (*S*)-**2** for a projected chiral synthesis of (*R*)-2-aminobutane (**3**) that we later accomplished[8] by a chemical route starting from commercially available (*R*)-2-aminobutanol (**4**) (Scheme 3). Surprisingly, reduction with baker's yeast of ethyl 4-chloroacetoacetate (**5a**) afforded the unexpected *S* configuration for the hydroxy ester **6a** and a relatively modest (60%) enantiomeric excess (ee). We used the compound **6a** to prepare (*S*)-(−)-4-hydroxypyr-

Scheme 2.

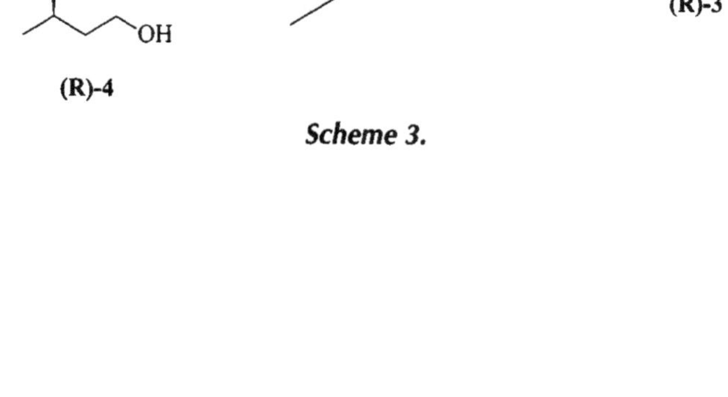

Scheme 3.

rolidin-2-one (**7**) (Scheme 4) and define the controversial configuration of this component of *Amanita muscaria*.[9]

The same result on the bioreduction of the compound **5** was obtained by Sih and co-workers, who addressed the reaction toward the production of the enantiomerically pure (*R*)-hydroxy ester **6a** (a valuable synthon for the preparation of the pharmacologically active (*R*)-carnitine) using a mutant of *Saccharomyces cerevisiae*.[10] The same authors gave an excellent example about the possibility of directing the biocatalytic process toward the desired stereoisomer by simple modifications of the substrate. The size of the ester moiety was made larger changing the ethyl into the octyl ester **5b** and, by incubation of this compound, the configurational outcome was reversed.[11] In this way, the optically pure (*R*)-hydroxy ester **6b** was prepared (Scheme 5). An alternative route to the use of a mutant for the preparation of (*R*)-ethyl ester **6a** could be the reduction by a different microorganism. We succeeded in this direction (Scheme 4) preparing enantiomerically pure (*R*)-ester **6a** by means of *Lactobacilli* strains.[12]

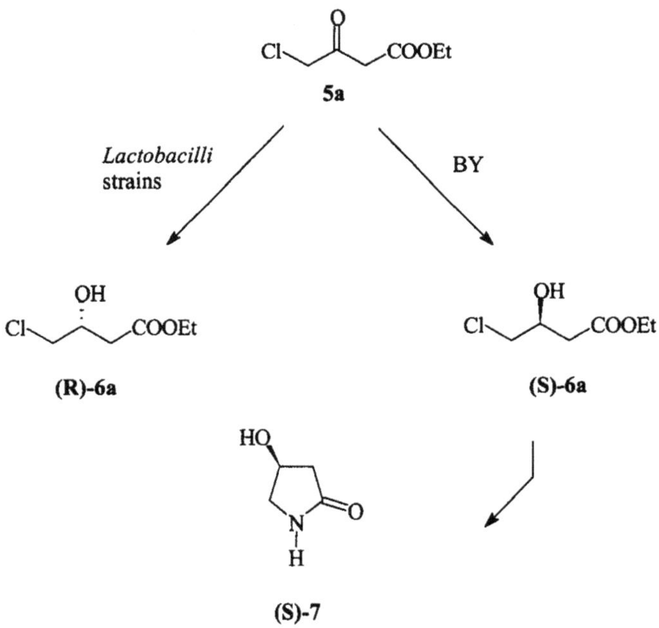

Scheme 4.

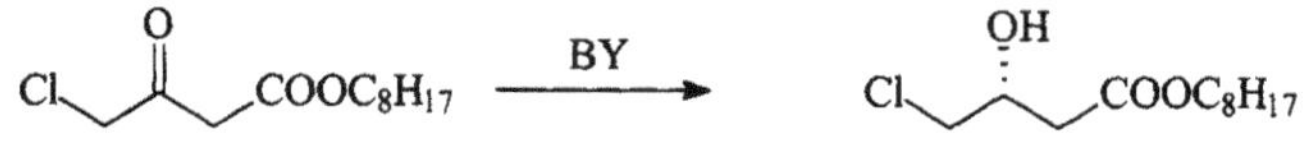

5b **(R)-6b**

Scheme 5.

The bioreduction of carbonyl compounds is still being actively investigated by several groups[13] and in this context we studied a few years ago[14] the incubation of some 3- and 4-oxoesters. For instance, ethyl levulinate **8** was reduced to >95% ee 4-hydroxy ester **10**. When 4-oxo-4-phenyl-butanoate (**9**) was subjected to the same reduction with *Saccharomyces cerevisiae* the main product was the (*S*)-(−)-lactone **11** (Scheme 6). Also the sodium salt of diethyl oxaloacetate **12** was reduced to optically pure diethyl (*S*)-malate **13** with fermenting baker's yeast and, among several other microorganisms tested, only *Lactobacillus fermentum* was able to afford (*R*)-**13** (80% ee, Scheme 7).[15] *Saccharomyces cerevisiae* reduced ketoacetals **14a,b** to the corresponding (*S*)-**15** with maximum 60% ee and for an enantioselective biotransformation of the substrates to both (*R*)- and (*S*)-hydroxy derivatives **15a,b** other microorganisms had to be used[16] (Scheme 8, Table 1).

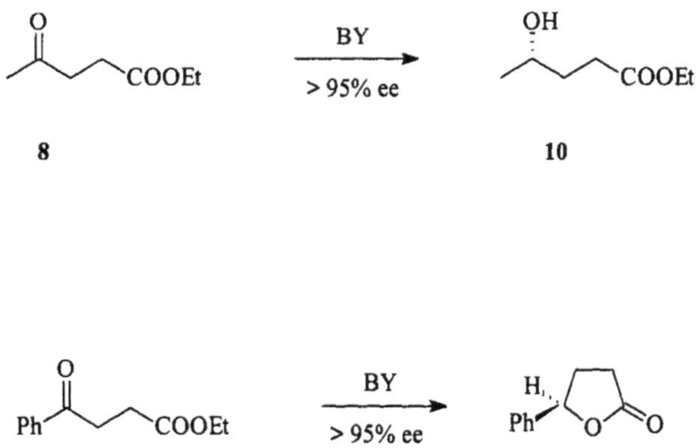

8 **10**

9 **11**

Scheme 6.

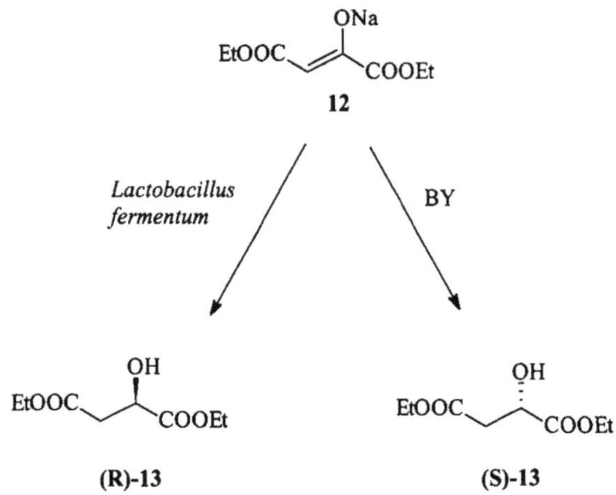

Scheme 7.

On the other hand, baker's yeast is particularly efficient in the reduction of hydroxy ketones **16a-c** to optically active (*R*)-1,2-diols **17a-c**[17] (Scheme 9). This is exemplified by the reduction of hydroxyacetone **16a** to (*R*)-1,2-propanediol (**17a**) as reported almost 50 years ago.[17a] The result, however, is not in agreement with Prelog's rule and it is conceivable that the dehydrogenase that catalyzes the reduction is different from the enzyme(s) that accept ketones as substrates.[18] On the other hand, pioneering works on these biocatalytic reductions have suggested that methyl ketones are the most suitable substrates for an enantioselective reduction.[19] We therefore reasoned that the benzyl ether of hydroxyacetone, 1-benzyloxyacetone (**18**), is a classical methyl ketone and could therefore be reduced by

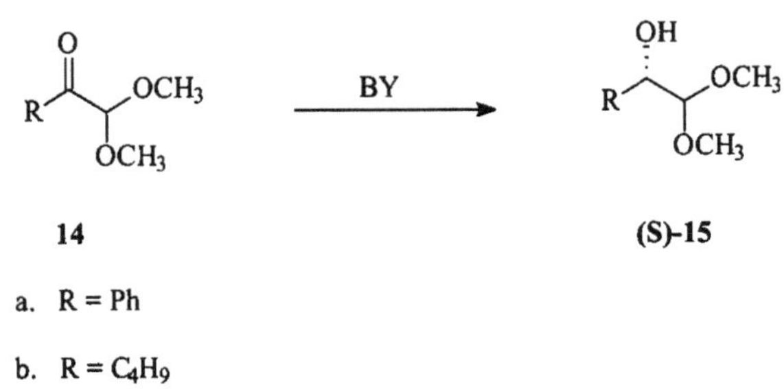

Scheme 8.

Table 1. Bioreduction of Compounds **14a** and **14b** with Different Yeasts

Microorganisms[a]	*Alcohol* **15a**		*Alcohol* **15b**	
	Yields (%, 48 h)	*ee*[b] *(%)*	*Yields (%, 48 h)*	*ee (%)*
Baker's yeast	45	46	26	60 (*S*)
Hansenula glucozyma CBS 5766	100	60	100	94 (*S*)
Hansenula minuta CBS 1708	100	96	50	20 (*R*)
Kloeckera saturna MIM	30	90	46	92 (*S*)
Sporobolomyces salmonicolor MAT,MM	18	84	40	64 (*S*)
Sporobolomyces alborubescens CBS 482	no reaction		15	94 (*S*)
Kluyveromyces fragilis CBS 397	no reaction		18	100 (*R*)
Kluyveromyces marxianus CBS 1553	53	92	60	100 (*R*)

Notes: [a]Abbreviations: MIM, Microbiologia Industriale, Milano; CBS, Centraalbureau voor Schimmelculture, Baarn; MAT, Microbiologia Agraria, Torino; MIP, Microbiologia Agraria, Perugia.
[b]Prevalent *S* configuration with all the strains.

fermenting baker's yeast to the (*S*)-hydroxy compound **19**, according to Prelog's rule. This was found to be true and we were able to prepare the optically pure (*S*)-alcohol **19** (Scheme 10). This product is the benzyl ether of the enantiomer of (*R*)-**17a**[20] and, therefore, using the same biocatalyst a simple manipulation of the substrate could again open the way to prepare both enantiomers of the product.

In another paper,[21] apart from a conventional baker's yeast-mediated reduction of two arylmethyl ketones to the corresponding (*S*)-carbinols, we showed that the biotransformation of the hydroxyketone **16b** afforded the corresponding (*R*)-diol **17b** in good yield and high optical purity only if the incubation condi-

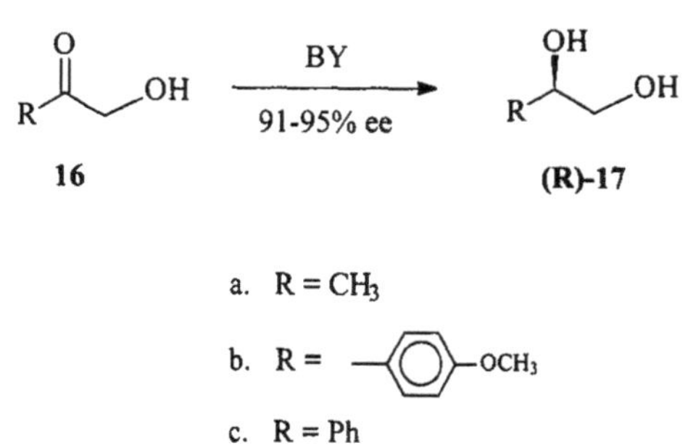

Scheme 9.

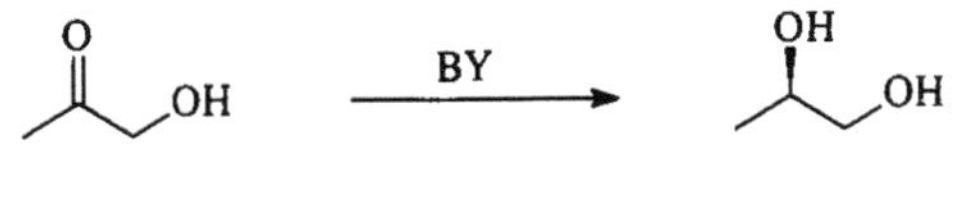

16a **(R)-17a**

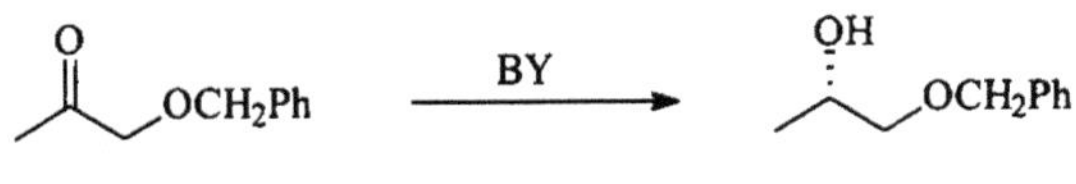

18 **(S)-19**

Scheme 10.

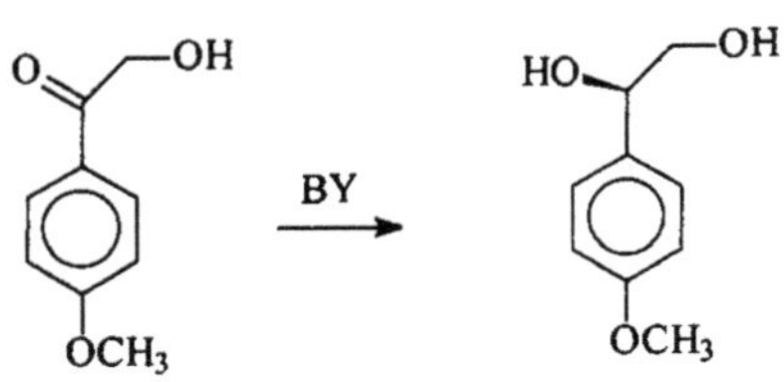

16b **(R)-17b**

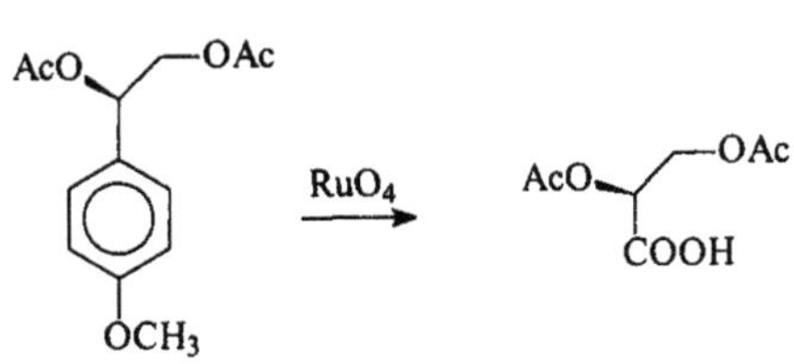

(R)-20 **(S)-21**

Scheme 11.

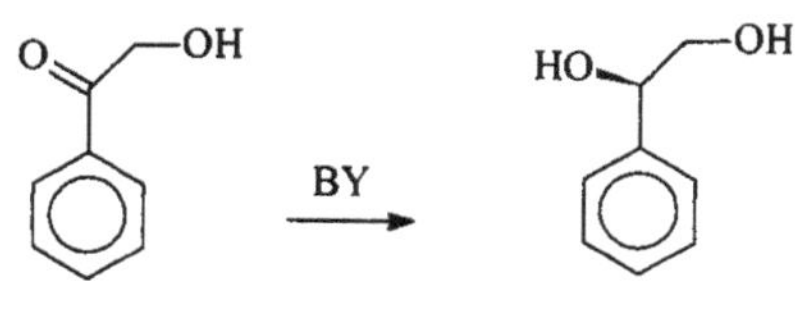

16c **(R)-17c**

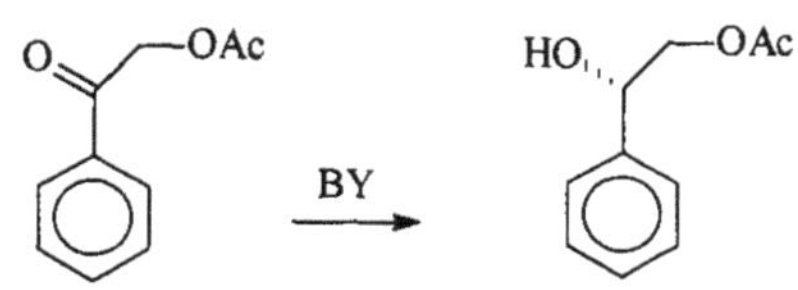

22 **(S)-23**

Scheme 12.

tions were carefully monitored. Ruthenium tetraoxide oxidation of the diacetate of the nearly optically pure (95% ee) (*R*)-diol **17b**, compound **20**, afforded 95% ee (*S*)-2,3-diacetoxypropanoic acid (**21**) (Scheme 11), previously used as an useful chiral synthon.[22]

Later, we found that a simple modification of the 2-hydroxyketone structure could again reverse the stereochemical outcome of the bioreduction (Scheme 12). In fact, phenacyl alcohol **16c** was reduced with baker's yeast to 92% ee (*R*)-diol **17c**,[23] whereas the acetate **22** afforded the (*S*)-monoacetate **23** (70% yield, 94% ee).[24] We extended this observation to other acetates of 2-hydroxyketones and invariably we found a reverse configuration for the product of reduction. However, from a synthetic point of view the acetate was not a good protecting group since during the incubation with baker's yeast we sometimes observed either hydrolysis of the substrate or migration of the acetate in the product. When the methyl ether of 2-hydroxyketones was incubated, the (*R*)-monomethoxy diol was formed.[25] The results are specifically illustrated using as substrates for the baker's yeast-mediated reduction of hydroxyketones, the hydroxyketone **24a**, the methyl ether **24b**, and the acetate **24c**. The corresponding diol **25a** and the monoprotected diols **25b,c** so obtained were (*R*)-**25a** (95% ee) and (*R*)-**25b** (72% ee) or the (*S*)-monoacetate **25c** (72% ee) (Scheme 13).

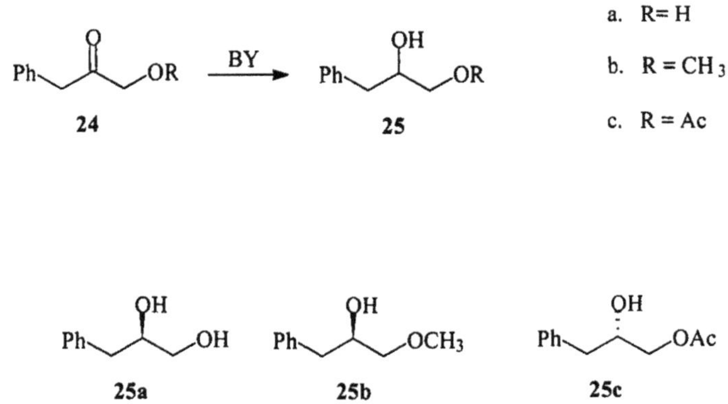

Scheme 13.

III. BIOHYDROGENATION OF DOUBLE BONDS WITH BAKER'S YEAST

A unique biocatalytic process that can be performed almost exclusively with microbiological systems is the biohydrogenation of unsaturated double bonds,[26] a reaction that belongs also to the repertoire of baker's yeast.[27] Especially intriguing in this regard was the report[28] on the baker's yeast-mediated transformation of ethyl 4,4-dimethoxy-3-methylbutenoate (**26**). It was postulated that during the fermentation the hydrolysis of the acetal group and the reduction of the intermediate aldehyde occurred. The saturated hydroxy ester so formed was then cyclized by an acidic treatment and the optically pure (*S*)-3-methyl-γ-butyrolactone (**27**) was isolated as the final product of the process (Scheme 14). This chiral synthon was commercially available when we used it as an intermediate for the synthesis of optically pure phenylsulfones **28a,b** according to Scheme 15. The compounds **28a,b** could constitute the chiral intermediates for the construction of steroid side chains.[29]

Later on, we decided to set up the incubation conditions for this biotransformation and studied in detail the reaction with special regard to the stereochemical aspects, trying to extend the method to similar unsaturated substrates.[30] The biohydrogenation afforded not easily reproducible results, with variable optical purities. More reproducible results in the preparation of the lactone **27** (90% ee) were obtained by us[31] using as substrate the potassium salt **29** instead of the ester **26** (Scheme 16).

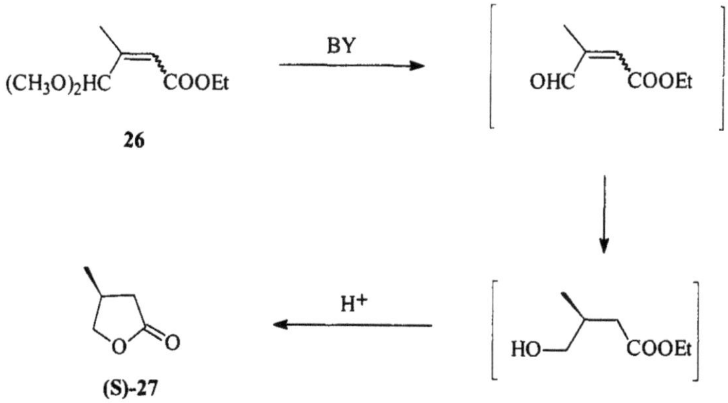

Scheme 14.

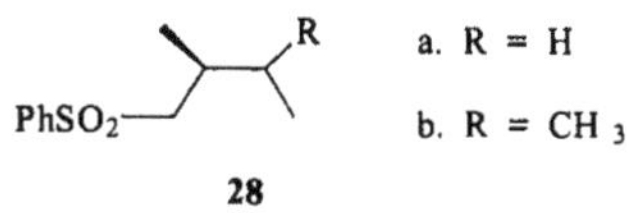

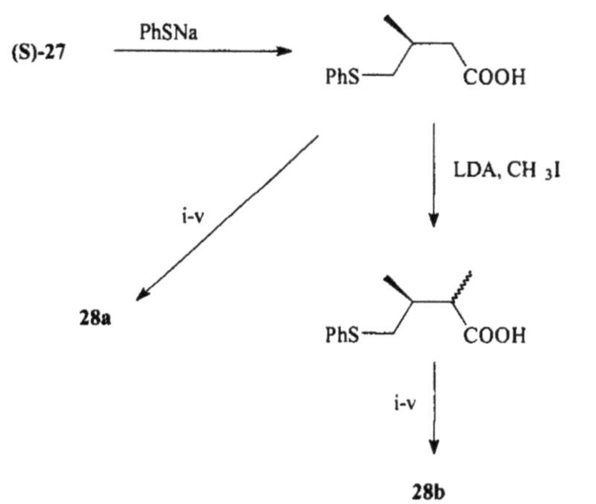

i. EtOH/H^+ ii. $LiAlH_4$ iii. CH_3SO_2Cl iv. $LiAlH_4$ v. MCPBA

Scheme 15.

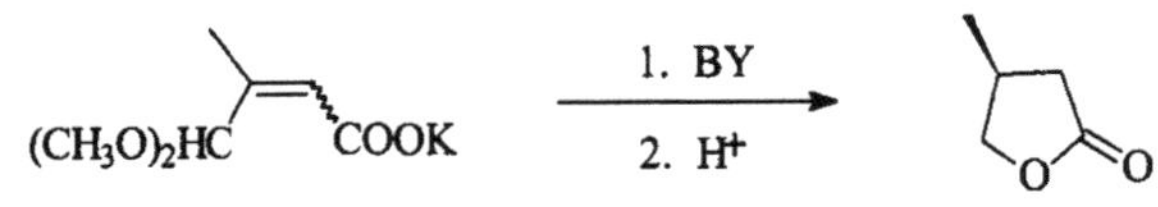

29 (S)-27

Scheme 16.

We also carried out the baker's yeast-mediated biohydrogenation of the ester **26** and isolated the (*S*)-saturated hydroxyester **30** (97% ee) that was further protected as methoxyethoxy ether and elaborated to the phenyl sulfone **31** that was the key chiral intermediate for the synthesis of (25*S*)-26-hydroxycholesterol (**32**)[32] according to Scheme 17. In this context, it seemed to us that the biohydrogena-

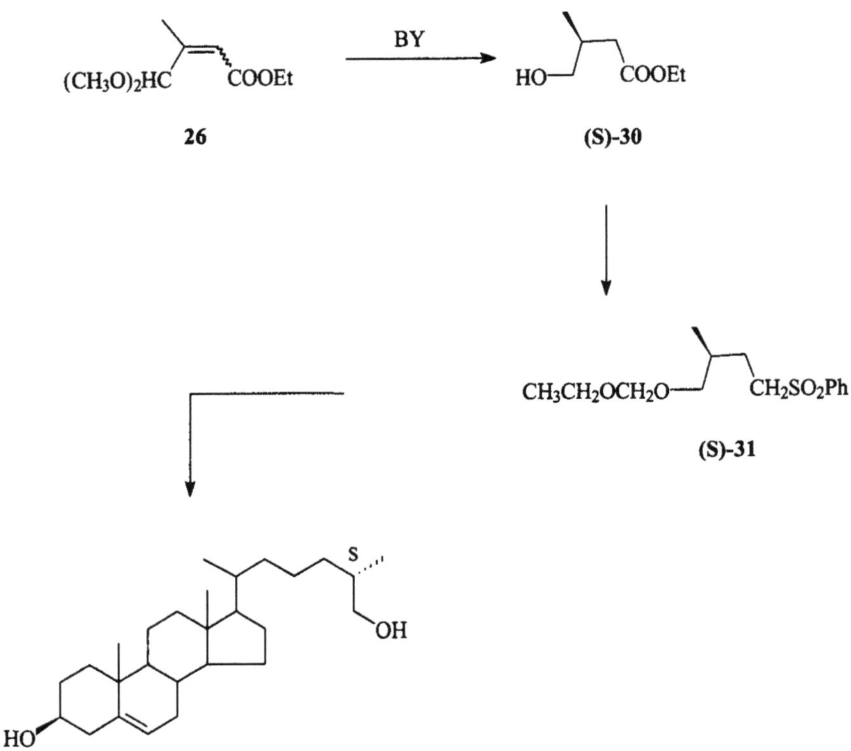

26 (S)-30 (S)-31 32

Scheme 17.

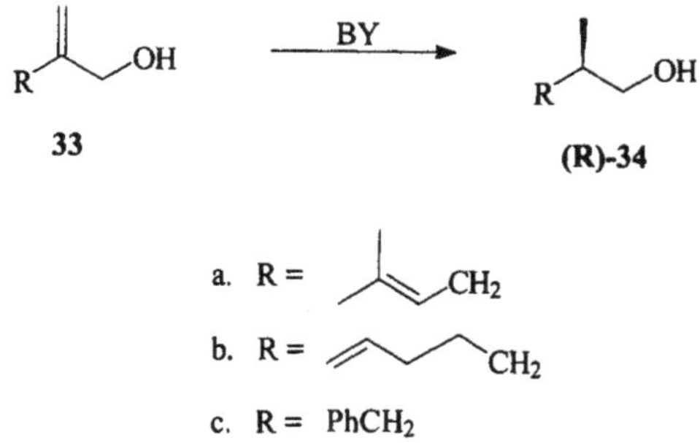

Scheme 18.

tion of compounds containing a methylene moiety could be an interesting biotransformation for the construction of a stereocenter characterized by a methyl group. Only a few similar processes had been described, mainly concerning α-methylene ketones.[33] We synthesized a few 2-substituted allyl alcohols containing the methylene moiety and a primary alcohol function, namely compounds **33a–c** and converted them to (*R*)-2-methyl alkanols **34a–c**[34] with ee between 92 and 98% (Scheme 18).

Recently, we have extended this biohydrogenation process to another interesting methylene alcohol synthesized from dihydroxyacetone **35**.[35] The compound **36** was converted with the aid of fermenting baker's yeast into 98% ee benzyl (*S*)-2-methyl-1,3-propanediol (**37**) (Scheme 19), a synthon that has been widely used for several chiral syntheses.[36] Using the same biocatalytic approach, the (*S*)-enantiomers of two fungicides have been recently prepared enantiomerically pure from 2-methyl-(4′-*tert*-butyl)cinnamaldehyde.[37]

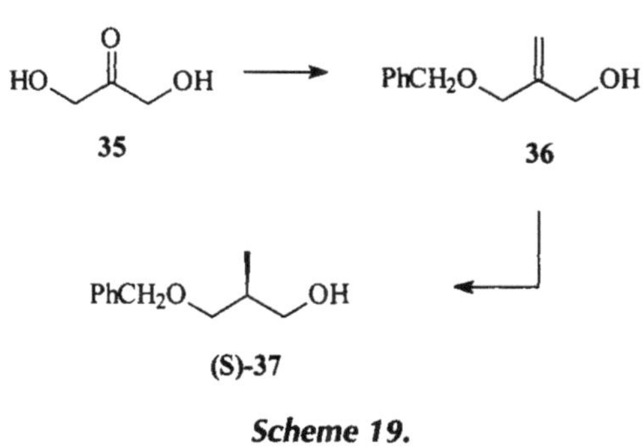

Scheme 19.

IV. ENZYME-CATALYZED STEREOSPECIFIC INTRODUCTION OF A DEUTERIUM LABEL

The stereospecific introduction of deuterium labeling into an organic compound has been used for a long time in bioorganic chemistry, either from a mechanistic or biosynthetic point of view.[38] A special mention should be dedicated to the conversion of a methyl group into a chiral moiety by introducing the proper substitutions of hydrogen atoms with the corresponding deuterium and tritium at a single carbon, which has been a historical breakthrough in enzymological studies.[39]

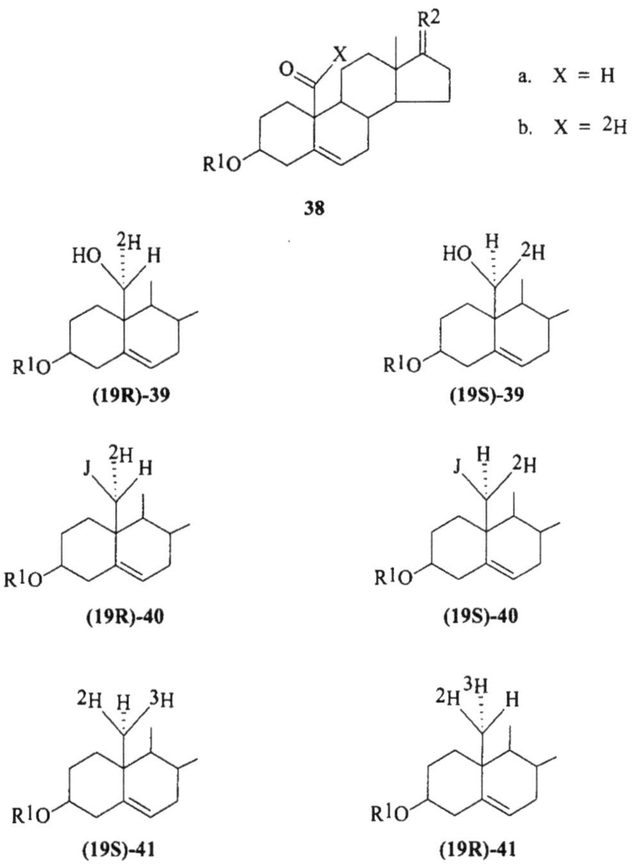

Scheme 20.

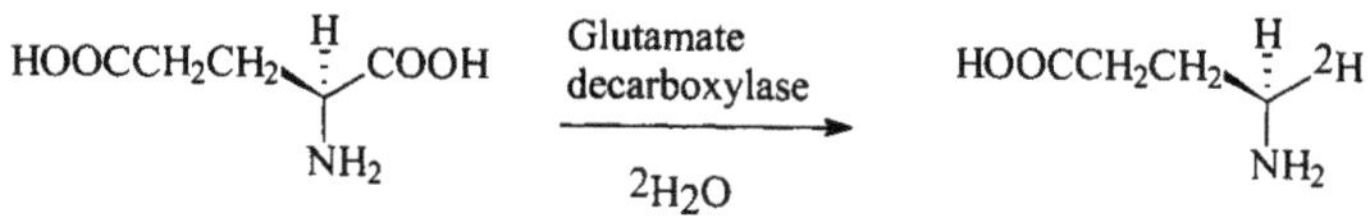

(S)-43 (R)-44

Scheme 21.

The construction of a chiral methyl group can be realized by chemical methods, as exemplified by the first synthesis of a 19-chiral methyl containing steroid.[40] In this case, taking advantage of the special stereochemical constriction for the angular position, the 19-aldehyde **38a** and its isotopomer **38b** are chemically reduced to the (*R*)- or (*S*)-deuteriated alcohols **39**. The reduction of the intermediate 19-iodomethyl derivatives **40** with tritiated lithium aluminum hydride affords the (*R*)- and (*S*)-19-[^{1}H, ^{2}H, ^{3}H]-steroids **41** (Scheme 20).

The stereospecific introduction of deuterium can be realized by an enzymatic reaction, provided that the stereochemical outcome is fully understood.[41] In the case of the decarboxylation of (*S*)-glutamate **43**, we demonstrated that the enzymatic reaction, catalyzed by the pyridoxal-phosphate-dependent decarboxylase from *Escherichia coli*, proceeded with retention of configuration.[42] Therefore, when the reaction was carried out on **43** in $^{2}H_2O$, (4*R*)-[4-^{2}H]-4-aminobutanoic acid (**44**) was prepared enantiotopically pure (Scheme 21).

Another decarboxylating enzyme, histidine decarboxylase from *Clostridium welchii*, was shown[43] to catalyze the reaction with retention of configuration. Thus, for instance, the proper enzyme transformed in $^{2}H_2O$ (*S*)-histidine **45** into (*R*)-deuteriated histamine **46** (Scheme 22).

Since both the decarboxylating enzymes were commercially available in a purified form, they could be used for the preparation of stereospecifically labeled compounds **44** and **46** on a preparative scale. From a general point of view, the above deuteriated compounds can be considered chiral synthons of established absolute stereochemistry and can be transformed into other polyfunctional chiral

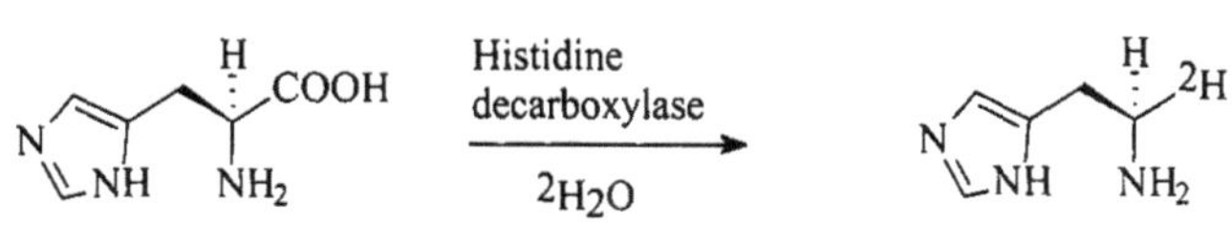

(S)-45 (R)-46

Scheme 22.

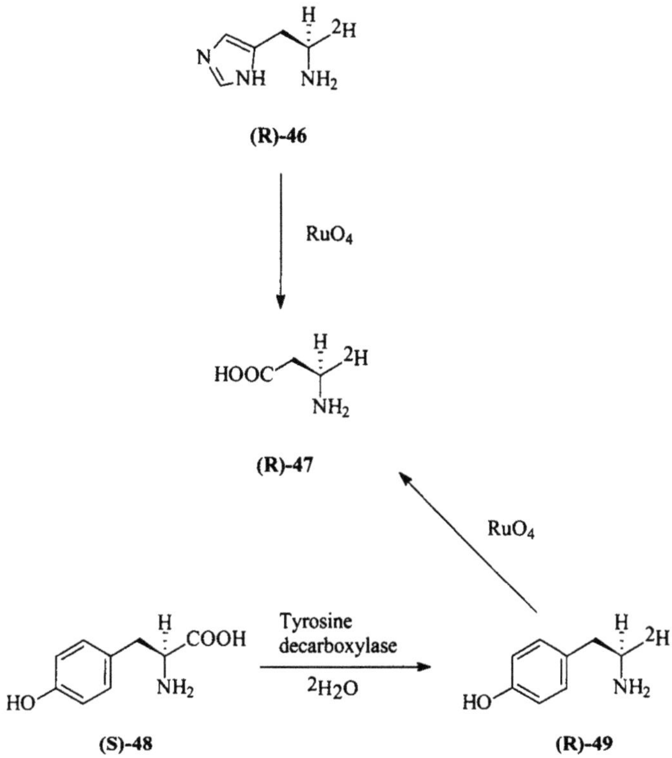

Scheme 23.

building blocks or incorporated into a target chiral molecule. In fact (Scheme 23), ruthenium oxidation of (2*R*)-[2-^2H]-histamine (**46**) transformed this compound into enantiotopically pure (3*R*)-[3-^2H]-β-alanine (**47**). The same result could be obtained by using another commercial decarboxylase, the tyrosine decarboxylase from *Streptococcus faecalis*, that catalyzes the decarboxylation of (2*S*)-tyrosine **48** into (2*R*)-[2-^2H]-tyramine (**49**).[44]

Furthermore, both the stereospecifically deuteriated glycines could be prepared by a chemo-enzymatic approach, starting from the (*S*)-glutamic acid **43** or the corresponding isotopomer, (2*R*,*S*)-[2-^2H]-**43**.[45] In fact, the enzymatic reaction with (*S*)-**43** in ^2H_2O afforded the (4*R*)-[4-^2H]-4-aminobutanoic acid **44**, whereas the (2*R*,*S*)-[2-^2H]-**43** in H_2O afforded the enantiomeric (4*S*)-[4-^2H]-4-aminobutanoic acid **44** along with the unreacted (2*R*)-[2-^2H]-**43** (Scheme 24).

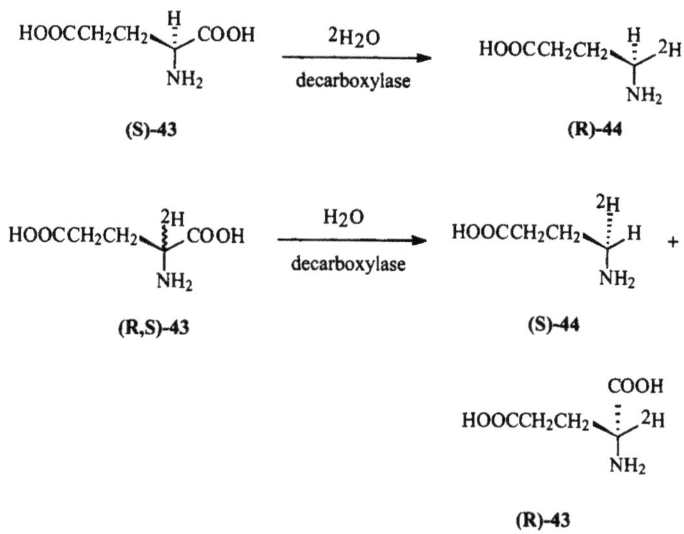

Scheme 24.

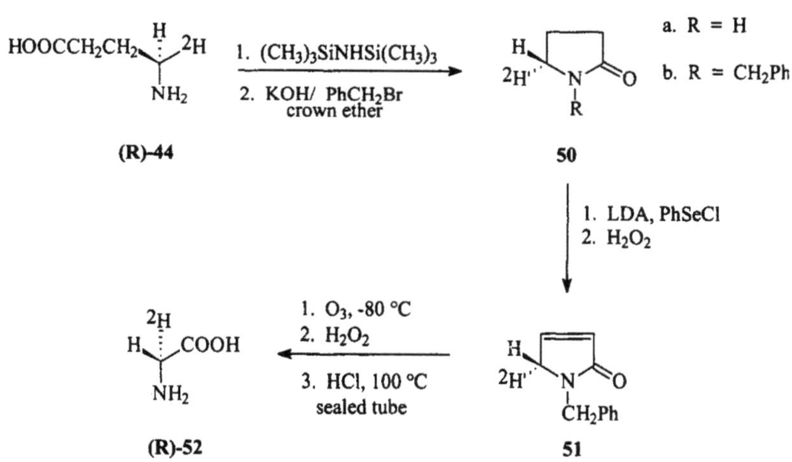

Scheme 25.

Starting from the isotopically and enantiomerically pure (4*R*)-[4-^{2}*H*]-4-aminobutanoic acid **44**, the preparation of the chiral glycine (*R*)-**52** was accomplished via the cyclization of (*R*)-**44** to the pyrrolidinone **50a**, which was protected as the benzyl pyrrolidinone **50b**. This was transformed into the 3-pyrrolinone **51** and this unsaturated compound was cleaved by ozonolysis. Following debenzylation of the intermediate oxalamide afforded enantiomerically pure (2*R*)-**52** (Scheme 25). Similarly, the (*S*)-[2-^{2}*H*]-glycines **52** could be prepared from (4*S*)-[4-^{2}*H*]-4-aminobutanoic acid **44**, obtained as previously described. Using a similar chemoenzymatic approach, we prepared (2*S*, 5*R*)-[5-^{2}*H*]-proline **53** from (4*R*)-[4-^{2}*H*]-4-aminobutanoic acid **44** by passing through to the 2-pyrrolidinone **50a** (Scheme 26).[46]

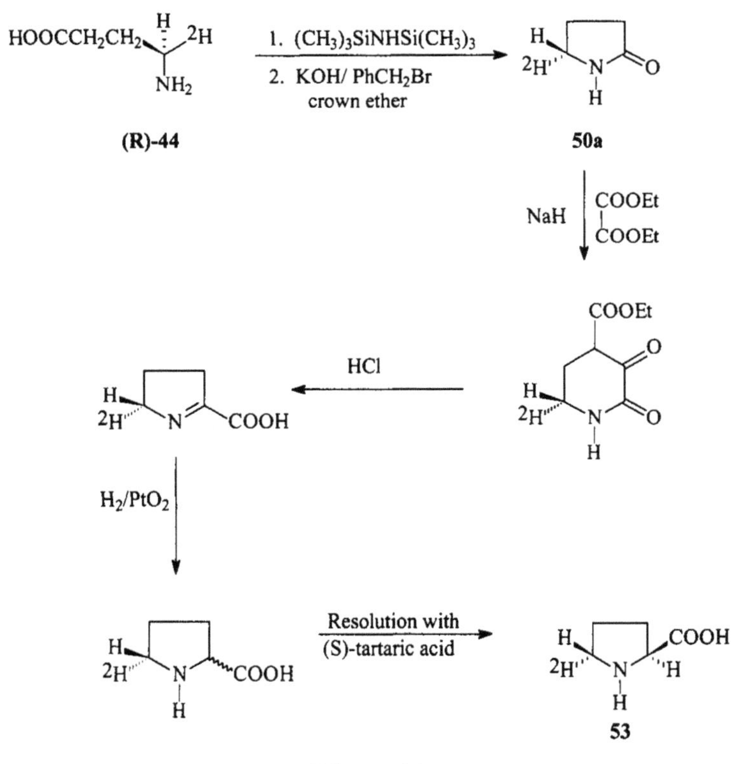

Scheme 26.

V. ENZYME-CATALYZED ASYMMETRIZATION OF PROCHIRAL COMPOUNDS

From the results on the enzymatic introduction of stereospecific labels, we learned the personal lesson that purified enzymes could be used as biological auxiliaries for the construction of stereogenic centers and thus started to extend the acquired experience to enzymatic asymmetrization of prochiral compounds.[47] The advantage of the use of enzymes on prochiral compounds is that a single enantiomer of the product of the enzymatic reaction can be obtained and, in an optimistic expectation, optically pure. A pioneering example of the application of this concept has been the asymmetrization of the prochiral diester **54** to the (*S*)-monoester **55** that was catalyzed by the hydrolytic enzyme pig liver esterase (PLE).[48] This compound contained two groups (carboxylic acid and ester) that

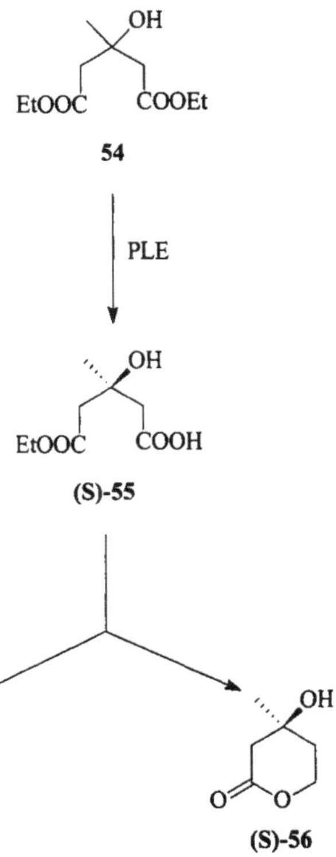

Scheme 27.

could be chemoselectively reduced to the corresponding hydroxy group, thus (*S*)-**55** could be converted by an enantiodivergent synthesis to both (*R*)- and (*S*)-mevalonolactone **56** (Scheme 27).

The asymmetrization of 3-hydroxyglutarate **57a** in the presence of the protease chymotrypsin (CHY) had been reported to proceed to the (*R*)-monoester **58a** with a high degree of enantioselectivity,[49] and optically pure **58a** was used as

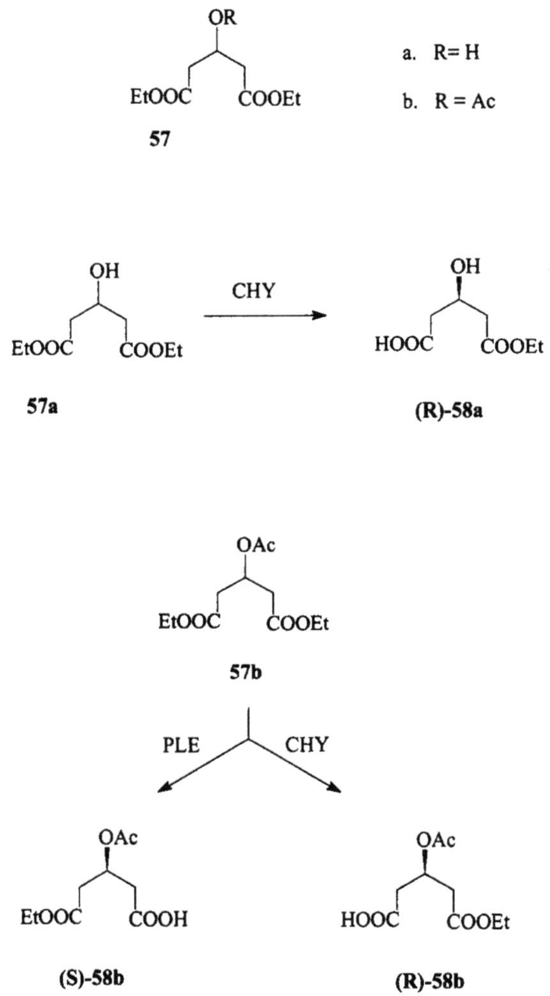

Scheme 28.

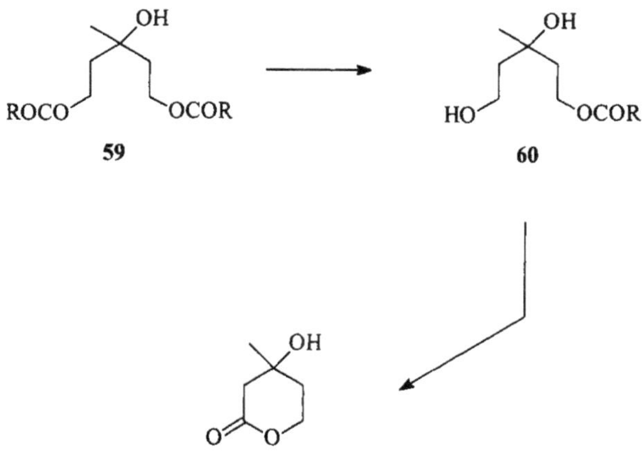

56

Scheme 29.

a chiral intermediate for the synthesis of several biologically active compounds.[49b,50] In an attempt to prepare optically pure **58a** we were unable to repeat the successful results reported in the literature. Few reports published later confirmed the difficulties encountered by us.[50] However, we found[51] that a simple modification of the substrate, i.e. acetylation of the 3-hydroxy group to give the ester **57b**, yielded a higher ee in the CHY-catalyzed hydrolysis (from 79 to 92%) to the (*R*)-monoester **58b**, and PLE afforded a 92% ee of the opposite enantiomer (*S*)-**58b** (Scheme 28).

However, this successful example of asymmetrization of a prochiral compound depends on the substrate and the enzyme used as biocatalyst. In fact, the hydrolysis of the prochiral diesters **59** to the monoesters **60**, that would be chemically transformed to mevalonolactone **56**,[52] proceeded with almost no enantioselection (Scheme 29).

Modest enantioselectivity was obtained with the hydrolysis of the racemic esters **61a-c** by several lipases (ee 27–47%), and only the relatively unusual enzyme, butyryl cholinesterase from electric eel, hydrolyzed the ester **61b** into the (*S*)-alcohol **62** with 62% ee.[53] From (*S*)-**62**, 52% ee, (*R*)-mevalonolactone **56** could be obtained (Scheme 30).

Another example of asymmetrization of a prochiral diester to a chiral monoester was found by us in the case of the diacetate of 2-methyl-1,3-propanediol **63**.[54] Here, the enzymatic reaction was carried out in the presence of the *Pseudomonas fluorescens* lipase (PFL) and from prochiral **63** a 90% ee (*R*)-

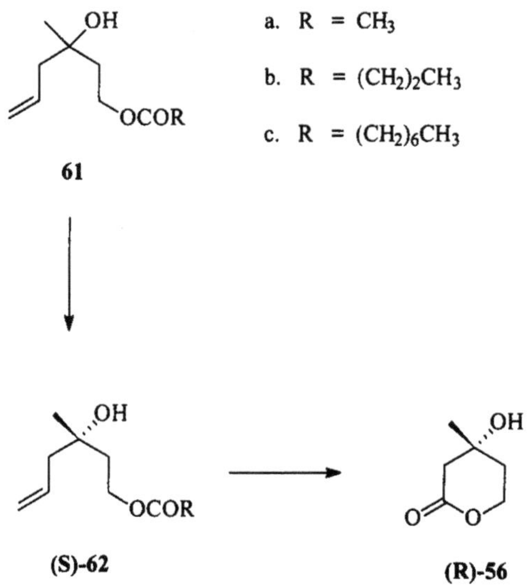

Scheme 30.

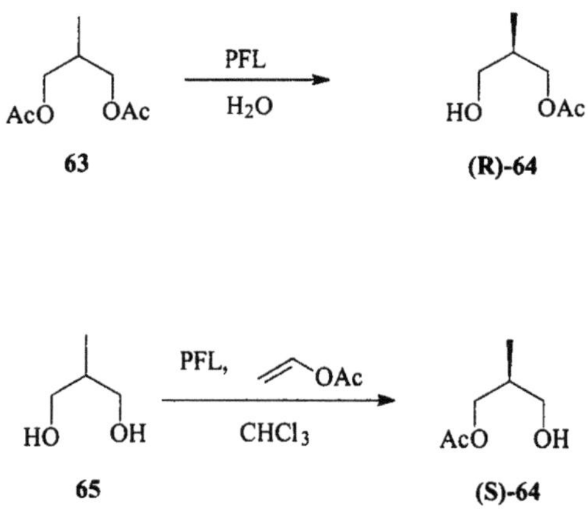

Scheme 31.

monoacetate **64** was obtained. Note that the acetylation of the diol **65** in the presence of the same enzyme, if carried out in organic solvent with the aid of vinyl acetate as acylating agent, afforded the (*S*)-monoacetate **54** with >98% ee (Scheme 31).

VI. ENZYMATIC RESOLUTION OF RACEMIC COMPOUNDS

If both enantiomers of a chiral compound are required, the enzymatic resolution becomes a very powerful tool in the hands of an organic chemist. In the case of aqueous hydrolysis a major drawback can be constituted by the solubility of the organic substrates, although many hydrolases can be used in aqueous medium containing a large variety of solvents without altering dramatically the catalytic activity of the biocatalyst.[55] This proved to be the case in the PLE-catalyzed hydrolysis of a few esters since it has been shown by several authors that the

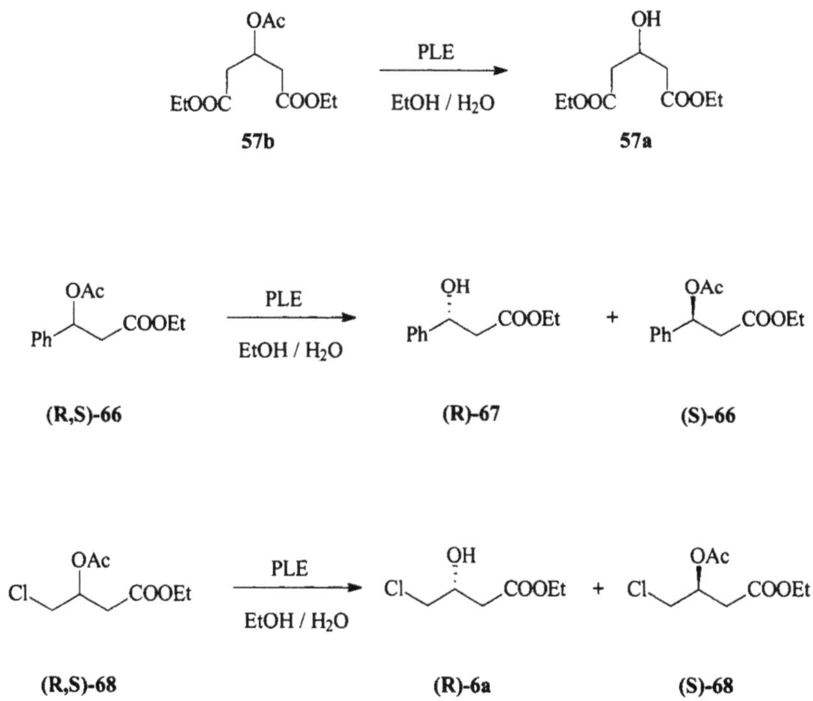

Scheme 32.

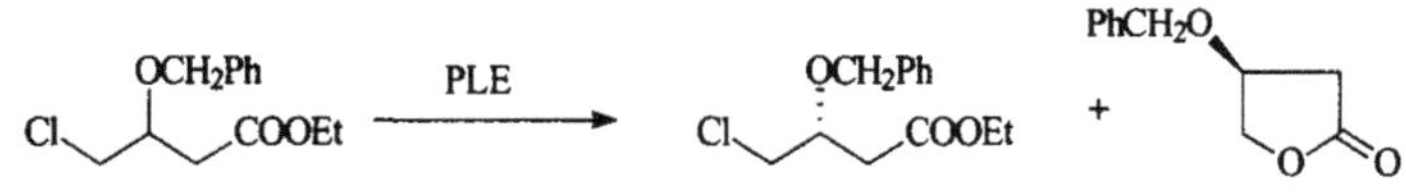

(R,S)-69 **(R)-69** **(S)-70**

Scheme 33.

activity and enantioselectivity of this enzyme is influenced by the presence of cosolvents in the aqueous medium.[56] Addition of ethanol has been beneficial in the PLE-catalyzed hydrolysis of substituted glutarates,[57] but in the case of the 3-acetoxy glutarate **57b** the main reaction was the hydrolysis of the acetate to afford 3-hydroxyglutarate (**57a**).[58] Taking advantage of this observation, the hydrolysis of the 3-acetoxy esters **66** and **68** proceeded in acceptable yields (lowered by the hydrolysis of the carboethoxy moiety) to the corresponding (*R*)-hydroxyesters **67** and **6a** and the (*S*)-acetates **66** and **68**[57] (Scheme 32).

By this route, the (*S*)-**68** could be prepared in 90% ee and the (*R*)-hydroxy ester **6a** reached 57% ee, at the best. It should be recalled that the (*R*)-enantiomer **6a** is the more valuable intermediate for an important drug, L-carnitine. We have already discussed a microbiological approach to this chiral synthon[12] and in the same paper we also showed that, among several other enzymes tested for the hydrolysis of the hydroxy ester **6a** or a few 3-derivatives, the best result still remained—the one shown in Scheme 32. Another interesting result was obtained carrying out the aqueous hydrolysis of the 3-benzyl ether (*R,S*)-**69** in the presence of PLE (Scheme 33). The (*S*)-acid was not recovered from the medium and the main products were the unreacted ester (*R*)-**69** (36% yield, 64% ee) and the (*S*)-lactone **70** (28% yield, 55% ee). Interestingly, when the PLE-catalyzed hydrolysis of **69** was carried out in the 20% ethanol, no reaction was observed after much longer times than that required for the reaction of (*R,S*)-**68** (48 vs. 2 h).

(*R*)-Malic acid is the unnatural stereoisomer of malic acid and a microbiological approach to its synthesis has previously been described by us.[15] In the same report, we had also studied the enzymatic resolution of (*R,S*)-diethyl malate **10** and the corresponding acetate and benzyl ether. The best result, in terms of yields

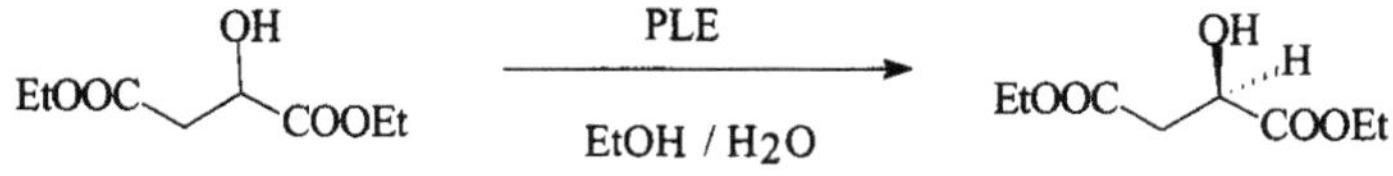

(R,S)-10 **(R)-10**

Scheme 34.

and enantioselectivity, were obtained with PLE in 20% ethanol at 0 °C (42% yield, 93% ee), recovering the unreacted (*R*)-malate **10** (Scheme 34).

VII. ENZYME-CATALYZED RESOLUTION IN ORGANIC SOLVENTS

Since the first publication on enzymatic catalysis in anhydrous solvents,[59] the chemical literature has faced an explosion of reports on applications of this new aspect in biocatalysis.[60] Lipases seem to be especially attractive in this respect due to their peculiar proteic structure, firmly established for a few of them by X-ray crystallography.[61] These enzymes can catalyze in organic solvents reactions virtually impossible in aqueous solution and their capabilities in nonconventional media differ from the performance in water, showing often higher stability, altered activity, and selectivity. From the practical point of view, several profound questions regarding how enzymatic activity compares to that in an aqueous medium can be ignored. An organic chemist should be astonished by the great simplicity of the experimental procedure of these enzymatic reactions that requires simple mixing the organic solution of the reagents in the presence of the powder of the enzymatic preparation, which is subsequently filtered off at the end of the reaction leaving the products in the solvent. The extensive applications in organic synthesis, where the structures are very different, is also highly impressive.

A few predictive rules about the stereochemical outcome of the lipase catalysis have already appeared,[62] although these are intended to be more as a guide to the use of the enzyme with respect to the substrates than a general model. In an organic solvent, a lipase is able to catalyze the reaction inverse to hydrolysis, i.e. esterification or transesterification, and, in order to minimize the reverse reaction, special acylating reagents have been proposed. We have carried out the transesterification reaction in an organic solvent, using vinyl acetate (VA) as acylating reagent[63] and PFL (renamed *Pseudomonas cepacia* lipase) as the most effective enzyme. Under these conditions, the vinyl alcohol that is formed irreversibly tautomerizes to acetaldehyde and prevents the back reaction (Scheme 35). We

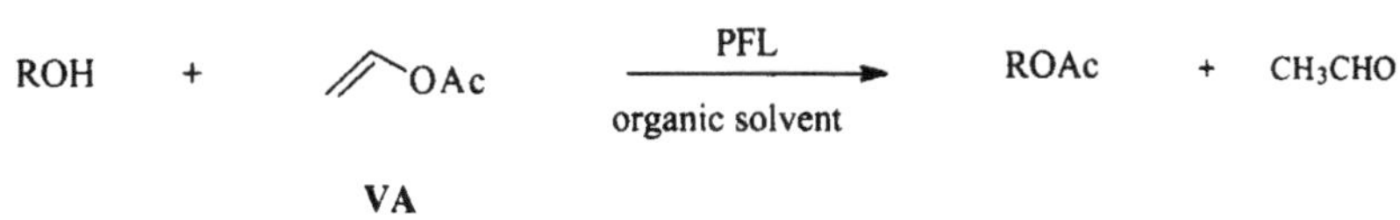

Scheme 35.

started to use this method shortly after its publication and achieved the first successful application to the asymmetrization of a prochiral diol, 2-methyl-1.3-propanediol (**65**) using chloroform as solvent.[54] We found that this reaction proceeded with the formation of the (*S*)-monoacetate **64** with the opposite stereochemical outcome of the hydrolysis of the diacetate **63** (see Scheme 31). In the same paper we observed that, using the same enzymatic transesterification process, the (*R,S*)-monoacetate **64** could afford 60% of the optically inactive diacetate **63** and 40% of unreacted monoester, (*S*)-**64** with >98% ee (Scheme 36).[54]

The same procedure was advantageously used for the resolution of racemic monoderivatives of the diol **65**, namely benzoate **71a** and mono ethers **71b,c** (Scheme 37). In this case, the unreacted (*S*)-**71a–c** and the acetates **72a–c** were formed in nearly optically pure form if the reaction was properly stopped at the desired conversion.[64] It should be mentioned that, due to the relative priorities of the groups, not all the acetates **72a–c** have the *R* configuration, but they are the

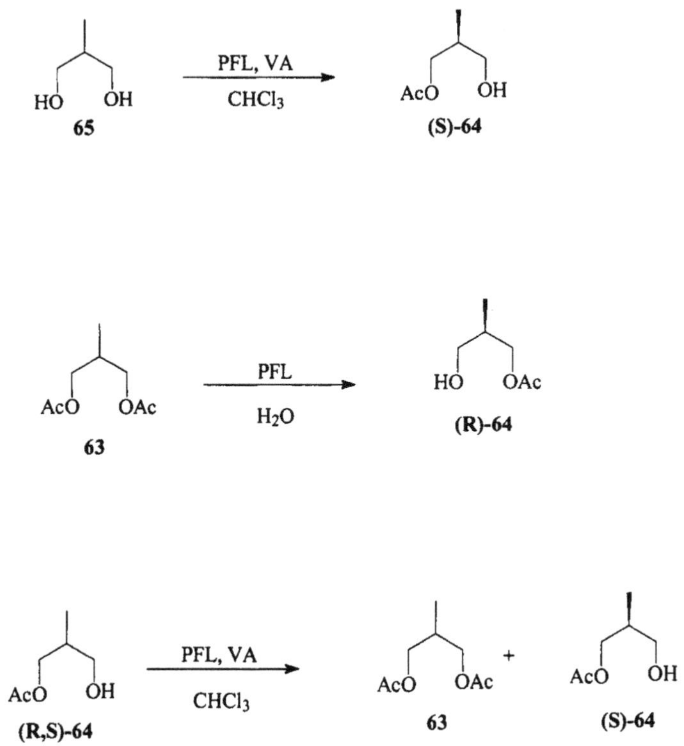

Scheme 36.

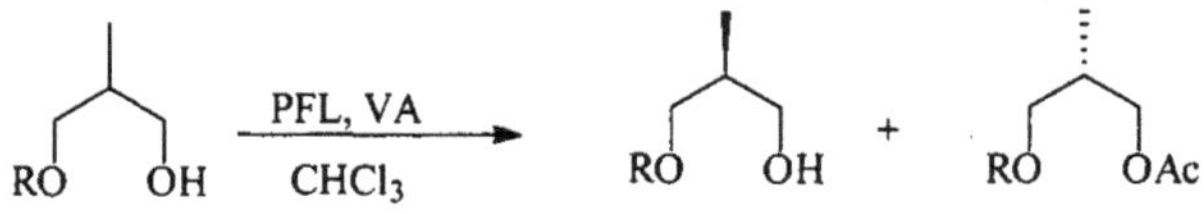

71 **71** **72**

a. R = COPh

b. R = CH_2Ph

c. R = $Si(CH_3)_2tBu$

Scheme 37.

acetates of the enantiomers of the unreacted (*S*)-alcohols **71a–c**. This will happen for other substrates as well and the original reference should be consulted for detailed considerations.

We have also prepared the phenyl selenyl alcohols **73a** and the (*R,S*)-phenylsulfide and sulfonyl derivatives **73b,c** that possess a stereogenic center within a general structure of a 2-methyl-substituted alcohol, and all the compounds were resolved (Scheme 38) into the optically pure (*R*)-alcohols **73a–c** and (*S*)-acetate **74a–c** with the same experimental protocol.[65,66] Simple chemical manipulations of (*R*)-**73a** furnished[65] the (*R*)-silyl ether **75** that was easily converted into the **71c** already prepared by resolution of the racemate, as previously described.[64] Similarly, the (*R*)-phenylsulfonyl alcohol **73c** is the synthetic equivalent of a C-5 isoprenoid synthon that could be used for the synthesis of (25*R*)-26-hydroxycholesterol (**32**) (Scheme 39).

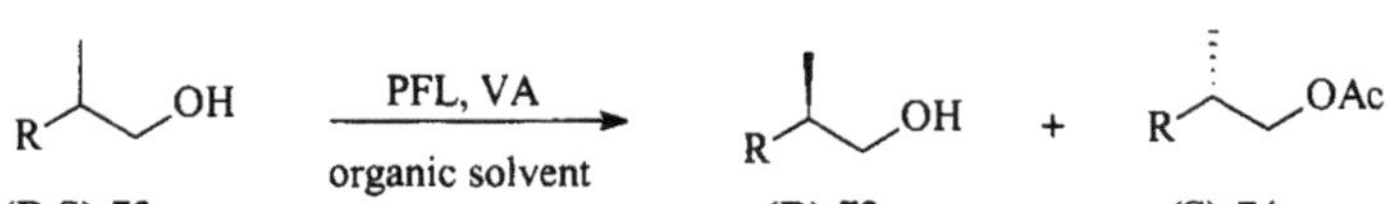

(R,S)-73 **(R)-73** **(S)-74**

a. R = $PhSeCH_2CH_2$

b. R = $PhSCH_2CH_2$

c. R = $PhSO_2CH_2CH_2$

Scheme 38.

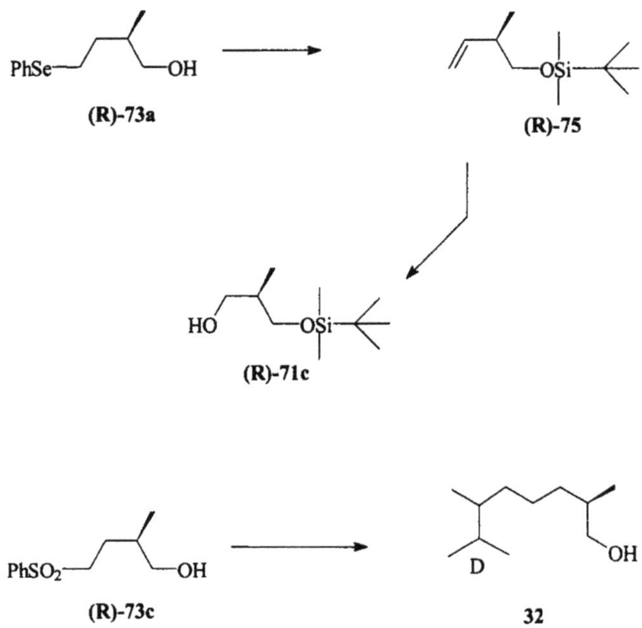

Scheme 39.

The above examples seemed to indicate that 2-methyl alkanols should be excellent substrates for the lipase PFL under the transesterification conditions, and the highest enantioselectivity could be achieved carrying out two separate reactions. If the unreacted alcohol is desired optically pure, the reaction has to be brought to 60% acetate, whereas for the acetate a similar result is obtained when the conversion is stopped at 40%. Following this protocol, we obtained excellent results with (*R/S*, 4*E*)-2-methyl-4-hexen-1-ol (**73d**)[67] and (*R,S*)-2-methyl-decanol (**73e**).[68] In both cases, the (*R*)-alcohols **73d,e** and the (*S*)-acetates **74d,e** were obtained optically pure (Scheme 40).

In light of the proposed model of the active site of PLE,[70] we tentatively assumed that, for the active site of PFL, two hydrophobic sites of different size (H_L and H_S) should also be responsible for the stereo–conformational transition state of the substrate approaching the acylated serine.[71] We also supposed that in the large hydrophobic site H_L a subsite H'_L was present where a small apolar substituent could be located and oriented in such a way that the stereochemical outcome of the products was determined (Scheme 41).

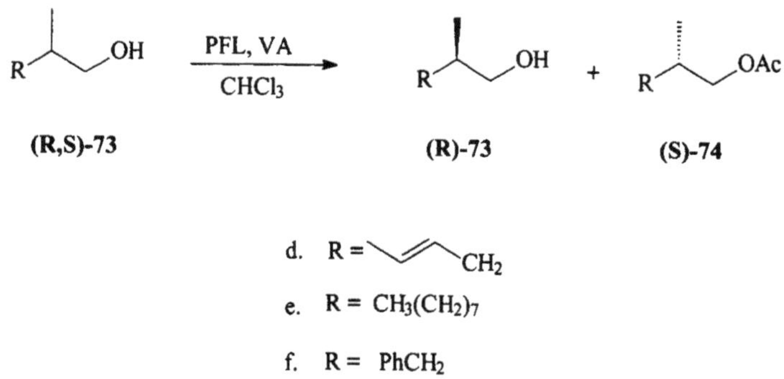

Scheme 40.

A few additional results allowed us to better define the structural requirements for our model as, for instance, the length of the side chain accepted in the large hydrophobic site H_L. We prepared the 4- and 1-benzyl ethers of 2-methyl-1,4-butanediol, compounds **76** and **77**, and found the same stereochemical outcome of the enzymatic reaction since the unreacted alcohols were (*R*)-**76** and (*R*)-**77** (98 and 40% ee, respectively). In both cases (*S*)-acetates **78** and **79** were formed (85 and 52% ee, respectively) and the overall enantioselectivity showed (Scheme 42) that in the 4-benzyl ether, bearing the alcohol function near the stereogenic center, the stereochemical discrimination of the two enantiomers was more efficient than in the case of the 1-benzyl counterpart.[72]

This result seemed to confirm that the 2-methyl-alkanol structure had the structural requisites to be conveniently accommodated in the PFL active site. Furthermore, we compared the results obtained with 3-phenyl-2-methyl-1-propanol

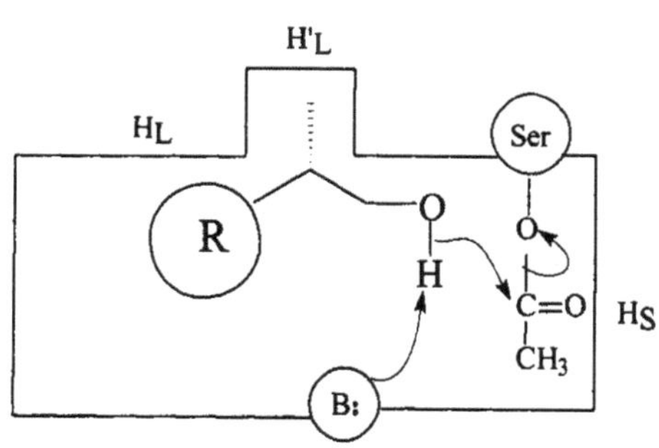

Scheme 41.

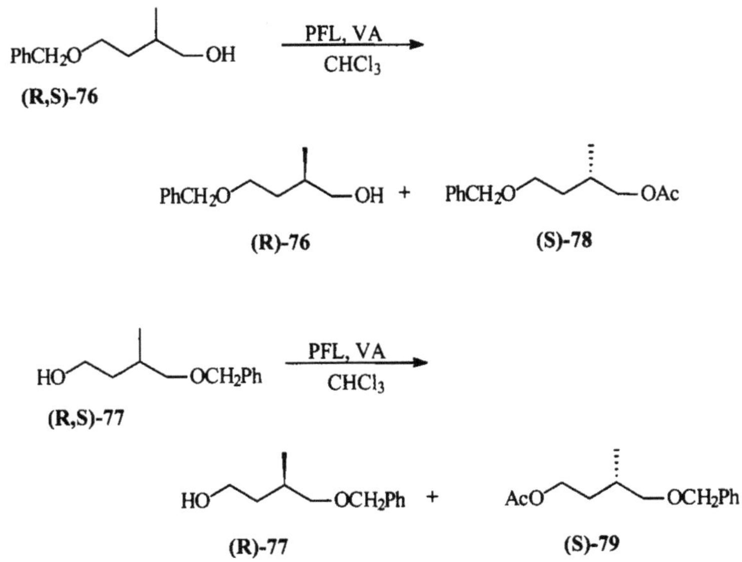

Scheme 42.

73f and its *nor*-analog, 2-phenyl-1-propanol **80**. In accord with data reported by other authors,[69] (*R,S*)-**73f** was efficiently and enantioselectively resolved into the (*R*)-alcohol **73f** and the (*S*)-acetate **74f**, whereas a low enantioselectivity was shown by the enzymatic process on 2-phenyl-1-propanol **80** [20% ee of the unreacted (*R*)-alcohol **80** and the (*S*)-acetate **81**]. As shown in Scheme 43, we proposed that a methylene group was critical in the separation from the stereogenic center and the remaining apolar part of the molecule located in the acylation to the H_L site.[73]

We were convinced that the subsite H'_L should be able to accept small apolar groups and assumed that a polar entity such as a hydroxy group should not be enantioselectively acylated. Therefore, we submitted to the enzymatic transacylation process two 1,2-diols corresponding to the structures **73f** and **80**[74] and also anticipated that conflicting acylation at the primary and secondary OH as well as a diacylation process could complicate the overall enzymatic reaction. The 1,2-acyl migration in these kind of diols could not be excluded either under a chemical or enzymatic reaction and might constitute an additional problem. In any event, by controlling the enzymatic acylation to the stage of the monoacetates, the results from the 1,2-diols **82** and **83** showed that both substrates were acylated with low enantioselectivity, leaving the unreacted (*S*)-diols with 30–40% ee. In

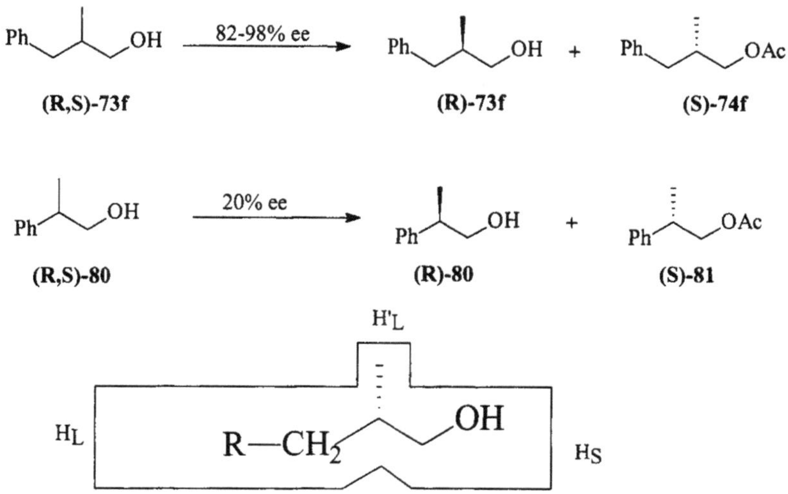

Scheme 43.

the case of the diol **83**, the corresponding (*R*)-monoacetate **85** was accompanied by substantial amounts of the 2-acetate and 1,2-diacetate (Scheme 44).

The 2-methyl ether derivatives **86** and **87** were acylated to the corresponding acetates **88** and **89** with a slight improvement of enantioselectivity (Scheme 45). In fact, from (*R,S*)-**86** the unreacted (*S*)-**86** (48% ee) and (*R*)-acetate **88** (53% ee)

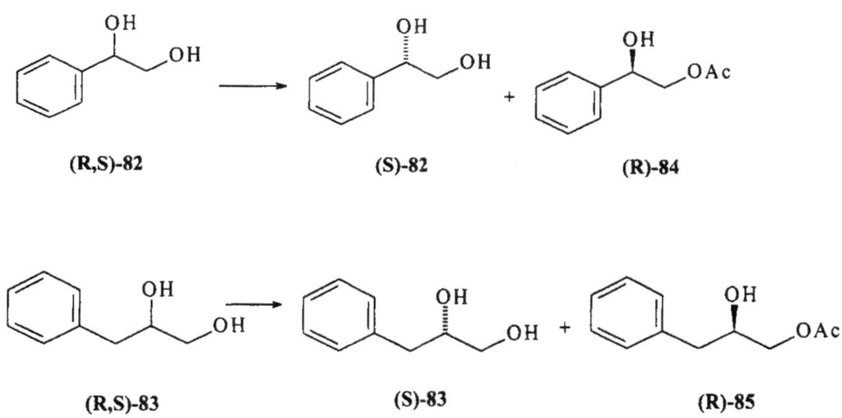

Scheme 44.

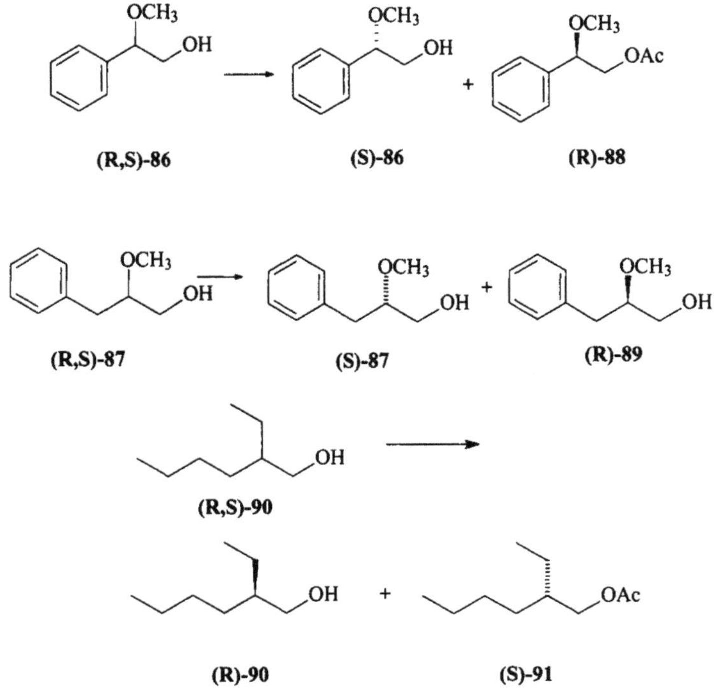

Scheme 45.

were formed, whereas from (*R,S*)-**87** the recovered (*S*)-alcohol had 63% ee and the (*R*)-acetate **89** was formed with 36% ee. This finding partially contradicted the assumption that an apolar group should be preferred to a polar one, but was consistent with the dimension of the subsite H'_L that could not accept an apolar moiety such as the ethyl group. In fact, for the (*R,S*)-2-ethyl hexanol **90** we found a modest enantioselectivity (50–60% ee) in the enzymatic acylation to the acetate **91**.[74]

We later confirmed the assumption about the dimension of the H'_L subsite, preparing a few 2-oxiranemethanols of general formula **92a–d** (Scheme 46). We found that in all cases the (*S*)-epoxyalcohols **92a–d** that remained unreacted were nearly optically pure. The enantiomeric acetates **93a–d** were recovered with excellent ee and the stereochemical outcome was consistent with the initial assumption,[75–77] considering, as previously mentioned, that the consequences of the priorities of the groups dictated the *S* configurations for both substrates and product.

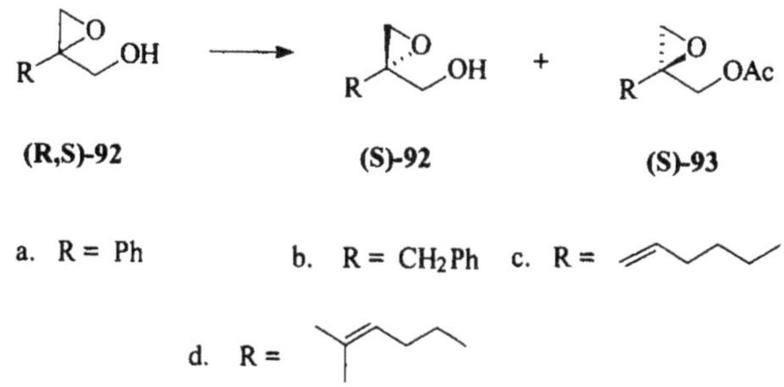

Scheme 46.

Interestingly, compound **92a** containing a phenyl group directly bound to the stereocenter showed the highest enantiomeric ratio *E*,[78] specially if compared to the structurally related 2-methyl alkanol, 2-phenyl-1-propanol **80**, and the secondary alcohol **94**, a more favorable substrate for the PFL-catalyzed transacylation.[62] For alcohol **94** a reverse stereochemical outcome was also observed, if compared to the primary alcohol analog **80** and to the substrate **73f** (Scheme 47). In Table 2 are summarized the values of *E* for compounds **73f**, **92a**, **80**, and **94**.

The high affinity of the oxirane moiety for the apolar subsite H_L' was confirmed, in our opinion, by the high degree of regio- and enantioselectivity in the

Table 2. PFL-Catalyzed Transesterification of Compounds **73f**, **80**, **92a** and **94**

Compound	*Time (h)*[a]	*ee (%)*	*E*	*Configuration*
73f	2	>98		*R*
			172	
74f	1	>98		*S*
80	8	20		*R*
			1.8	
81	3	20		*S*
92a	3	90		*S*
			200	
93a	0.5	>98		*S*
94	9[b]	96		*S*
			95	
95	7.5[c]	>98		*R*

Notes: [a]Corresponding to 40% conversion in the case of acetates and 60% in the case of alcohols.
[b]50% conversion.
[c]40% conversion.

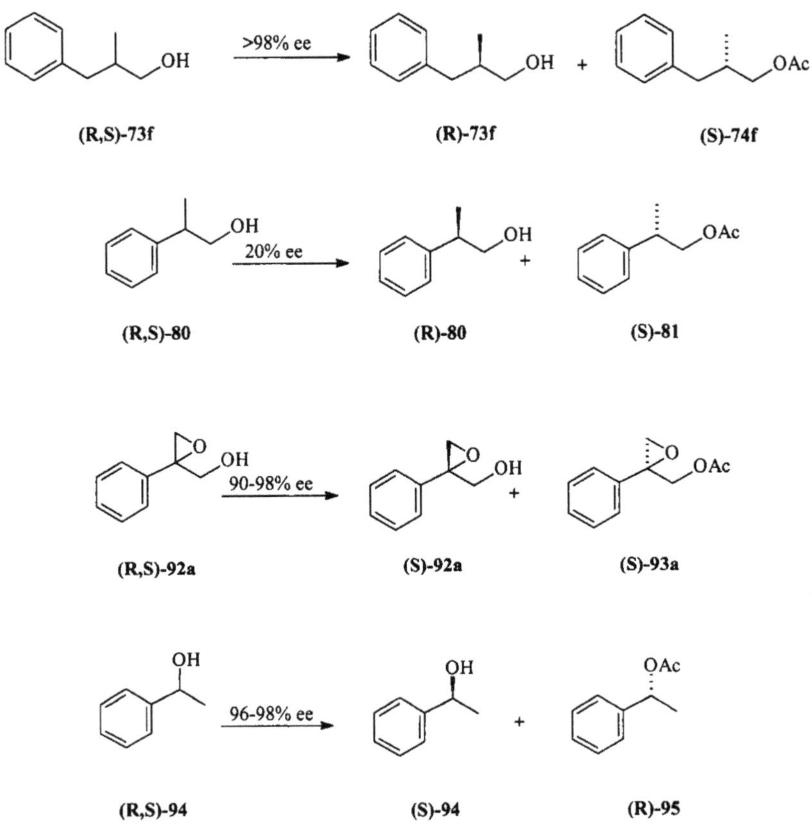

Scheme 47.

transesterification reaction of 2-substituted 1,4-butanediols **96a–c**. We studied the regioselectivity of the reaction through the relative ratio of the 1-acetates **97a–c** versus the 4-acetates **98a–c**.[79] The highest regioselectivity was registered with the substrate **96c** (97:3) and in this case, when the starting diol remained as unreacted material (40%), a 86% ee was evaluated for (*S*)-**96c** (Scheme 48). In diol **96a**, the freely rotating C-2 methyl group seems to induce less regioselection than the two more rigid methylene and oxirane groups (in **96b** and **96c**, respectively). Finally, the fact that the enzymatic process acylates the more hindered C-1 hydroxyl seems to suggest a preferred conformation of the molecules, in such a way that in the apolar subsite the C-2 small substituents find the proper place (Scheme 49).

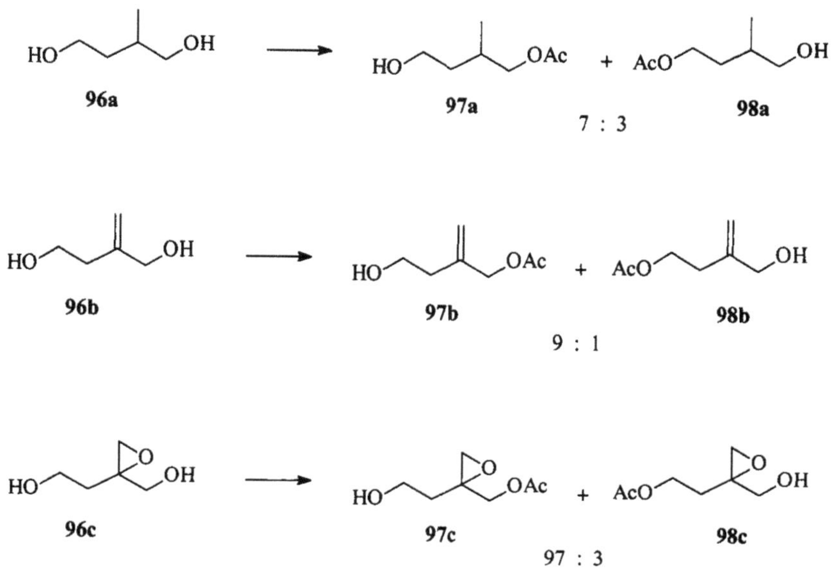

Scheme 48.

It is also worth mentioning that chiral oxiranemethanols can be versatile building blocks and a few of them are available by the Sharpless asymmetric epoxidation procedure.[80] We have prepared enzymatically two chiral oxiranemethanols for the synthesis of important natural products in enantiomerically pure form. In fact, (*S*)-**92c** and (*S*)-**93c** were prepared because they could be chiral synthons for both isomers of frontalin:[76] (*S*)-**92c** is the intermediate for the synthesis of (1*S*, 5*R*)-frontalin **99**, and (*S*)-**92d** is a good chiral synthon for the preparation of (*R*)-

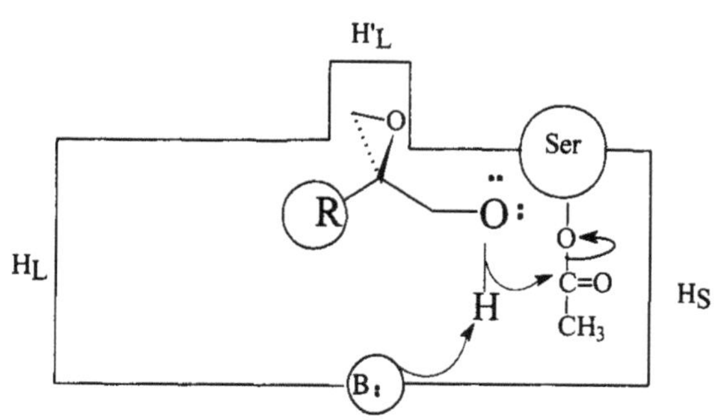

Scheme 49.

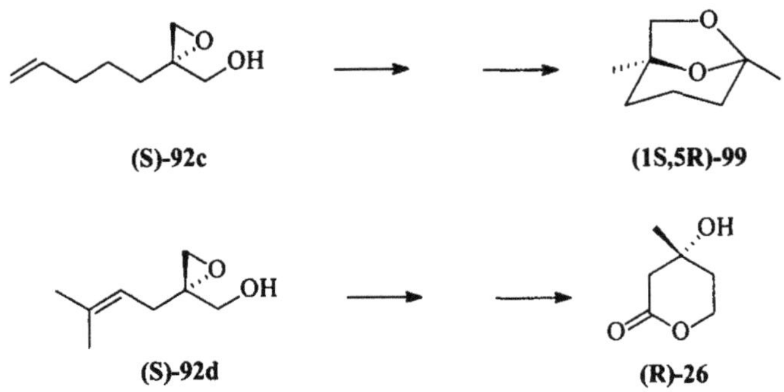

Scheme 50.

mevalonolactone **26** (98% ee), a compound that we tried to synthesize by other chemico–enzymatic approaches[53,77] (Scheme 50).

Strictly related to the previous structures of oxiranemethanols were the 2-methyl-1,2-diols **100a–c**, and the enzymatic resolution of these racemic diols seemed interesting for several reasons. First, we and others[81] noticed that 1,2-diols were not enantioselectively resolved by lipase-catalyzed reactions. Second, the diols could represent a more advanced intermediate for synthesis. Finally, the structural characteristics might be compatible with the supposed active site model of PFL. Therefore, the racemic diols **100a–c** were prepared and enzymatically acylated under the usual conditions to the corresponding 1-acetates **101a–c** (Scheme 51). Again, the resolution of (*R*,*S*)-diol **100a** was almost not enantioselective and the result was superimposable to that obtained from the 2-methyl analog **80**, whereas the process was highly enantioselective for the diols **100b** and **100c** (compare with the results from compounds **73f**, **92b**, and **92d**). The stereochemical outcome was confirmed since the unreacted diols and the acetates were in the *R*- and *S* configuration, respectively.[82] Although our results were

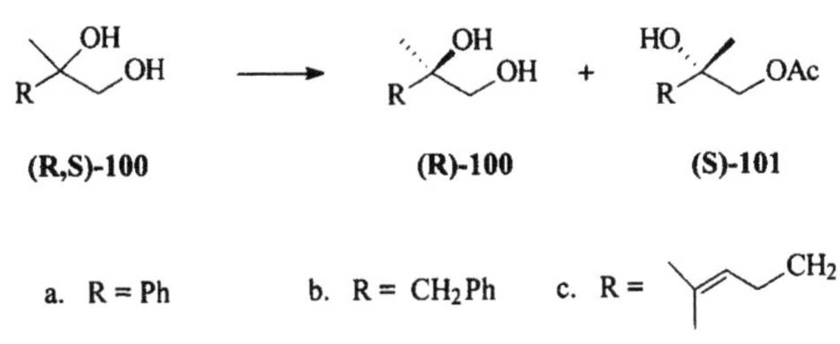

Scheme 51.

rather consistent with our initial assumption, a few ambiguous results indicate that we have unveiled only a small part of the scenario of the active site of the enzyme. The same stereochemical model described for several lipases seems to rationalize mainly the results from secondary alcohols.[62] The same authors have recently tried to extend their model to primary alcohols,[83] but in this case the enzyme enantio preference does not seem to be so easily extrapolated as for secondary alcohols.

In any event, we have demonstrated that primary alcohols substituted at position 2 with small apolar groups can be good substrates for regio- and enantioselective transesterifications catalyzed by PFL in organic solvents. Recently, we have shown that the above moiety can also be part of a steroid structure and the results are still in agreement with the previous assumptions. In fact, the epimeric mixture of the C-22 alcohol **102** can be resolved by PFL (Scheme 52) and the (20*S*)-acetate **103** and the (20*R*)-alcohol **102** are formed in the reaction.[84]

However, when an aromatic ring is present in 2-methyl-substituted alkanols, we have noticed that another factor may become critical, e.g. the distance of the aromatic group from the stereogenic center. This can be partially explained by the presence of one or more aromatic amino acids present at the active site of the enzyme that might influence the stereochemical outcome in organic solvents thanks to electronic interactions. A methylene group between the stereogenic center and the aromatic moiety seems to be the optimum; this has also been recently confirmed for a series of 2-methyl-naphthyl alkanols.[85] For these compounds the problem of regioselectivity also has to be considered because the naphthyl alkanols may exist as 1- or 2-regioisomers. The results obtained from compounds **104a,b** and **106a,b** show that the 2-regioisomers **106** react faster than the 1-isomer **104** and that the highest *E* has been obtained for the alcohols structurally related to **73f**, i.e. distant by a methylene group from the stereogenic center. In fact, **104b** affords the (*R*)-alcohol **104b** and the (*S*)-acetate **105b** with >98% ee (Table 3). In all cases, the unreacted (*R*)-alcohols **104a,b** and **106a,b** are obtained together with (*S*)-acetates **105a,b** and **107a,b** (Scheme 53).

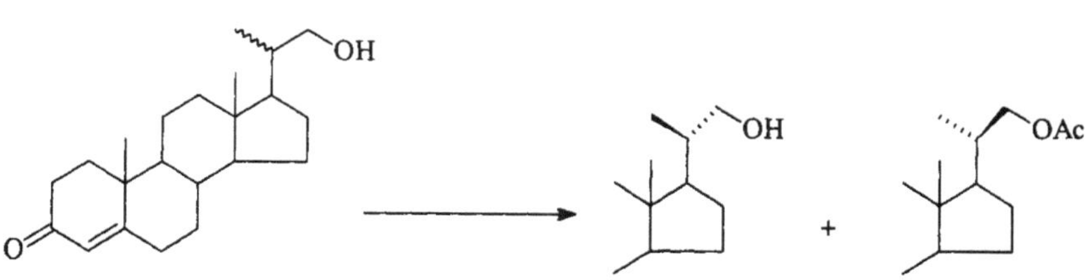

(20 R,S)-102 **(20 R)-102** **(20S)-103**

Scheme 52.

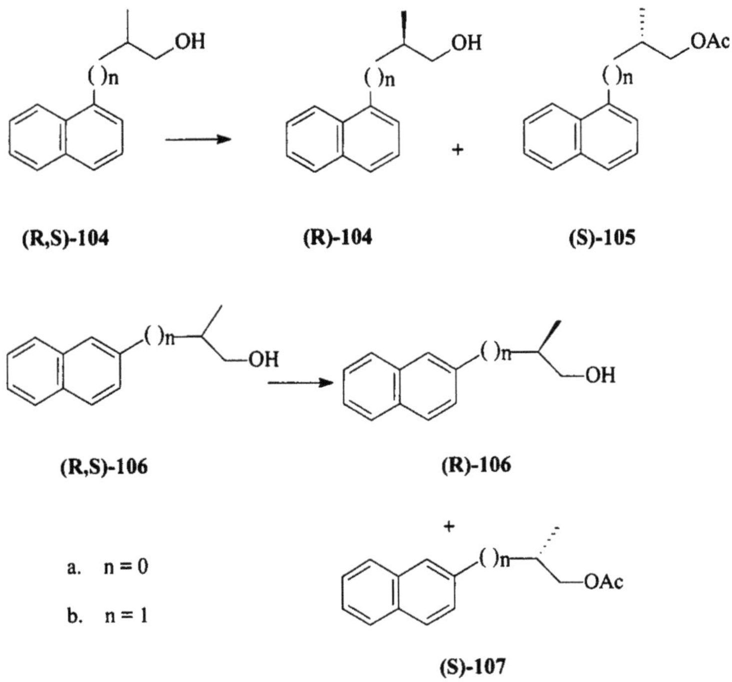

Scheme 53.

Table 3. PFL-Catalyzed Transesterification of Naphtyl Alcohols **104** and **106**

Substrate	*Product*	*Time (h)*	*Yield (%)*	*ee (%)*[a]	*Config.*[b]	*E*
104a	(–)-**105a**	86	35	72	*S*	9
104a	(+)-**104a**	97	43	70	*R*	
106a	(–)-**107a**	16	32	62	*S*	6
106a	(+)-**106a**	22	31	65	*R*	
104b	(+)-**105b**	4	30	>98	*S*	150
104b	(+)-**104b**	23	34	>98	*R*	
106b	(+)-**107b**	3	40	92	*S*	45
106b	(+)-**106b**	7	40	85	*R*	

Notes: [a]The acetates were converted into the alcohols by reaction with $LiAlH_4$ and the ee established by NMR.
[b]Configurations assigned by optical rotation or by NMR.

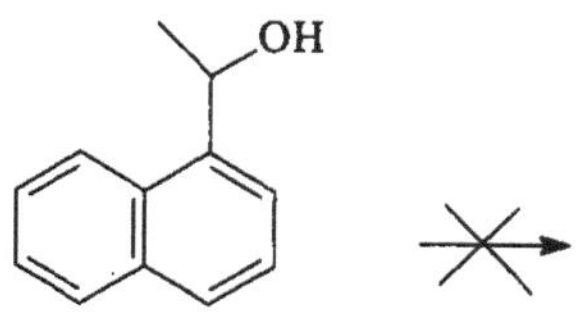

(R,S)-108

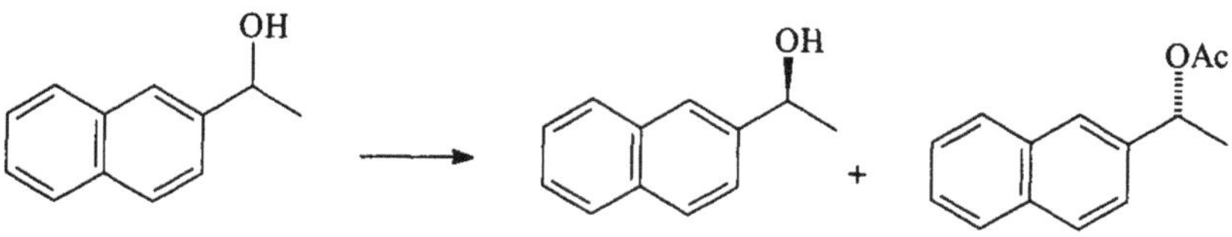

(R,S)-109 **(S)-109** **(R)-110**

Scheme 54.

It is interesting to observe that the secondary naphthyl alcohols **108** (a 1-isomer) does not react under the conditions of the enzymatic reaction, whereas the 2-isomer **109** affords nearly optically pure (*S*)-**109** and (*R*)-**110** (Scheme 54). A similar reversal of configuration going from primary to secondary alcohols has been observed by others[83] and ourselves.[73]

Finally the peculiar influence of the aromatic portion of the substrate on the enzymatic reaction has been confirmed by the results obtained from the acylation of the 2-substituted-1,4-butanediols **111a–c**. The ratio between the acetates **112a–c** and **113a–c** was strongly influenced by the aromatic moiety of the substrate[86] since for the 2-benzyl **111b** there was no regioselection, whereas **111a** and **111c** furnished opposite results (Scheme 55). The ratio **112a/113a** was 3:7 with the 4-acetate being preferred, and for the 2-*O*-benzyl ethers **112c/113c** the result was again in favor of the 1-acetate (7:3).

We studied in more detail how the stereochemistry of the substrate could influence the regioselectivity of the reaction, preparing optically pure (*R*)- and (*S*)-**111c**. We found that, whereas (*S*)-**111c** was regioselectively acylated at the C-1 position [(*S*)-**112c**:**113c**, 9:1], the (*R*)-**111c** was acylated with less regioselection and with predominant C-4 acylation [(*R*)-**112c**:**113c**, 35:65] (Scheme 56).

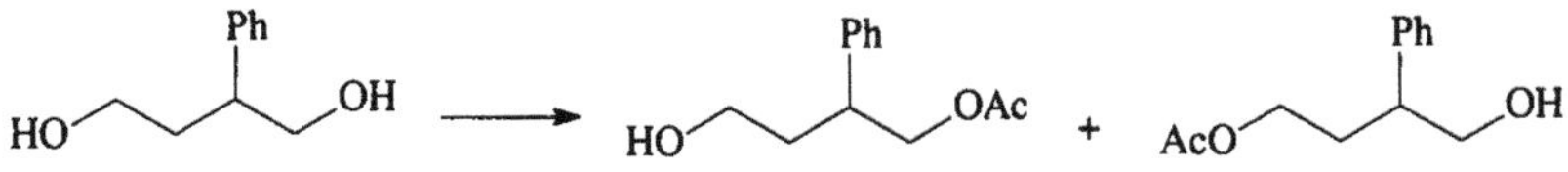

111a 112a 113a

3 : 7

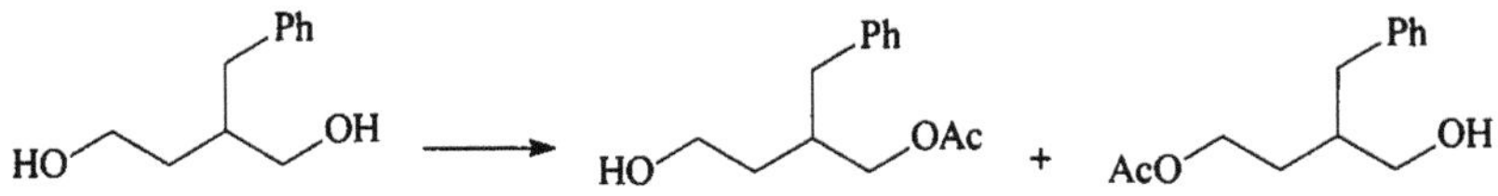

111b 112b 113b

1 : 1

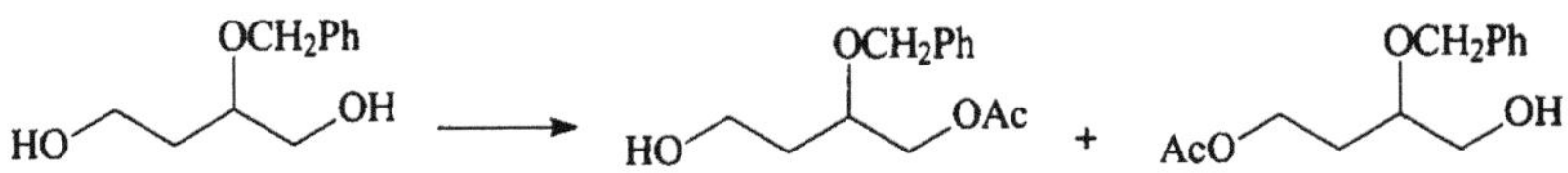

111c 112c 113c

7 : 3

Scheme 55.

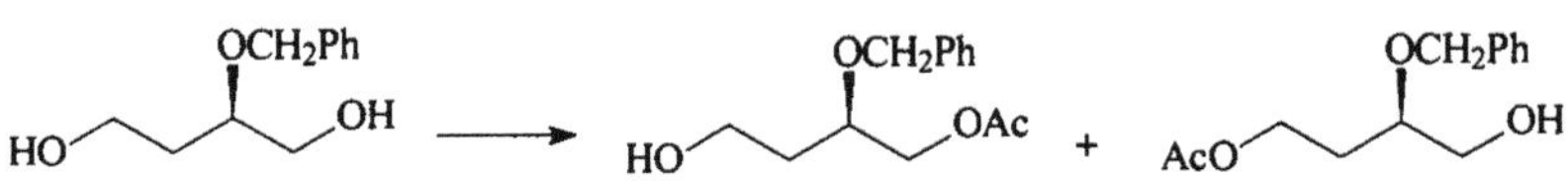

(R)-111c (R)-112c (R)-113c

35 : 65

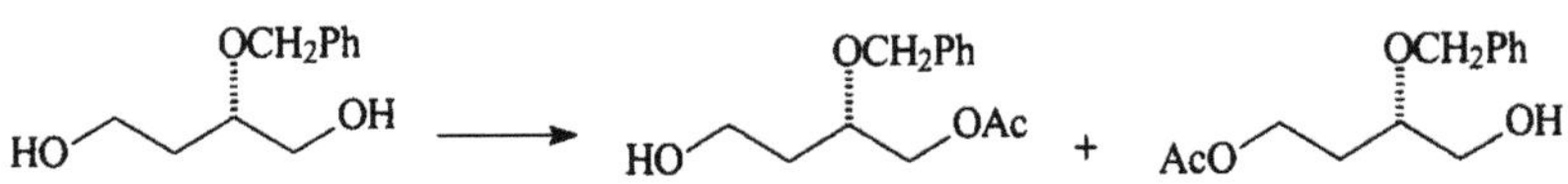

(S)-111c (S)-112c (S)-113c

9 : 1

Scheme 56.

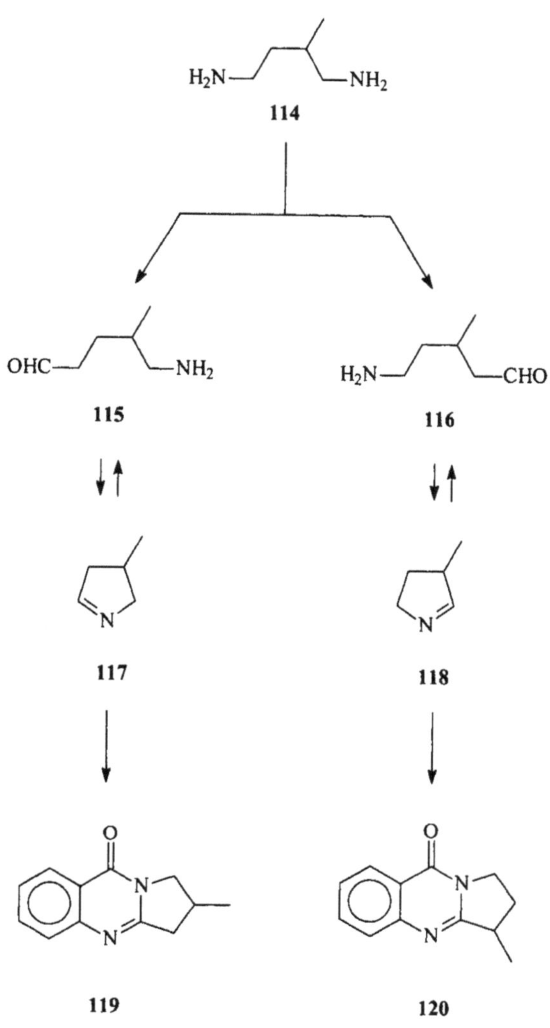

Scheme 57.

We had carried out similar studies on the regio- and enantioselectivity of the oxidation of 2-methyl-1,4-diaminobutane (2-methylputrescine) **114** by diamino oxidase from pea seedlings (PDAO) or pig kidney (KDAO). In this case the ratio between the two products of oxidation, the isomeric amino aldehydes **115** and **116** that were in equilibrium with the corresponding pyrrolines **117** and **118**, was

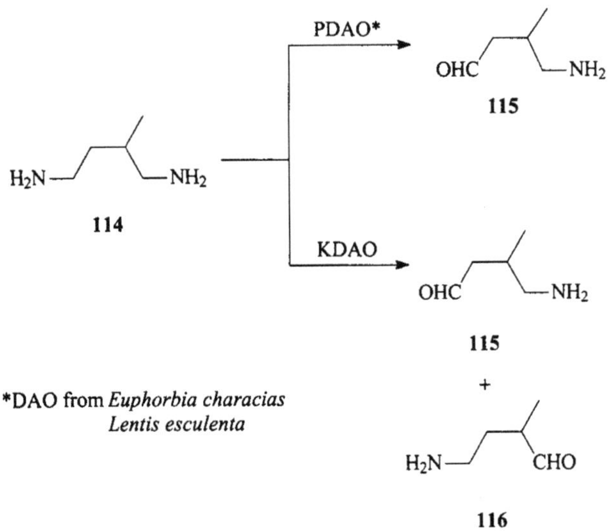

Scheme 58.

established through the products of reaction with 2-aminobenzaldehyde, i.e. the isomeric methylquinazolones **119** and **120** (Scheme 57). The results were that PDAO catalyzed the regioselective oxidation of **114** to the aminoaldehyde **115** (determined as **119**), whereas KDAO showed virtually no regioselectivity[87] since an equimolar ratio of **119** and **120** was formed (Scheme 58). We synthesized the optically pure (*R*)- and (*S*)-**114** and found[88] that PDAO regioselectively catalyzed

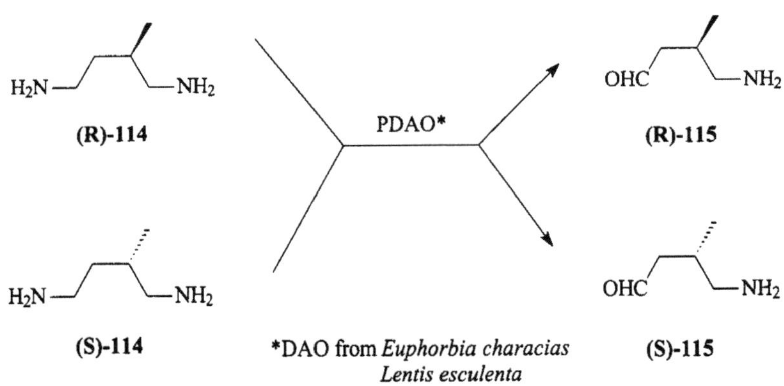

Scheme 59.

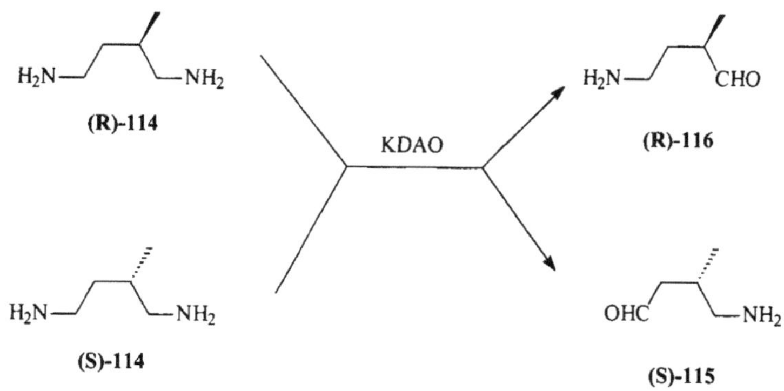

Scheme 60.

the oxidation of the C-4 amino group to the aminoaldehyde **115** (Scheme 59). The results obtained with KDAO on (*R*)- and (*S*)-**114**[88] explained the lack regioselectivity previously reported. In fact, (*R*)-**114** was regioselectively oxidized at the C-1 amine affording almost exclusively (*R*)-**116**, whereas from (*S*)-**114** only the (*S*)-**115** was formed (Scheme 60).

VIII. CONCLUSIONS

We hope that the results presented in this report may witness that biocatalysis can be an efficient tool in the hands of organic chemists for the synthesis of enantiomerically pure compounds. A judicious choice of the biocatalyst most suitable to the desired reaction can determine the success of the biotransformation. This is at the time still an empirical practice, although many examples are currently available and one can decide to use a biocatalyst by correlation with successful applications. Additionally, a few models are being elaborated for a kind of structure–activity relationship.

It is also very encouraging that biochemical technologies leading to optically pure compounds are attracting interest for industrial applications.[89,90] In this field, the economical rationale will be decisive and will dictate the choice of the most efficient biocatalyst with a compromise relating to the price of the process, whereas academic research may hold in perspective fascinating new horizons going from biologically and chemically modified enzymes to production of less accessible enzymes through the techniques of molecular biology.

ACKNOWLEDGMENTS

We thank Ministero dell'Università e Ricerca Scientifica e Tecnologica and the Italian National Research Council (CNR) for financial support.

REFERENCES AND NOTES

1. Seebach, D.; Hungerbühler, E. In *Modern Synthetic Methods 1980*; Scheffold, R., Ed.; Salle & Sauerländer: Frankfurt am Main, 1980, p. 91.
2. A rich literature has flourished on this topic; for a review accounting the progress on these researches up to 1991, see: Santaniello, E.; Ferraboschi, P.; Grisenti, P.; Manzocchi, A. *Chem. Rev.* **1992**, *92*, 1071. For more recent books and reviews, see: (a) Mori, K. *Synlett* **1995**, 1097. (b) *Enzyme Catalysis in Organic Synthesis*; Drauz, K.; Waldmann, H. Eds.; VCH: Weinheim, 1995.
3. Prelog, V. *Pure Appl. Chem.* **1964**, *9*, 119.
4. For exhaustive reviews, see: (a) Servi, S. *Synthesis* **1990**, 1. (b) Csuk, R.; Glänzer, B.I. *Chem. Rev.* **1991**, *91*, 49.
5. The incubation conditions have to be carefully controlled in order to ensure >95% ee: Wipf, B.; Kupfer, E.; Bertazzi, R.; Leuenberger, H.G.W. *Helv. Chim. Acta* **1983**, *66*, 485.
6. For a few significant examples of applications of **2** to asymmetric syntheses, see for instance: (a) Mori, K. *Tetrahedron* **1981**, *37*, 1341. (b) Nicolau, K.C.; Pavia, M.R.; Seitz, S.P. *J. Am. Chem. Soc.* **1981**, *103*, 1224. (c) Kramer, A.; Pfander, H. *Helv. Chim. Acta* **1982**, *65*, 293. (d) Seidel, W.; Seebach, D. *Tetrahedron Lett.* **1982**, *23*, 159.
7. Seebach, D.; Sutter, M.A.; Weber, R.H.; Züger, M.F. *Organic Syntheses* J. Wiley: New York, 1985; Vol. 63, p. 1.
8. Santaniello, E.; Casati, R.; Milani, F. *J. Chem. Soc., Perkin Trans. 1* **1985**, 919.
9. Santaniello, E.; Casati, R.; Milani, F. *J. Chem. Res. (S)* **1984**, 132.
10. Zhou, B.-N.; Gopalan, A.S.; VanMiddlesworth, F.; Shieh, W.-R; Sih, C.J. *J. Am. Chem. Soc.* **1983**, *105*, 5925.
11. Trincone, A.; Nicolaus, B.; Lama, L.; Marsiglia, F.; Gambacorta, A. *Biotechnol. Lett.* **1991**, *13*, 31.
12. Aragozzini, F.; Valenti, M.; Santaniello, E.; Ferraboschi, P.; Grisenti, P. *Biocatalysis* **1992**, *5*, 325.
13. (a) Hunt, J.R.; Carter, A.S.; Murrell, J.C.; Dalton, H.; Hallinan, K.O.; Crout, D.H.G.; Holt, R.A.; Crosby, J. *Biocatalysis Biotranf.* **1995**, *12*, 159. (b) Koul, S.; Crout, D.H.G.; Errington, W.; Tax, J. *J. Chem. Soc., Perkin Trans. 1* **1995**, 2969. (c) Zymanczykduda, E.; Lejczak, B.; Kafarski, P.; Grimaud, J.; Fischer, P. *Tetrahedron* **1995**, *51*, 11809. (d) Mochizuki, N.; Hiramatsu, S.; Sugai, T.; Ohta, H.; Morita, H.; Itokawa, H. *Bioscience Biotechnol. Biochem.* **1995**, *59*, 2282. (e) Nakamura, K.; Kondo, S.; Kawai, Y.; Hida, K.; Kitano, K.; Ohno, A. *Tetrahedron: Asymmetry* **1996**, *7*, 409.
14. Manzocchi, A.; Casati, R.; Fiecchi, A.; Santaniello, E. *J. Chem. Soc., Perkin Trans. 1* **1987**, 2753.
15. Santaniello, E.; Ferraboschi, P.; Grisenti, P.; Aragozzini, F.; Maconi, E. *J. Chem. Soc., Perkin Trans. 1* **1991**, 601.
16. Ferraboschi, P.; Santaniello, E.; Tingoli, M.; Aragozzini, F.; Molinari, F. *Tetrahedron: Asymmetry* **1993**, *4*, 1931.

17. (a) Levene, P.A.; Walti, A. In *Organic Syntheses*; J. Wiley: New York, 1943, Coll. Vol. II, p. 545. (b) Guetté, J.P.; Spassky, N. *Bull. Soc. Chim. Fr.* **1972**, 4217. (c) Barry, J.; Kagan, H. B. *Synthesis* **1981**, 453.
18. Several dehydrogenases are present in baker's yeast that can accept a substrate with different stereochemical outcome; see: (a) Shieh, W.R.; Gopalan, A.S.; Sih, C.J. *J. Am. Chem. Soc.* **1985**, *107*, 2993. (b) Heidlas, J.; Engel, K.H.; Tressl, R. *Eur. J. Biochem.* **1988**, *172*, 633.
19. MacLeod, R.; Prosser, H.; Fikentscher, L.; Lanyi, J.; Mosher, H.S. *Biochemistry* **1964**, *3*, 838.
20. Manzocchi, A.; Fiecchi, A.; Santaniello, E. *Synthesis* **1987**, 1007.
21. Ferraboschi, P.; Grisenti, P.; Manzocchi, A.; Santaniello, E. *J. Chem. Soc., Perkin Trans. 1* **1990**, 2469.
22. (a) Takayama, H.; Ohmori, M.; Yamada, S. *Tetrahedron Lett.* **1980**, *21*, 5027. (b) Bhatia, S.K.; Hajdu, J. *Tetrahedron Lett.* **1987**, *28*, 3767.
23. Early reports on this biotransformation had already shown the feasibility of the reaction: Ridley, D.D.; Stralow, M. *J. Chem. Soc., Chem Commun.* **1975**, 400.
24. Manzocchi, A.; Fiecchi, A.; Santaniello, E. *J. Org. Chem.* **1988**, *53*, 4405.
25. Ferraboschi, P.; Grisenti, P.; Manzocchi, A.; Santaniello, E. *Tetrahedron* **1994**, *50*, 10539.
26. Simon, H.; Bader, J.; Günther, H.; Neumann, S.; Thanos, J. *Angew. Chem., Int. Ed. Engl.* **1985**, *24*, 539.
27. (a) Fuganti, C.; Ghiringhelli, D.; Grasselli, P. *J. Chem. Soc., Chem. Commun.* **1975**, 846. (b) Gramatica, P.; Manitto, P.; Poli, L. *J. Org. Chem.* **1985**, *50*, 4625.
28. Leuenberger, H.G.W.; Boguth, W.; Barner, R.; Schmid, M.; Zell, R. *Helv. Chim. Acta* **1979**, *62*, 455.
29. Ferraboschi, P.; Santaniello, E. *Synth. Commun.* **1984**, *14*, 1199.
30. Ferraboschi, P.; Grisenti, P.; Casati, R.; Fiecchi, A.; Santaniello, E. *J. Chem. Soc., Perkin Trans. 1* **1987**, 1743.
31. Ferraboschi, P.; Grisenti, P.; Fiecchi, A.; Santaniello, E. *Org. Prep. Proc. Int.* **1989**, *21*, 371.
32. Ferraboschi, P.; Fiecchi, A.; Grisenti, P.; Santaniello, E. *J. Chem. Soc., Perkin Trans. 1* **1987**, 1749.
33. (a) Utaka, M.; Onoue, S.; Takeda, A. *Chem. Lett.* **1987**, 971. (b) Sakai, T.; Matsumoto, S.; Hidaka, S.; Imajo, N.; Tsuboi, S.; Utaka, M. *Bull. Chem. Soc. Jpn.* **1991**, *64*, 3473.
34. Ferraboschi, P.; Casati, S.; Santaniello, E. *Tetrahedron: Asymmetry* **1994**, *5*, 19.
35. Ferraboschi, P.; Santaniello, E., unpublished results.
36. Branca, Q.; Fischli, A. *Helv. Chim. Acta* **1977**, *60*, 925.
37. Majeric, M.; Avdagic, A.; Hamersak, Z.; Sunjic, V. *Biotechnol. Lett.* **1995**, *17*, 1189.
38. Floss, H.G.; Tsai, M.D. *Adv. Enzymol. Relat. Areas Mol. Biol.* **1979**, *50*, 243.
39. (a) Cornforth, J.W.; Redmond, J.W.; Eggerer, H.; Buckel, W.; Gutschow, C. *Nature* **1969**, *221*, 1212. (b) Lüthy, J.; Rétey, J.; Arigoni, D. *Nature* **1969**, *221*, 1213.
40. Caspi, E.; Santaniello, E.; Patel, K.; Arunachalam, T. *J. Chem. Soc., Chem. Commun.* **1984**, 379.
41. Whitlock, H.W., Jr. In *Applications of Biochemical Systems in Organic Chemistry*; Jones, J.B.; Sih, C.J.; Perlman, D., Eds.; J. Wiley: New York, 1976, p. 1045.
42. Santaniello, E.; Galli Kienle, M.; Manzocchi, A.; Bosisio, E. *J. Chem. Soc., Perkin Trans. 1* **1979**, 1677.
43. Santaniello, E.; Manzocchi, A.; Biondi, P.A. *J. Chem. Soc., Perkin Trans. 1* **1981**, 307.
44. Bellau, B.; Burba, J. *J. Am. Chem. Soc.* **1960**, *82*, 5751.
45. Santaniello, E.; Casati, R.; Manzocchi, A. *J. Chem. Soc., Perkin Trans. 1* **1985**, 2389.
46. Gramatica, P.; Manitto, P.; Manzocchi, A.; Santaniello, E. *J. Label. Compounds Radiopharm.* **1981**, *18*, 955.

47. Schoffers, E.; Golebiowski, A.; Johnson, C.R. *Tetrahedron* **1996**, *52*, 3769.
48. Huang, F.C.; Lee, L.F.H.; Mittal, R.S.D.; Ravikumar, P.R.; Chan, J.A.; Sih, C.J.; Caspi, E.; Eck, C.R. *J. Am. Chem. Soc.* **1975**, *97*, 4144.
49. (a) Cohen, S.G.; Khedouri, J. *J. Am. Chem. Soc.* **1961**, *83*, 4228. (b) Brooks, D.W.; Palmer, J.T. *Tetrahedron Lett.* **1983**, *24*, 3059.
50. (a) Gopalan, A.S.; Sih, C.J. *Tetrahedron Lett.* **1984**, *25*, 5235. (b) Rosen, T.; Heathcock, C.H. *Tetrahedron* **1986**, *42*, 4909.
51. Santaniello, E.; Chiari, M.; Ferraboschi, P.; Trave, S. *J. Org. Chem.* **1988**, *53*, 1567.
52. Santaniello, E.; Casati, R.; Ceriani, L.; Ferraboschi, P.; Grisenti, P. *Chem. Phys. Lipids* **1988**, *49*, 97.
53. Santaniello, E.; Canevotti, R.; Casati, R.; Ceriani, L.; Ferraboschi, P.; Grisenti, P. *Gazz. Chim. Ital.* **1989**, *119*, 55.
54. Santaniello, E.; Ferraboschi, P.; Grisenti, P. *Tetrahedron Lett.* **1990**, *31*, 5657; for a correction, see: *Tetrahedron Lett.* **1991**, *32*, 430.
55. Boland, W.; Frössl, C.; Lorenz, M. *Synthesis* **1991**, 1049.
56. (a) Björkling, F.; Boutelje, J.; Gatenbeck, S.; Hult, K.; Norin, T. *Tetrahedron Lett.* **1985**, *26*, 4957. (b) Björkling, F.; Boutelje, J.; Gatenbeck, S.; Hult, K.; Norin, T.; Szmulik, P. *Bioorg. Chem.* **1986**, *14*, 176.
57. Sabbioni, G.; Jones, J.B. *J. Org. Chem.* **1987**, *52*, 4565.
58. Santaniello, E.; Ferraboschi, P.; Grisenti, P.; Manzocchi, A.; Trave, S. *Gazz. Chim. Ital.* **1989**, *119*, 581.
59. Zaks, A.; Klibanov, A.M. *Science* **1984**, *224*, 1249.
60. Koskinen, A.M.P.; Klibanov, A.M., Eds.; *Enzymatic Reactions in Organic Media*; Chapman & Hall: London, 1996.
61. Derewenda, Z.S.; Sharp, A.M. *Trends Biochem. Sci.* **1993**, *18*, 20.
62. Kazlauskas, R.J.; Weissfloch, A.N.E.; Rappaport, A.T.; Cuccia, L.A. *J. Org. Chem.* **1991**, *56*, 2656.
63. (a) Degueil-Castaing, M.; De Jeso, B.; Drouillard, S.; Maillard, B. *Tetrahedron Lett.* **1987**, *28*, 953. (b) Wang, Y.-F.; Lalonde, J.J.; Momongan, M.; Bergbreiter, D.E.; Wong, C.-H. *J. Am. Chem. Soc.* **1988**, *110*, 7200.
64. Grisenti, P.; Ferraboschi, P.; Manzocchi, A.; Santaniello, E. *Tetrahedron* **1992**, *48*, 3827.
65. Ferraboschi, P.; Grisenti, P.; Santaniello, E. *Synlett* **1990**, 545.
66. Ferraboschi, P.; Grisenti, P.; Manzocchi, A.; Santaniello, E. *J. Org. Chem.* **1990**, *55*, 6214.
67. Ferraboschi, P.; Brembilla, D.; Grisenti, P.; Santaniello, E. *Synlett* **1991**, 310.
68. Ferraboschi, P.; Grisenti, P.; Manzocchi, A.; Santaniello, E. *J. Chem. Soc., Perkin Trans. 1* **1992**, 1159.
69. Delinck, D.L.; Margolin, A.L. *Tetrahedron Lett.* **1990**, *31*, 6797.
70. Toone, E.J.; Werth, M.J.; Jones, J.B. *J. Am. Chem. Soc.* **1990**, *112*, 4946.
71. Deleuze, H.; Langrand, G.; Millet, H.; Baratti, J.; Buono, G.; Triantaphylides, C. *Biochim. Biophys. Acta* **1987**, *911*, 117.
72. Grisenti, P.; Ferraboschi, P.; Casati, S.; Santaniello, E. *Tetrahedron: Asymmetry* **1993**, *4*, 997.
73. Ferraboschi, P.; Casati, S.; De Grandi, S.; Grisenti, P.; Santaniello, E. *Biocatalysis* **1994**, *10*, 279.
74. Ferraboschi, P.; Santaniello, E., unpublished results.
75. Ferraboschi, P.; Brembilla, D.; Grisenti, P.; Santaniello, E. *J. Org. Chem.* **1991**, *56*, 5478.
76. Ferraboschi, P.; Casati, S.; Grisenti, P.; Santaniello, E. *Tetrahedron: Asymmetry* **1993**, *4*, 9.
77. Ferraboschi, P.; Casati, S.; Grisenti, P.; Santaniello, E. *Synlett* **1994**, 754.

78. (a) Chen, C.-S.; Fujimoto, Y.; Girdaukas, G.; Sih, C.J. *J. Am. Chem. Soc.* **1982**, *104*, 7294. (b) Chen, C.-S.; Wu, S.-H.; Girdaukas, G.; Sih, C.J. *J. Am. Chem. Soc.* **1987**, *109*, 2812. (c) Sih, C.J.; Wu, S.-H. *Topics in Stereochem.* **1989**, *19*, 63.
79. Ferraboschi, P.; Grisenti, P.; Manzocchi, A.; Santaniello, E. *Tetrahedron: Asymmetry* **1994**, *5*, 691.
80. Pfenninger, A. *Synthesis* **1986**, 89.
81. (a) Theil, F.; Weidner, J.; Ballschuh, S.; Kunath, A.; Schick, H. *J. Org. Chem.* **1994**, *59*, 388. (b) Theil, F. *Catalysis Today* **1994**, *22*, 517.
82. Ferraboschi, P.; Casati, S.; Grisenti, P.; Santaniello, E. *Tetrahedron: Asymmetry* **1994**, *5*, 1921.
83. Weissfloch, A.N.E.; Kazlauskas, R.J. *J. Org. Chem.* **1995**, *60*, 6959.
84. Ferraboschi, P.; Molatore, A.; Verza, E.; Santaniello, E. *Tetrahedron: Asymmetry* **1996**, *7*, 551.
85. Ferraboschi, P.; Casati, S.; Manzocchi, A.; Santaniello, E. *Tetrahedron: Asymmetry* **1995**, *6*, 1521.
86. Ferraboschi, P.; Casati, S.; Verza, E.; Santaniello, E. *Tetrahedron: Asymmetry* **1995**, *6*, 1027.
87. Santaniello, E.; Manzocchi, A.; Biondi, P. A.; Simonic, T. *Experientia* **1982**, *38*, 782.
88. Santaniello, E.; Manzocchi, A.; Biondi, P.A.; Secchi, C.; Simonic, T. *J. Chem. Soc., Chem. Commun.* **1984**, 803.
89. Collins, A.N.; Sheldrake, G.N.; Crosby, J., Eds. *Chirality in Industry*; J. Wiley: Chichester, 1992.
90. Sheldon, R.A. *Chirotechnology: Industrial Synthesis of Optically Active Compounds*; M. Dekker: New York, 1993.

THE APPLICATION OF MICROBIAL METHODS TO THE SYNTHESIS OF CHIRAL FINE CHEMICALS

Nicholas J. Turner and Stanley M. Roberts

Advances in Asymmetric Synthesis
Volume 2, pages 285–307

ISBN 1-55938-797-1

I. INTRODUCTION

The use of enzymes and whole cells as catalysts in organic synthesis has expanded enormously over the past 10 years. The increasing need to synthesize new pharmaceutical compounds in single enantiomer form, rather than as racemates, has stimulated the search for enantioselective methods in organic synthesis. Since enzymes are chiral, the reactions they catalyze are often stereoselective, and hence they are potentially of great value as catalysts for asymmetric synthesis.[1,2] Significantly, many more biotransformations are now being operated on a large scale in industry, helping to dispense some of the popular myths concerning issues of scale-up.

In practice, two types of enzymic transformation have proved to be of greatest practical utility, namely hydrolysis[3] (and the reverse reaction—esterification)[4] and reduction.[5] Hydrolytic reactions are generally carried out using isolated enzymes (e.g. lipases, esterases, acylases), whereas reductions can be effected using either whole cells (e.g. baker's yeast) or isolated enzymes (e.g. alcohol dehydrogenases). In the latter case, if the reductions are to be carried out on a medium to large scale, it is necessary to recycle the cofactor NADH/NADPH to avoid using it in stoichiometric quantities.

In the following review, examples of our recent work will be presented. Attention will be focused initially on novel hydrolytic and redox reactions. This will be followed by an overview of some of the other enzymatic transformations that are possible; finally some considerations of future developments will be presented.

II. HYDROLYSIS REACTIONS

A. Lipases and Esterases

Lipases and esterases are without doubt the most widely used biocatalysts in organic synthesis. This is due to a number of important factors: (1) they are robust catalysts that are easy to use, requiring no additional cofactors; (2) they are commercially available in partially purified form; and (3) they operate on a wide range of substrates with very high enantioselectivity. The substrates for lipases and esterases are esters. These can be esters of complex chiral acids (e.g. R^*CO_2Me or R^*CO_2Et) or esters of complex chiral alcohols (e.g. R^*OAc, R^*OCOBu^n) (Figure 1). As a general rule, esterases catalyze the hydrolysis of the former type of substrate, whereas lipases are active against the second class of ester. From a stereochemical viewpoint, the reactions can be classified as kinetic

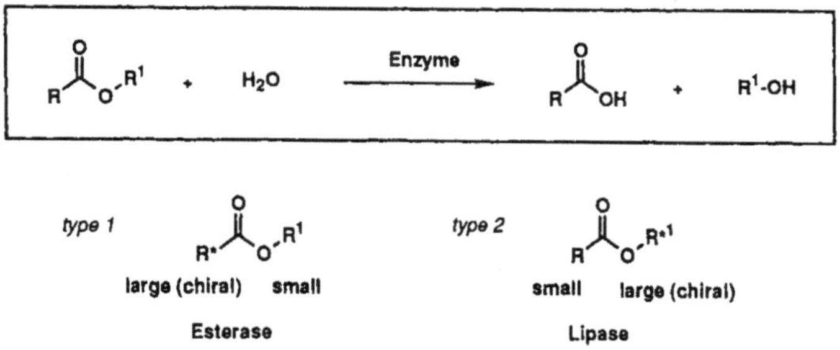

Figure 1. Lipase/esterase catalyzed hydrolysis of esters.

resolutions (which account for by far the largest group of reported reactions) or asymmetric transformations, the latter making use of either prochiral or *meso*-substrates (Figure 2).

An example of a kinetic resolution is given in Figure 3. The hydroxylactone (**1**) is readily prepared in racemic form from cyclopentadiene and glyoxylic acid[6] and the butanoate ester (**2**) is formed in a standard fashion. Hydrolysis of the racemate in aqueous buffer catalyzed by *Pseudomonas fluorescens* (*cepacia*) lipase (PFL) takes place with almost complete selectivity to produce optically active ester and optically active alcohol (Figure 3). A simple workup procedure

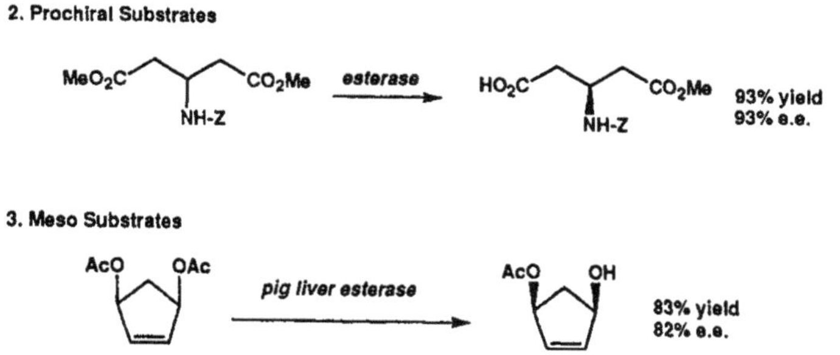

Figure 2. Strategies for using hydrolases to prepare chiral intermediates.

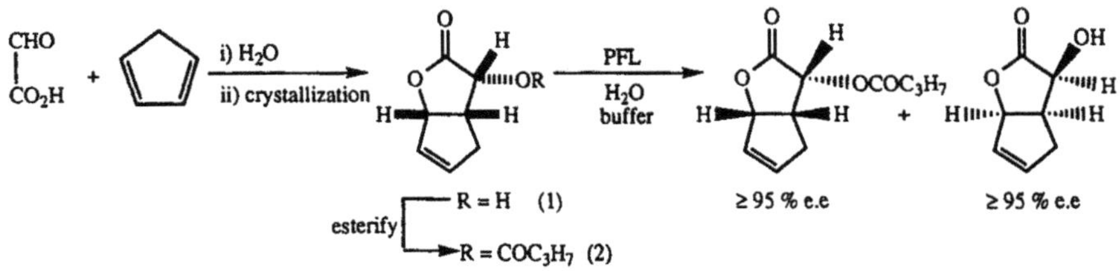

Figure 3. Preparation and resolution of 4-hydroxy-2-oxabicyclo[3.3.0]hept-7-en-3-one.

was utilized: the optically active ester was extracted into hexane while the hydroxylactone was retained in the aqueous layer. The latter compound was then extracted from the aqueous layer into ethyl acetate. A workup procedure avoiding column chromatography was mandatory since the process was run on a multi-kilogram scale. The alcohol was used to prepare the HIV-reverse transcriptase inhibitor carbovir (**3**), while the ester was employed to synthesize the δ-lactone

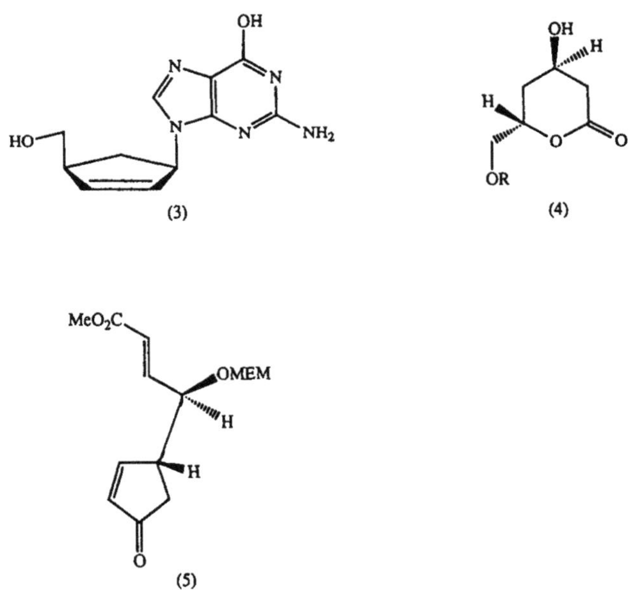

Figure 4. Structure of carbovir, a hypocholestemic precursor and a synthon for brefeldin.

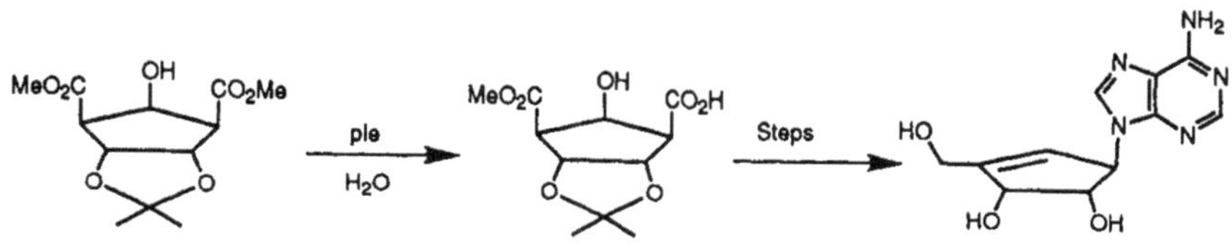

(6) (7) (8)

Figure 5. Resolution of a *meso*-diester.

(**4**) *en route* to hypocholestemic agents (Figure 4).[7] In addition, the cyclopentenone derivative (**5**) was available from the same pots of optically active materials. This compound was used in an efficient synthesis of brefeldin (*vide infra*).

The use of both products from the kinetic resolution is noteworthy. However, in a more general circumstance a kinetic resolution generally leads to ca. 50% of the desired chiron and ca. 50% of unwanted material. It is often more efficient to plan to use a *meso*-diester in a synthetic pathway, allowing the enzyme to produce a chiral half-ester. For instance, the dimethyl ester (**6**) may be synthesized from norbornadiene (Figure 5): hydrolysis catalyzed by pig liver esterase produces the ester (**7**) (95% ee) in almost quantitative yield.[8] The ester (**7**) was converted into neplanocin (**8**),[9] an interesting natural product whose biosynthesis is being studied.[10]

Similarly, the diacetate (**9**), prepared from cyclopentadiene, is hydrolyzed by porcine pancreatic lipase to produce the optically pure alcohol (**10**) in 100% yield

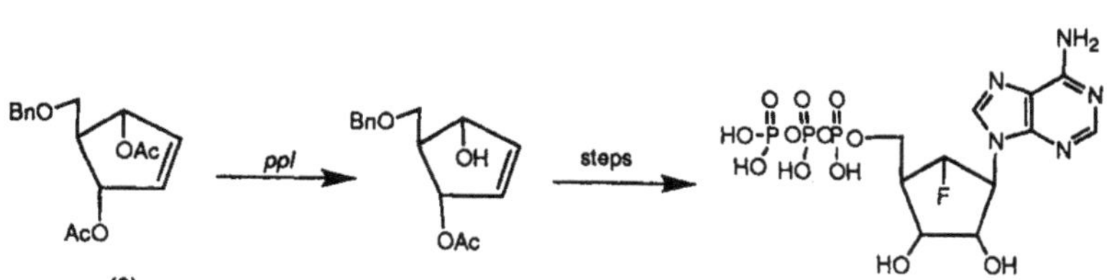

(9) (10) (11)

Bn= CH2Ph

100% yield
>98%e.e.

Figure 6. Resolution of a *meso*-diacetate and the preparation of an ATP mimic.

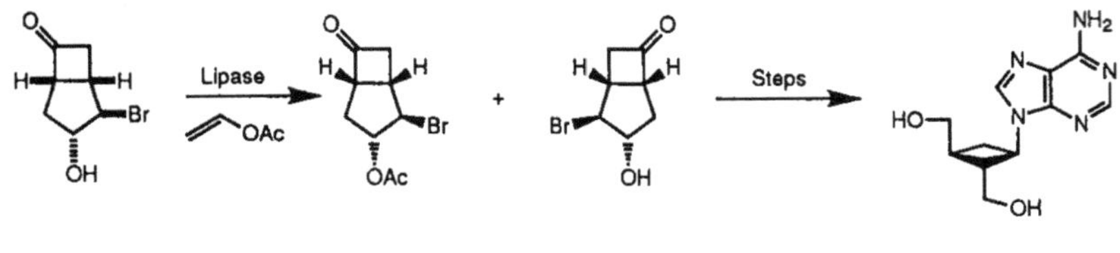

(±)-bromohydrin (12) (13) (14)

Figure 7. Resolution of 2-bromo-3-hydroxybicyclo[3.2.0]heptan-6-one and the preparation of carbocyclic oxetanocin.

(Figure 6). The alcohol (**10**) was taken through a multistep sequence to produce the fluorocarboxylic analog (**11**) of ATP.[11]

Although there are now many lipases and esterases that are commercially available,[12] in practice only relatively few different types have been used for most of the reactions reported. An initial screen against the substrate of interest will normally include porcine pancreatic lipase, *Pseudomonas fluorescens* lipase, *Candida cylindracea* lipase, *Mucor miehei* lipase, and pig-liver esterase. This combination of catalysts display a wide substrate specificity together with attendant high stereoselectivity.

Some time ago, an important observation was made by the Klibanov group regarding the use of hydrolytic enzymes under low water conditions.[13] They demonstrated that lipases and esterases could be used under near-anhydrous conditions in organic solvents to catalyze the *synthesis of esters*. Recently, we have used *Mucor miehei* lipase under low-water conditions to produce a synthon for carbocyclic oxetanocin in optically pure form.

Thus commercially available racemic bicyclo[3.2.0]hept-2-ene-6-one was reacted with HOBr to give the bromohydrin (**12**) and this alcohol was enantioselectivity acetylated using the *Mucor* enzyme in vinyl acetate to give the ester (**13**) and unreacted alcohol; the optically pure alcohol was transformed, over 10 steps, into the antiviral target (**14**) (Figure 7).[14]

In another example. again connected to our on-going program concerning anti-AIDS chemotherapeutics, the allylic alcohol (**15**) was acetylated with high enantioselectivity using *M. miehei* lipase in vinyl acetate. Both products from the enzyme-catalyzed acetylation (after chromatography) were reacted further, being converted into the diphosphorylphosphonates (**16**) and (**17**). Surprisingly the enantiomer (**17**) with the absolute configuration unlike the natural nucleosides was a more effective and indeed a potent HIV-reverse transcriptase inhibitor (Figure 8).[15]

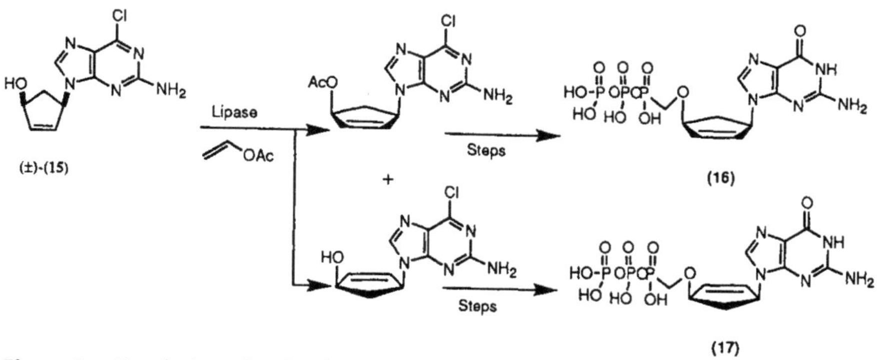

Figure 8. Resolution of a chiral secondary alcohol and the synthesis of inhibitors of HIV-reverse transcriptase.

We have been equally successful in our search for an efficient synthesis of the non-proteinogenic chiral amino acid, L-(*S*)-*tert*-leucine (**18**) in homochiral form (Figure 9). L-(*S*)-*tert*-leucine (**18**) is widely used as both a chiral auxiliary for asymmetric synthesis and also as a component of therapeutic peptides, owing the fact that its introduction into a peptide sequence often confers resistance to proteolytic cleavage. The key feature of the biotransformation (a type of transesterification) is that the unreactive enantiomer of the substrate oxazolinone (**19**) undergoes epimerization in situ to the reactive enantiomer, thereby effecting greater than 50% conversion of the starting material (94% yield, 99.5% ee). Subsequent two-step hydrolysis of the product from the reaction yields L-(*S*)-*tert*-leucine in homochiral (99.5% ee) form.[16] In collaboration with Chiroscience (Cambridge, UK) we have shown that the biotransformation can be carried out on a scale of 250 g per liter.

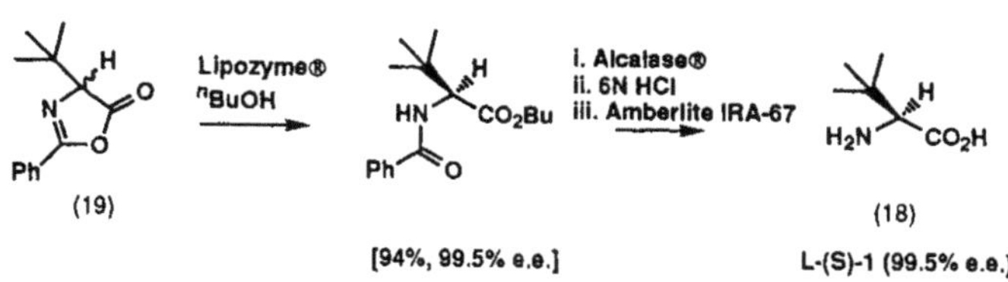

Figure 9. Synthesis of L-(*S*)-*tert*-leucine via a lipase catalyzed dynamic resolution process.

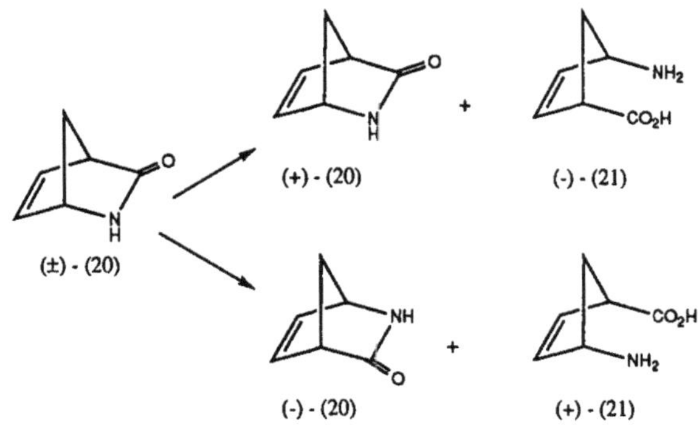

Figure 10. Resolution of 2-aza-bicyclo[2.2.1]hept-5-en-3-one.

B. Acylases

Enzymes which cleave amide bonds (acylases) often do so in a selective manner and they have been used on a large scale to produce 6-aminopenicillanic acid from fermented penicillins.[17] The same class of enzymes is also employed to prepare optically pure amino acids from racemic *N*-acyl derivatives.[18]

The use of acylases for the production of nonnatural amino acids is exemplified by the preparation of the γ-amino acid (**21**) from the γ-lactam (**20**). In fact either enantiomer of the unsaturated amino acid (**21**) can be formed depending on the organism (and the in-situ acylase) used for the biotransformation (Figure 10). The optically pure amino acids and optically pure lactams are available on a very large scale and have been put to use in the synthesis of carbovir and surrogate compounds as well as pharmalogically active analogs of γ-aminobutyric acid (GABA).[19]

C. Nitrile Hydrolases

Whereas the literature is now replete with many reports of the application of lipases and esterases (and to a lesser extent acylases) in organic synthesis, by contrast there are relatively few examples of the use of enzymes for the hydrolysis of nitriles.[20] This situation is somewhat surprising given the relatively harsh conditions required to hydrolyze nitriles to amides and carboxylic acids (2M NaOH or 6M HCl) in the laboratory. Recently, in addition to our work described below, the groups of Professor Ohta at Keio University, Japan[21] and Professor Griengl at Graz University, Austria[22] have begun to explore the synthetic application of nitrile hydrolases. Our work at Exeter has focused mainly on the use of an

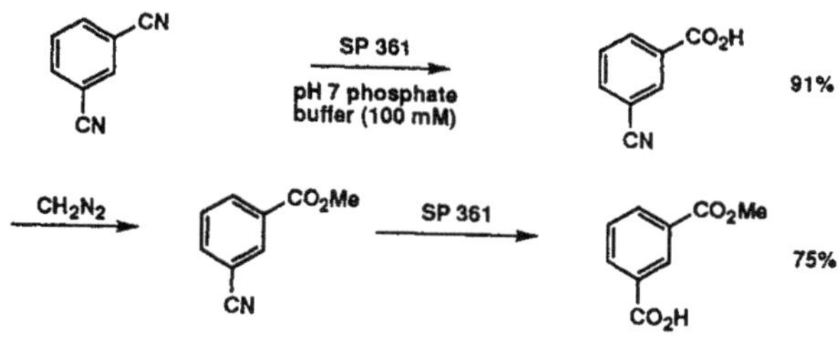

Figure 11. Selective hydrolysis of aromatic nitriles under mild conditions.

immobilized whole cell biocatalyst (SP 361), generously provided by Novo Industri of Denmark. The *Rhodococcus* cells have been immobilized on an ion exchange resin while still retaining the activity of the constituent nitrile hydrolase enzymes. The cells contain two enzymes: one that is able to convert nitriles to amides (hydratase) and one that can convert amides to carboxylic acids (amidase). Early on[23] we were able to show that this immobilized catalyst was very effective in catalyzing the hydrolysis of a wide range of nitriles under exceptionally mild conditions (pH 7.5, 35 °C). This presented unique opportunities for effecting the selective hydrolysis of nitrile groups in the presence of acid and base sensitive functionality (Figure 11). More recently our attention has been directed towards stereoselective transformations. We have shown that certain substituted aryl propionic acids can be prepared in optically active form by hydroly-

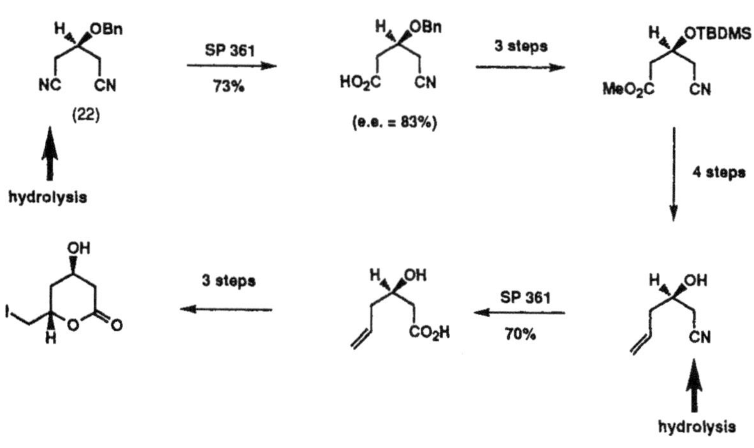

Figure 12. Asymmetric hydrolysis of 3-benzyloxyglutaronitrile.

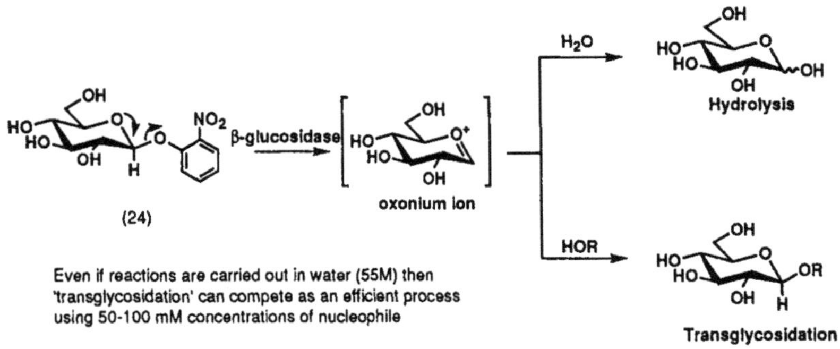

Figure 13. Mechanism of action of glycosidases.

sis of the corresponding racemic nitrile.[24] In addition, asymmetric hydrolysis of a range of 3-hydroxyglutaronitriles (e.g. **22**) has been shown to yield optically active nitrile acids[23] that are useful synthons for the preparation of biologically active δ-lactones[25] (Figure 12).

D. Glycosidases

Finally in the section on hydrolytic enzymes we turn to the glycosidases. These enzymes catalyze the hydrolysis of glycosidic linkages, most often those linking carbohydrates in oligosaccharides. The mechanism of hydrolysis is represented diagrammatically in Figure 13. Using *o*-nitrophenyl-β-D-glucoside (**24**) as a representative substrate, with β-glucosidase as the enzyme, cleavage of the exocyclic C–O bond occurs, resulting in the generation of an enzyme-stabilized oxonium ion intermediate. Thereafter, attack by a molecule of water results in overall hydrolysis. However, by adding a second nucleophile to the reaction, usually an alcohol, it is possible to intercept the glycosyl cation and divert the reaction towards glycoside synthesis. Examples are given below of both *hydrolysis* and *transglycosidation* modes.

Morphine-6-glucuronide (**25**) is a potent analgesic that possesses approximately 10 times greater activity than morphine.[26] Indeed there is some evidence to suggest that morphine may be a prodrug of morphine-6-glucuronide, the former undergoing rapid glucuronidation in vivo in the liver. Treatment of morphine-3,6-diglucuronide (**26**) with β-glucuronidase from *Patella vulgata* results in highly selective hydrolysis of the 3-glucuronide without any cleavage at the 6 position (Figure 14).[27]

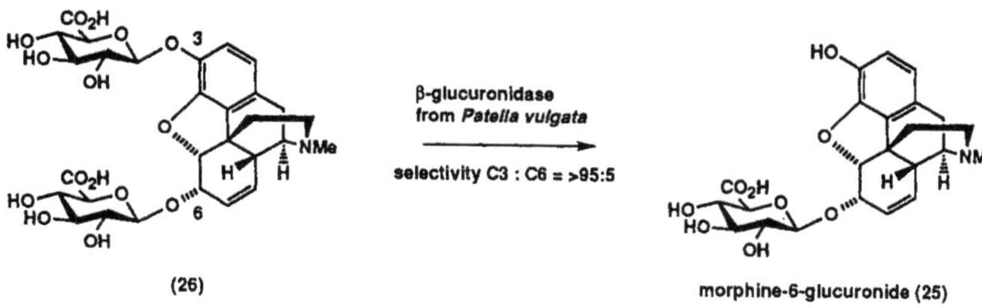

Figure 14. Synthesis of morphine-6-glucuronide.

With regard to transglycosidation, we have recently developed a method for the stereospecific attachment of carbohydrates to amino acid derivatives.[28] In order to avoid using large excesses of the acceptor nucleophile[29] we have chosen to reverse the ratio of the donor to acceptor and instead add ~6 equivalents of the donor, over a period of 24 h, via syringe pump. Using the method, we are able to obtain up to 30% yields of the corresponding glycosides (Figure 15). By matching the appropriate *o*-nitrophenylglycoside with the appropriate glycosidase (e.g. *o*-nitrophenyl β-D-glucoside with β-glucosidase, *o*-nitrophenyl-β-D-xyloside with β-xylosidase, etc.) it is possible to synthesize a range of analogs by this method.[30]

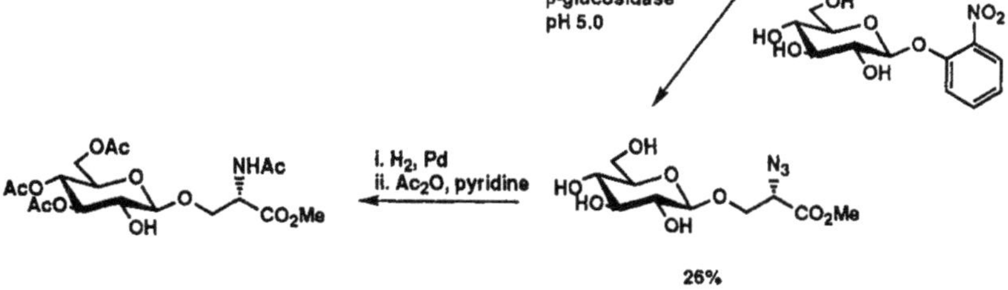

Figure 15. Glycosidase catalyzed synthesis of glycoamino acid derivatives.

III. REDUCTION REACTIONS

Microbial methods have been widely applied to the reduction of a variety of functional groups (Figure 16). Most attention[31] has been directed towards the asymmetric reduction of carbonyl groups, including 1,2- and 1,3-dicarbonyl compounds, although increasingly the reduction of other functional groups (e.g. C=C, NO_2, CO_2H) is being examined. Since almost all biochemical reductions require either NADH or NADPH as an essential cofactor, then *either* a whole cell system must be used in which cofactor recycling occurs naturally, *or* if an isolated enzyme is used then a secondary cofactor regeneration system must be employed.

A study of the fastest and most effective way of preparing the (*S*)-alcohol (**27**) using a whole cell system (baker's yeast) or an isolated enzyme or, indeed, an alternative chemical method (including classical resolution) led to the conclusion that the process shown in Figure 17 was the easiest to employ on a 1–10 g scale.[32]

The alcohol (*S*)-**27** was converted into the organometallic species (**28**) (Figure 18), which was linked to the enone (**5**) and then transformed into the antifungal

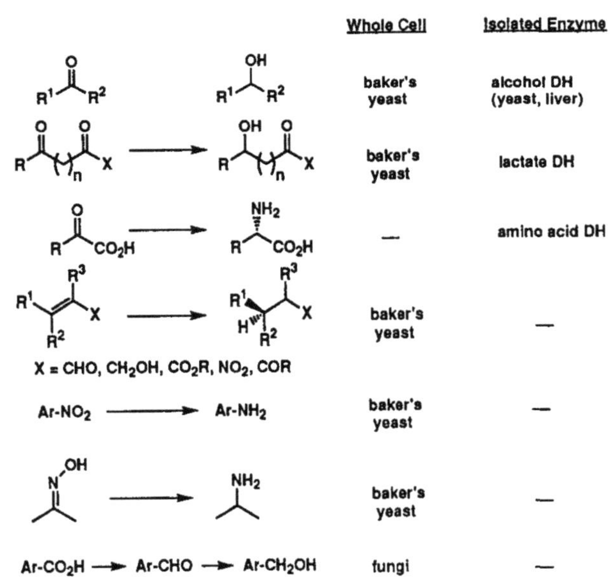

Figure 16. Overview of some biocatalytic reductions.

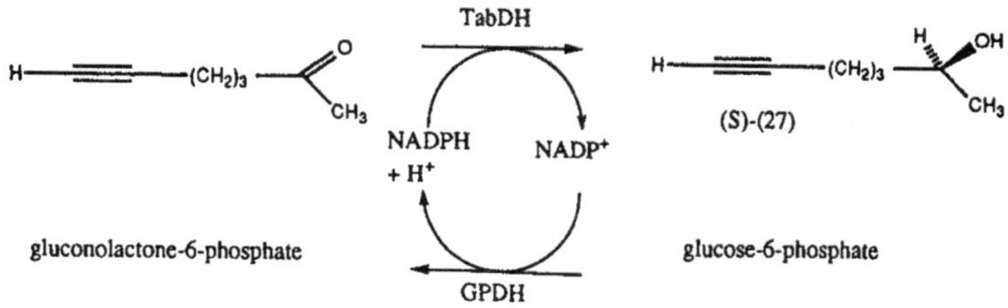

TabDH = Thermoanaerobium brockii dehydrogenase

GPDH = Glucose-6-phosphate dehydrogenase

Figure 17. Synthesis of (*S*)-6-hydroxyhept-1-yne.

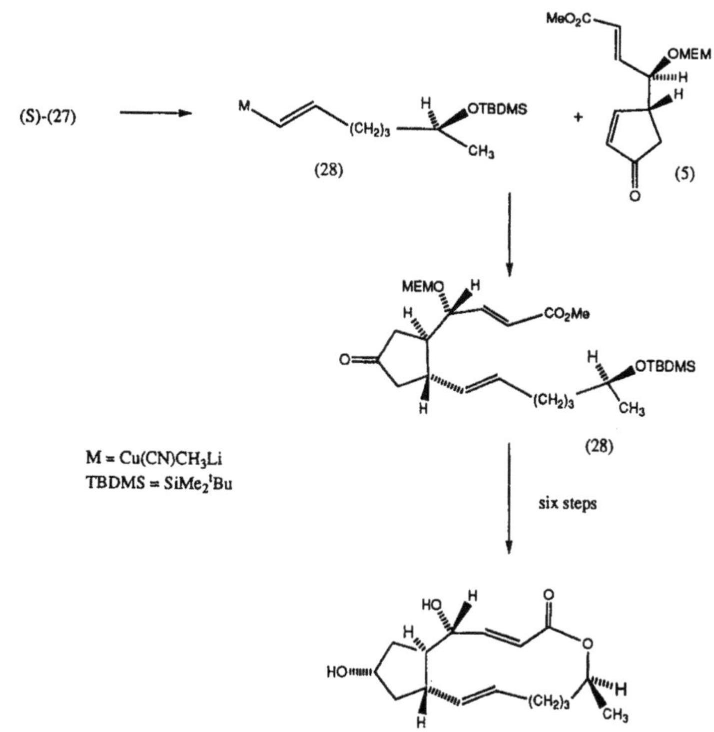

M = Cu(CN)CH$_3$Li

TBDMS = SiMe$_2$^{t}Bu

Figure 18. Synthesis of Brefeldin.

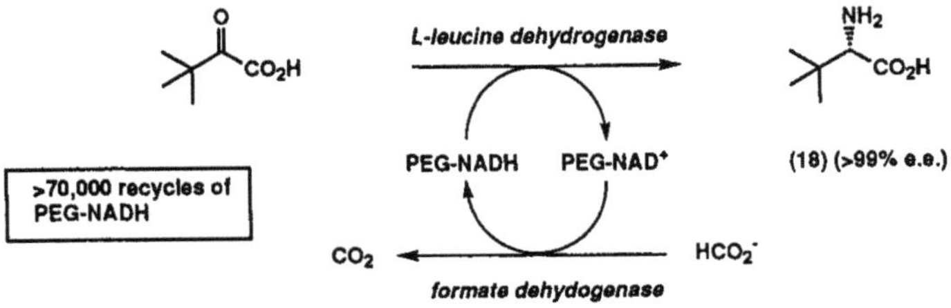

Figure 19. Synthesis of L-(*S*)-*tert*-leucine (Degussa AG).

agent brefeldin (**29**).[33] A large-scale application of related methodology is illustrated by the synthesis of L-*tert*-leucine (**18**) carried out by Degussa (Figure 19).[34] Regeneration of the cofactor NADH is achieved, in this instance, by adding a second enzyme, formate dehydrogenase, together with formate. The NADH can then be added in catalytic quantities and recycled many times (>70,000 recycles).

The reduction of β-keto esters using fermenting baker's yeast (*Saccharomyces cerevisae*) is a widely known and much exploited reaction.[35] For example, ethyl acetoacetate (ethyl 3-oxobutanoate) (**30**) is reduced by baker's yeast to yield ethyl (3*S*)-hydroxybutanoate (**31**) with modest enantiomeric excess (Figure 20). Many variations of this reaction exist in which the substrate β-ketoester is substituted at either the 2- or 4-positions or both. We have carried out another variation of this reduction by using a racemic β-keto-δ-hydroxyester (**32**) as the substrate (Figure 21).[36] The yeast effects a simultaneous kinetic resolution and asymmetric reduction leading to the *syn*-diol (**33**) with very high diastereomeric excess (>95%) and moderate enantiomeric excess (87%). The diol (**33**) was subsequently transformed into the δ-lactone (**34**) which is the enantiomer of the lactone found in the anti-hypocholestemic agents, mevinolin and compactin.

The reduction of aromatic nitro compounds to the corresponding anilines is usually carried out using metal-based reagents (e.g. Sn/HCl or Zn/HCl). Since 1914 it has been known that certain yeasts are also able to catalyze this reaction[37] although the transformation has remained largely unexploited. Our initial investi-

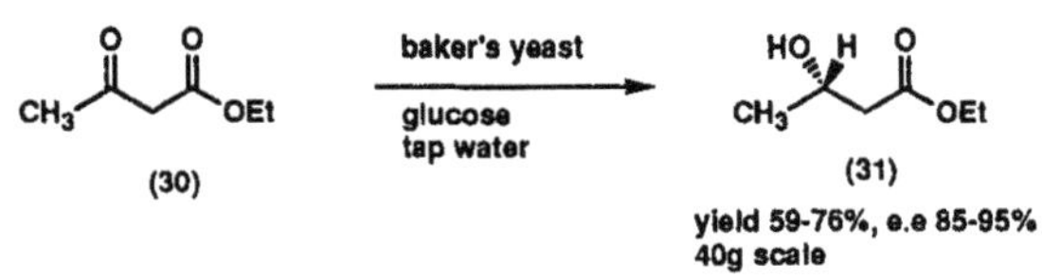

Figure 20. Baker's yeast reduction of ethylacetoacetate.

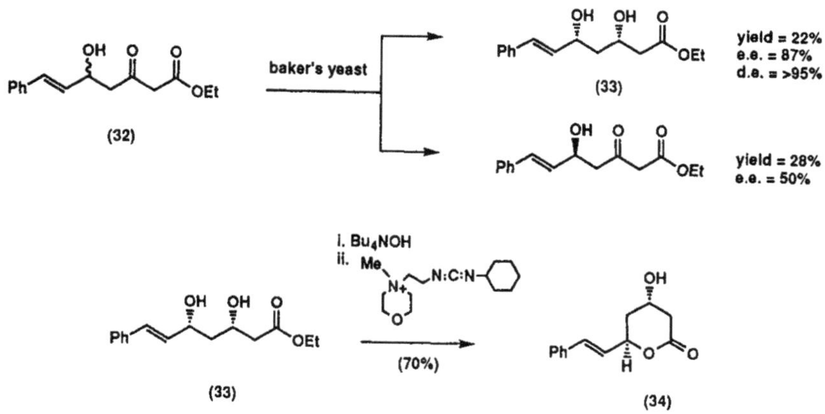

Figure 21. Asymmetric synthesis of δ-lactones.

gations have revealed that the reduction proceeds best with electron-withdrawing substituents on the aromatic ring. More significantly it has been possible, in some cases, to achieve the regioselective reduction of unsymmetrical dinitroarenes (e.g. **35** to **36**). The site of reduction can be predicted by a consideration of the relative steric and electronic factors operating[38] (Figure 22).

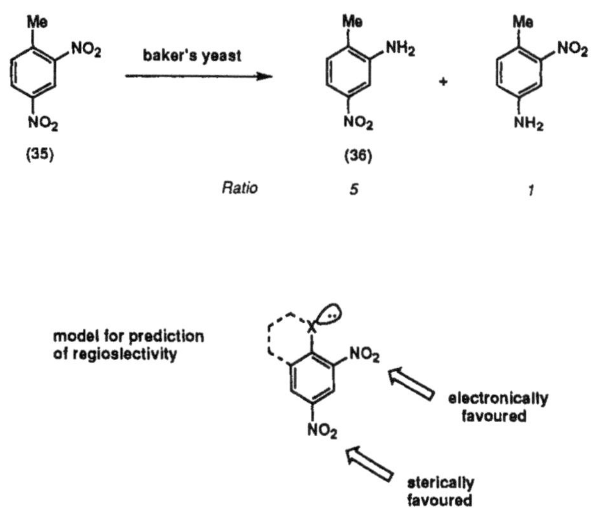

Figure 22. Regioselective reduction of dinitroarenes.

IV. OTHER ENZYME-CATALYZED TRANSFORMATIONS

In addition to hydrolysis and reduction, many other enzyme-catalyzed transformations are currently being investigated by research groups both in universities and in industry. Important advances are being made with oxidations (especially Baeyer–Villiger type oxidations of ketones to lactones and hydroxylation processes), carbon–carbon bond synthesis, and oligosaccharide synthesis. Emphasis is being placed on the development of biocatalytic systems that will complement, and thereby increase, the various synthetic methodologies that are currently available.

For example, the stereocontrolled Baeyer–Villiger reaction has been studied in detail. Bicyclic and monocyclic substrates have been employed (Figures 23 and 24): a whole cell system (*Acinetobacter calcoaceticus* NCIMB 9871) was utilized to effect the enantioselective oxidation of the dihalonorbornanone (**37**) to give optically pure lactone (**38**) and recovered optically pure ketone. The latter compound was converted into the fluoro-carbocyclic analog (**39**) of the anti-AIDS agent AZT (Figure 23) using a chemical Baeyer–Villiger reaction in the

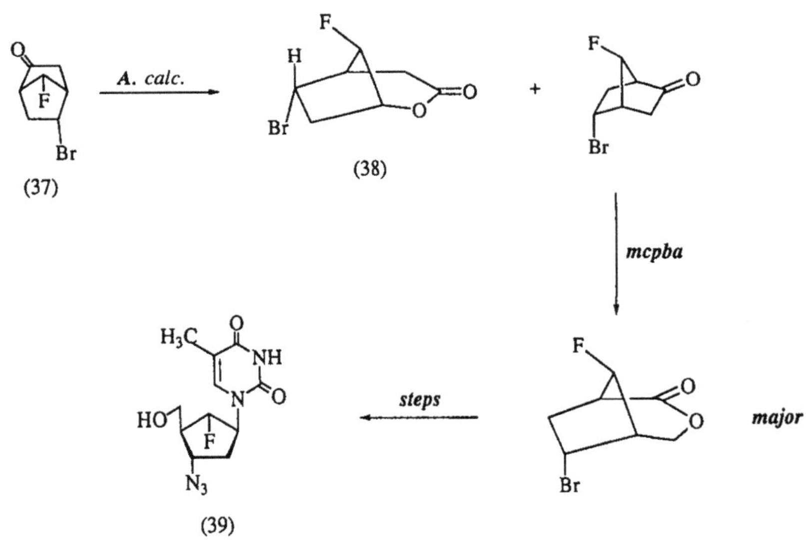

Figure 23. Whole-cell Baeyer–Villiger oxidation of 5-bromo-7-fluorobicyclo [2.2.1]heptan-2-one.

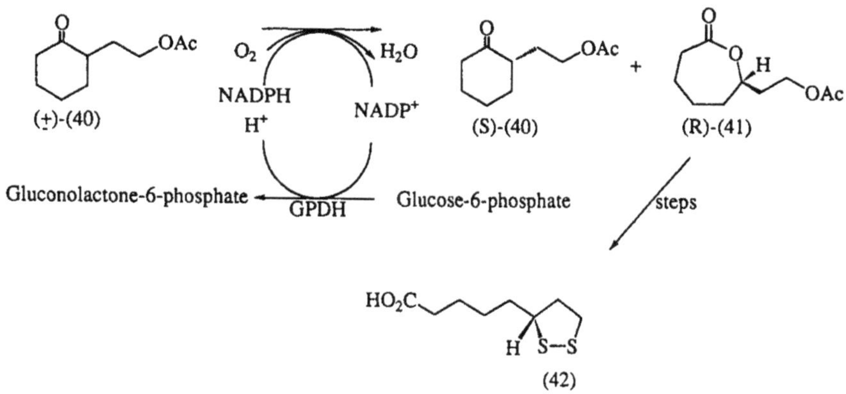

Figure 24. Enzyme-catalyzed Baeyer–Villiger oxidation of 2-(acetoxyethyl) cyclohexan-1-one and the synthesis of lipoic acid.

first step.[39] The differing regioselectivity observed between the biological and the chemical Baeyer–Villiger reactions is noteworthy.

Oxidation of the monocyclic compound (**40**) was effected using a monooxygenase (MO2) from *Pseudomonas putida* NCIMB 10007 (Figure 24). The enantioselective oxidation produced the lactone (**41**) which was transformed into lipoic acid (**42**) using standard methodology. In this bio-Baeyer–Villiger reaction a partially purified enzyme was used as a catalyst requiring the use of cofactor recycling (cf. Figure 17).[40]

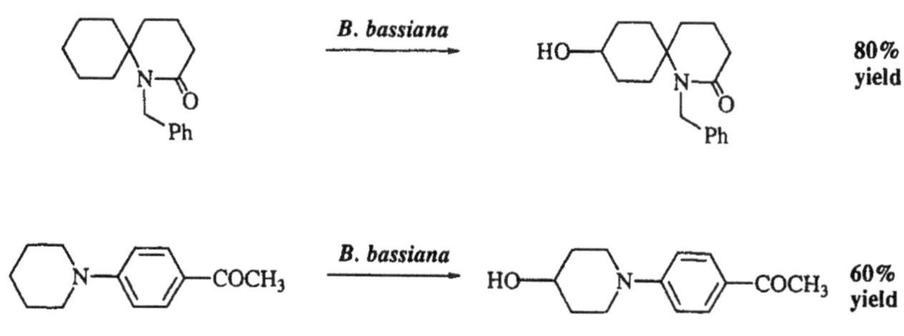

Figure 25. Examples of remote hydroxylation of heterocyclic compounds using *B. bassiana*.

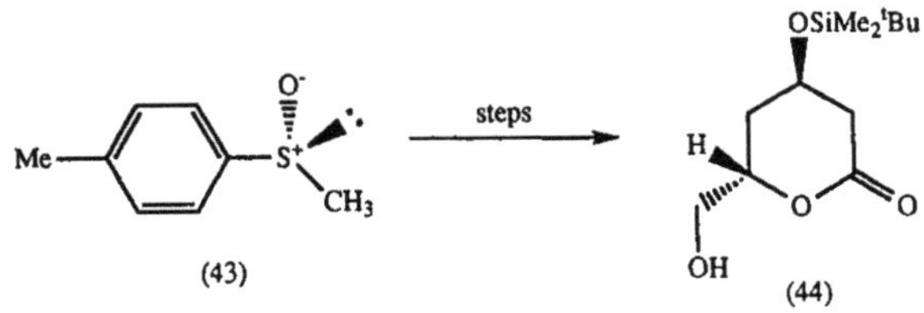

(43) (44)

Figure 26. Use of methyl tolylsulfoxide to prepare a precursor of hypocholestemic agents.

Hydroxylation reactions have been particularly well studied in connection with the modification of steroids and alkaloids.[41] The site of hydroxylation is often difficult to predict when other, novel substrates are employed, though high yields of products can be obtained in transformations that are difficult to emulate using nonbiological methods (Figure 25).[42] The microorganism *Beauvaria bassiana* which is featured in Figure 25 is often used in hydroxylation reactions of this type.

Heteroatom oxidations using biocatalysts are also well known; given the use of sulfoxides in organic synthesis it is not surprising that the stereoselective oxidation of sulfides has attracted particular attention. While the Kagan–Sharpless reaction provides one possibility,[43] the employment of biocatalysts is an alternative strategy. The use of baker's yeast is particularly easy for an inexperienced

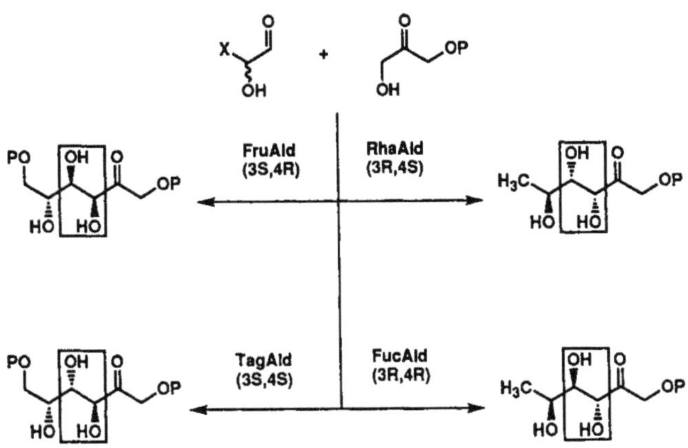

Figure 27. Diastereoselective aldolase reactions.

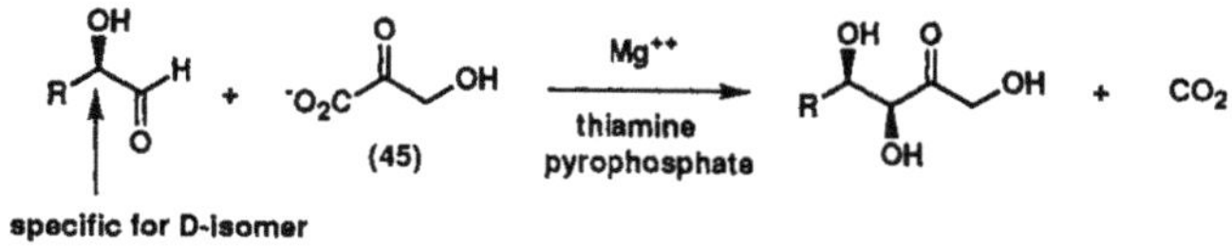

Figure 28. Transketolase catalyzed synthesis of C–C bonds.

scientist and this organism has been shown to convert methyl *para*-tolylsulfide into the (*R*)-sulfoxide.[43] The latter compound has been used to prepare the δ-lactone (**44**) (Figure 26), another variant of the building blocks used to synthesize hypocholestemic agents.[44]

Enzyme-catalyzed carbon–carbon bond synthesis has been studied in some detail, particularly with respect to the aldolases. These enzymes catalyze many versions of the asymmetric aldol reaction involving condensation of a carbonyl-containing substrate with an aldehyde (Figure 27).[45] Unfortunately, most aldolases require the nucleophilic carbonyl-containing substrate to be phosphorylated and this has tended to limit their widespread application. With this in mind, we have recently begun to study the related reaction catalyzed by transketolase (Figure 28). This enzyme uses hydroxypyruvate[45] as one of the substrates and a wide range of aldehydes as the electrophile. The enzyme has been cloned and overexpressed in *Escherichia coli*.[46] The level of expression of the enzyme in *E. coli* is so high that a 1000-liter fermentation produced enough enzyme to catalyze the formation of 1 ton of product in 1 h! Since the enzyme is known to be selective for α-hydroxyaldehydes of D-configuration we have recently devised a synthesis of D-hydroxyaldehydes, in homochiral form, using D-amino acids as precursors[47] (Figure 29).

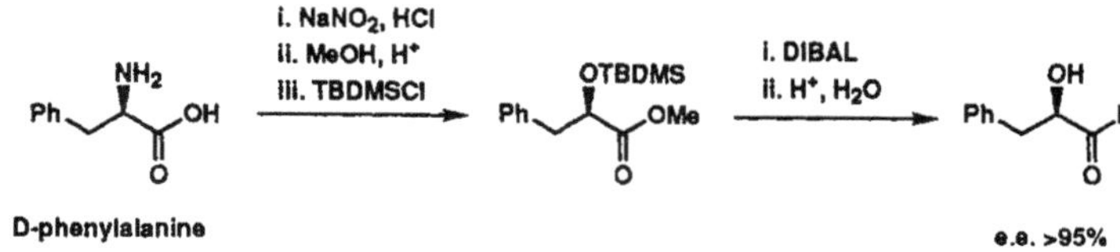

Figure 29. Synthesis of homochiral D-α-hydroxyldehydes.

V. FUTURE PROSPECTS

The use of certain biocatalytic reactions in synthetic sequences has now become well established and to some extent routine. This allows the synthetic organic chemist to add the repetoire of reactions performed by enzymes to more conventional methods when considering the best way to access a target structure. Even within the area of biocatalysis the range of different enzyme-catalyzed reactions means that an array of pathways to the desirable fine chemical is available. The routes to mevinic acid lactone mentioned in this article illustrate this point (Figure 30).

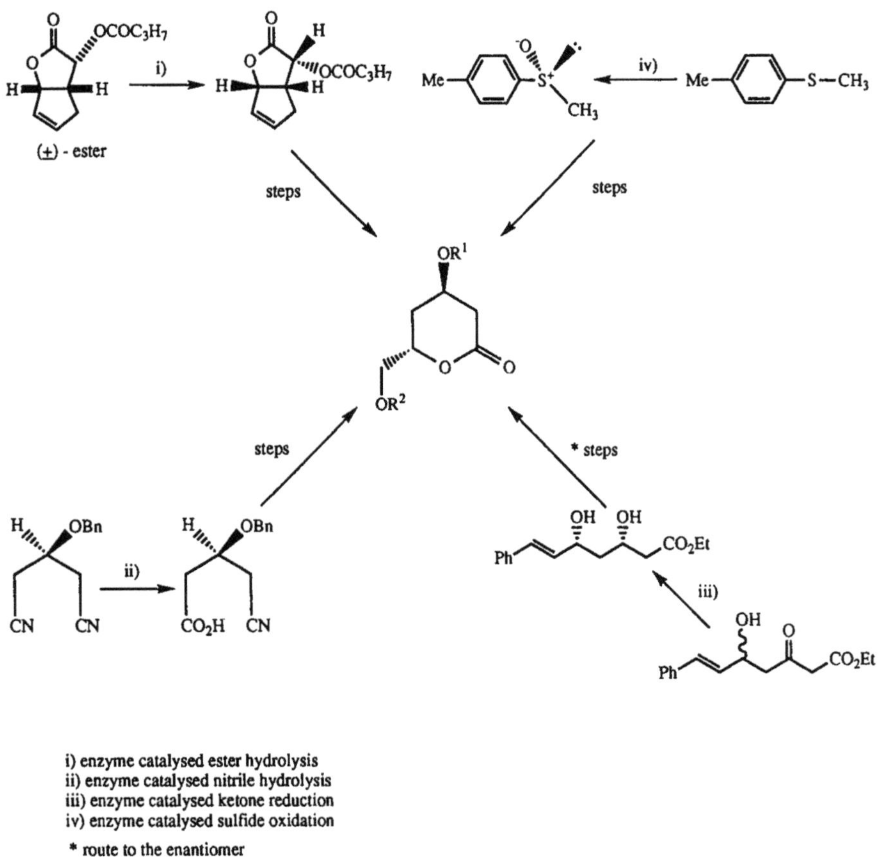

Figure 30. Biotransformations leading to a well-established synthon for the mevinic acid-type hypocholestemic agents.

Undoubtedly the emphasis in the future will be to expand the range of synthetically usefully reactions such that a number of different oxidations, reductions, C–C bond forming reactions etc. can be reliably used. This will necessitate further research, particularly in the areas of (1) increased screening of microbial sources for novel enzymic activities, (2) overexpression of proteins in microbial hosts, (3) a better understanding of the substrate specificity and stereoselectivity of useful enzymes, and (4) efficient cofactor recycling systems.

Chemists involved in biotransformations have become increasingly aware of the importance of collaborating with scientists from other disciplines in order to progress the development of novel biocatalysts. It is now typical for chemists to embark on these projects in association with other groups who have complementary expertise in microbiology enzymology, molecular genetics, and biochemical engineering. By using this approach, it is possible from the beginning of the project to identify most of the important parameters that are crucial to the successful development of a particular biotransformation.

REFERENCES AND NOTES

1. For recent books see: (a) Roberts, S.M.; Turner, N.J.; Willetts, A.J.; Turner, M.K. *Introduction to Biocatalysis Using Enzymes and Microorganisms*; Cambridge University Press: New York, 1995. (b) Wong, C.-H.; Whitesides, G.M. *Enzymes in Synthetic Organic Chemistry*; Pergamon Press: Oxford, 1994. (c) Faber, K. *Biotransformations in Organic Chemistry*, 2nd ed.; Springer-Verlag: 1994.
2. For recent general reviews of enzyme-catalyzed organic synthesis see: (a) Turner, N.J. *Nat. Prod. Rep.* **1994**, *11*, 1. (b) Santaniello, E.; Ferraboschi, P.; Grisenti, P.; Manzocchi, A. *Chem. Rev.* **1992**, *92*, 1071.
3. Zhu, L.-M.; Tedford, M.C. *Tetrahedron* **1990**, *46*, 6587.
4. Klibanov, A.M. *Acc. Chem. Res.* **1990**, *23*, 114.
5. Turner, N.J. *Chem. Ind.* **1994**, 592.
6. McCague, R.; Olivo, H.F.; Roberts, S.M. *Tetrahedron Lett.* **1993**, *34*, 3785.
7. McCague, R.; MacKeith, R.A.; Olivo, H.F.; Roberts, S.M.; Taylor, S.J.C.; Xiong, H. *Bioorg. Med. Chem.* **1994**, *2*, 387.
8. Cotterill, I.C.; Cox, P.B.; Drake, A.F.; LeGrand, D.M.; Hutchinson, E.J.; Latouche, R.; Pettman, R.B.; Pryce, R.J.; Roberts, S.M.; Ryback, G.; Sik, V.; Williams, J.O. *J. Chem. Soc., Perkin Trans. 1* **1991**, 3071.
9. Hill, J.M.; Hutchinson, E.J.; Le Grand, D.M.; Roberts, S.M.; Thorpe, A.J.; Turner, N.J. *J. Chem. Soc., Perkins Trans. 1* **1994**, 1483.
10. Hill, J.M.; Jenkins, G.N.; Rush, C.P.; Turner, N.J.; Willetts, A.J.; Buss, A.D.; Dawson, M.J.; Rudd, B.A.M. *J. Am. Chem. Soc.* **1995**, *116*, 5391.
11. (a) Le Grand, D.M.; Roberts, S.M. *J. Chem. Soc., Chem. Commun.* **1993**, 1284. (b) Payne, A.N.; Roberts, S.M. *J. Chem. Soc., Perkin Trans. 1* **1992**, 2633.
12. See, for example: *Preparative Biotransformations*; Roberts, S.M.; Wiggins, K.; Casy, G.; Phythian, S., Eds.; Wiley: New York, 1992.
13. Cambou, B.; Klibanov, A.M. *J. Am. Chem. Soc.* **1984**, *106*, 2687.

14. Cotterill, I.C.; Roberts, S.M. *J. Chem. Soc., Perkin Trans. 1* **1992**, 2585.
15. Merlo, V.; Roberts, S.M.; Storer, R.; Bethell, R.C. *J. Chem. Soc., Perkin Trans. 1* **1994**, 1477.
16. Turner, N.J.; Winterman, J.R.; McCague, R.; Parratt, J.S.; Taylor, S.J.C. *Tetrahedron Lett.* **1995**, *36*.
17. See, for example: Davies, H.G.; Green, R.H.; Kelly, D.R.; Roberts, S.M. *Biotransformations in Preparative Organic Chemistry*; Academic Press: London, 1989, pp. 69–70.
18. See, for example, ref. 17, pp. 65–68.
19. Taylor, S.J.C.; McCague, R.; Wisdom, R.; Lee, C.; Dickson, K.; Ruecroft, G.; O'Brien, F.; Littlechild, J.; Bevan, J.; Roberts, S.M.; Evans, C.T. *Tetrahedron: Asymmetry* **1993**, *4*, 1117.
20. For a review on the synthetic applications of nitrile hydrolases see: Crosby, J.; Moilliet, J.; Parratt, J.S.; Turner, N.J. *J. Chem. Soc., Perkin Trans. 1* **1994**, 1679.
21. Kakeya, H.; Sakai, N.; Sugai, T.; Ohta, H. *Tetrahedron Lett.* **1991**, *32*, 1343.
22. de Raadt, A.; Klempier, N.; Faber, K.; Griengl, H. *J. Org. Chem.* **1993**, *58*, 3179.
23. Cohen, M.A.; Sawden, J.; Turner, N.J. *Tetrahedron Lett.* **1990**, *31*, 7223.
24. Cohen, M.A.; Parratt, J.S.; Turner, N.J. *Tetrahedron: Asymmetry* **1992**, *3*, 1543.
25. (a) Crosby, J.; Parratt, J.S.; Turner, N.J. *Tetrahedron: Asymmetry* **1992**, *3*, 1547. (b) Beard, T.; Cohen, M.A.; Parratt, J.S.; Turner, N.J.; Crosby, J.; Moilliet, J. *Tetrahedron: Asymmetry* **1993**, *4*, 1085. (c) Maddrell, S.; Turner, N.J., unpublished results.
26. Osborne, R.J.; Joel, S.P.; Trew, D.; Slevin, M.L. *The Lancet* **1988**, 828.
27. Brown, R.T.; Carter, N.E.; Scheinmann, F.; Turner, N.J. *Tetrahedron Lett.* **1995**, *36*.
28. Turner, N.J.; Webberley, M.C. *J. Chem. Soc., Chem. Commun.* **1991**, 1349.
29. Toone, E.J.; Simon, E.S.; Bednarski, M.D.; Whitesides, G.M. *Tetrahedron* **1989**, *45*, 5365.
30. Baker, A.; Turner, N.J.; Webberley, M.C. *Tetrahedron: Asymmetry* **1994**, *5*, 2517.
31. Lemière, G.L. In *Enzymes as Catalysts in Organic Synthesis*; Schneider, M.P.; Reidel: Dordrecht, 1986, p. 17.
32. Casy, G.; Gorins, G.; McCague, R.; Olivo, H.F.; Roberts, S.M. *J. Chem. Soc., Chem. Commun.* **1994**, 1085.
33. Carnell, A.J.; Casy, G.; Gorins, G.; Kompany-Saeid, A.; McCague, R.; Olivo, H.F., Roberts, S.M.; Willetts, A.J. *J. Chem. Soc., Perkin Trans. 1* **1994**, 3431.
34. Schutte, H.; Hummel, W.; Tsai, H.; Kula, M.R. *Appl. Microbiol. Biotechnol.* **1985**, *22*, 306.
35. For example see: Servi, S. *Synthesis* **1990**, 1.
36. Johnson, D.V.; Turner, N.J., unpublished results.
37. Neubert, C.; Welde, E. *Biochem. Z.* **1914**, *60*, 427.
38. Davey, C.L.; Powell, L.W.; Turner, N.J.; Wells, A. *Tetrahedron Lett.* **1994**, *35*, 7867.
39. (a) Levitt, M.S.; Newton, R.F.; Roberts, S.M.; Willetts, A.J. *J. Chem. Soc., Chem. Comm.* **1990**, 619. (b) Highcock, R.M.; Hilpert, H.; Myers, P.L.; Roberts, S.M.; Storer, R. *J. Chem. Soc., Perkin Trans. 1* **1991**, 1127.
40. Adger, B.; Bes, M.T.; Grogan, G.; McCague, R.; Pedragosa-Moreau, S.; Roberts, S.M.; Villa, R.; Wan, P.W.A.; Willetts, A.J. *J. Chem. Soc., Perkin Trans. 1* **1995**.
41. Izuka, H.; Naito, A. *Microbial Conversion of Steroids and Alkaloids*; Springer-Verlag: Berlin, 1981.
42. Floyd, N.; Munyemana, F.; Roberts, S.M.; Willetts, A.J. *J. Chem. Soc., Perkin Trans. 1*, **1993**, 881.
43. Kagan, H.B.; Dunach, E.; Nemeck, C.; Pitchen, P.; Samuel, O.; Zhao, S.H. *Pure Appl. Chem.* **1985**, 1911.
44. Beecher, J.; Brackenridge, I.; Roberts, S.M.; Tang, J.; Willetts, A.J. *J. Chem. Soc., Perkin Trans. 1* **1995**.

45. Fessner, W.D.; Sinerius, G.; Schneider, A.; Dreyer, M.; Schulz, G.E.; Badia, J.; Aguilar, J. *Angew. Chem. Int. Ed. Engl.* **1991**, *30*, 555.
46. Hobbs, G.; Lilly, M.D.; Turner, N.J.; Ward, J.M.; Woodley, J.M.; Willetts, A.J. *J. Chem. Soc., Perkin Trans. 1* **1993**, 165.
47. Humphrey, A.J.; Turner, N.J., unpublished results.

INDEX

www.ingramcontent.com/pod-product-compliance
Ingram Content Group UK Ltd.
Pitfield, Milton Keynes, MK11 3LW, UK
UKHW020239290926
15470UKWH00002B/636

9 781559 387972